WINGS OF WAR

By the same author

Five Up: A Chronicle of Five Lives
The Sport of Princes: Reflections of a Golfer
Flying Colours: The Epic Story of Douglas Bader

WINGS OF WAR

Airmen of All Nations Tell Their Stories
1939–1945

Edited by Laddie Lucas

Hutchinson
London Melbourne Sydney Auckland Johannesburg

Hutchinson & Co. (Publishers) Ltd

An imprint of the Hutchinson Publishing Group

17–21 Conway Street, London W1P 6JD

Hutchinson Group (Australia) Pty Ltd
30–32 Cremorne Street, Richmond South, Victoria 3121
PO Box 151, Broadway, New South Wales 2007

Hutchinson Group (NZ) Ltd
32–34 View Road, PO Box 40–086, Glenfield, Auckland 10

Hutchinson Group (SA) Pty Ltd
PO Box 337, Bergvlei 2012, South Africa

First published 1983

Set in Linotron Sabon

Printed in Great Britain by Redwood Burn, Trowbridge, Wilts.,
and bound by Wm Brendon & Son Ltd, Tiptree, Essex

British Library Cataloguing in Publication Data

Wings of war.
1. World War, 1939–1945—Aerial operations
I. Lucas, Laddie
940.54'4 D785

ISBN 0 09 154280 4

Contents

Part Three In the Teeth of Adversity

Part Four Retaliation

Part Five Gathering Onslaught

Part Six Maximum Effort

Part Seven Aerial Dividend

Acknowledgements

The editor acknowledges the remarkable help he has received, at home and overseas, in gathering this collection of writings. So many have contributed to it in one way or another that to mention one individual would be, almost invariably, to overlook another thus causing unintentional hurt.

There are, however, some collective thanks which can, and should, be expressed without risk of injustice.... To all those contributors whose names appear in the text that follows; their cooperation and assistance have often far exceeded reasonable expectations.... To Air Commodore Henry Probert and his staff at the Air Historical Branch of the Ministry of Defence in London; this work could not have been undertaken without their continuing help which has always been kindly and considerately offered. Through it, British posts overseas were alerted to this project at the start; thereafter the aid which our Air Advisers and Attachés provided in the various capitals was a telling act in securing the international character of the collection. All responded readily in enabling contacts to be located and none more so (because the scope was wider for each) than Air Commodore Martin Chandler in Bonn, Air Commodore John Parker in Paris, Wing Commander Keith Harding in Oslo, Wing Commander Olly Knight in Prague, Wing Commander Peter Cooke in Harare and, across the Atlantic, Air Commodore Ron Dick in Washington.

In the same way, Alec Douglas and his staff at the Directorate of History, Defence Headquarters, in Ottawa, have made valuable suggestions to support the Canadian representation. This backing was, in turn, supplemented by the interest shown by the Wartime Pilots' and Observers' Association in Winnipeg, whose President, Don Aylard, went out of his way to ensure the Association's help....

Then, from South Africa, the extent of the Republic's significant contribution to victory forty and more years ago is reflected in the response the editor has received from old wartime comrades. Among them, Charles Barry, a fine photo reconnaissance pilot in his day, was instrumental in setting things going in Johannesburg at the outset, thus opening the way for the offer of historically important material from additional sources.

Finally, the editor warmly acknowledges the willing support he has received from those who have provided the English translations where they were needed. To Alexander Bell and his talented family, to Peter Clapham and Charles Pretzlik, both of whom made their marks in the Royal Air Force's offensive Mosquito operations, and Penny Garton, who handled the Italian copy, special thanks are due.

To all these, and other kind friends, the editor offers his salute.

Explanatory Note

This anthology is designed to typify the air war of 1939–45. It embraces the writings of most of the combatant countries. It balances the views of enemy and ally. Its principal aim is not so much to provide an extensive historical record as to offer a readable sequence of experiences and anecdotes, opinion and fact, which combine to form an authoritative impression of the great aerial campaigns. Forty years on, some of the personal descriptions of achievement – and disaster – still seem barely credible.

Most of the operational theatres and flying roles are covered. The editor has not, however, disguised his preference for the unusual and unconventional selection. His concern, in the main, has been to marshal together a collection of first-hand, individual accounts of particular incidents and themes rather than to labour a more commonplace and comprehensive approach.

Much of the writing is new. There was a heartening readiness on the part of a number who had made their special, and often distinguished, contributions to the fighting to set down, before it was too late, their own version of the controversies, topics and operations in which they had earlier been involved. In some cases they have made assessments of contemporary leaders which, in the past, they might have felt inclined to avoid. Time changes attitudes.

It is, however, regrettable – and in some ways surprising – that the Soviet Union has chosen not to contribute to the collection. There was so much for a selected few of her surviving airmen to relate – much, indeed, that the rest of the world would have liked to know of the heroics over the Eastern front.

Through the British staff in Moscow an opportunity was opened to the Soviet defence authorities to embellish a noble record of some forty years ago. They have elected to remain silent, thereby recalling a stance with which the Allied air forces became all too sadly familiar in war.

For the rest, views conflict; opinions diverge; emotions fluctuate; old enmities die hard. Yet, bridging the divide, is the priceless bond of flying which has drawn men intimately together in the face of challenge and which these writings often portray. The link strengthens, not weakens, with the years. It is a human, close and continuing thing. Transcending battle, it represents, for those who have known it, one of life's treasures.

It is, therefore, not surprising that, with these ties holding friendships together, the response, worldwide, to the editor's request for material has been strong and spontaneous. Contributions of one kind or another have far outstripped the limits of space.

By this means much personal and original writing has been obtained which otherwise would have remained unknown.

In this way, too, the record has been served.

Flight

How can they know that joy to be alive
  Who have not flown?
To loop and spin and roll and climb and dive,
  The very sky one's own,
The urge of power while engines race,
  The sting of speed,
The rude winds' buffet on one's face,
  To live indeed.

How can they know the grandeur of the sky,
  The earth below,
The restless sea, and waves that break and die
  With ceaseless ebb and flow;
The morning sun on drifting clouds
  And rolling downs –
And valley mist that shrouds
  The chimneyed towns?

So long has puny man to earth been chained
  Who now is free,
And with the conquest of the air has gained
  A glorious liberty.
How splendid is this gift He gave
  On high to roam,
The sun a friend, the earth a slave,
  The heavens home.

This poem was written in 1938 by Flight Cadet B.P. Young, aged twenty, during his last year at the Royal Air Force College, Cranwell, where he won the Sword of Honour. Born and educated in South Africa, Brian Young was shot down in France on 13 May 1940, in a Hurricane of 615 Squadron during the first days of the German breakthrough in the West.

After months in hospital he returned to operations in 1942 as the captain of a Sunderland flying boat in the Battle of the Atlantic. As a flight commander in 422 (Royal Canadian Air Force) Squadron of Coastal Command he often flew, operationally, the aircraft which visitors to London have quite recently seen moored on the River Thames, within sight of Tower Bridge on the route from Westminster Pier to Greenwich.

Young retired from the Royal Air Force in 1973 with the rank of Air Vice-Marshal.

PART ONE

Beginnings

The interim between the two world wars coincided with the period that 'Johnnie' Johnson has fairly called 'the golden age of flying'. It is right, then, as we prepare to browse through some of the writings of the 1939–45 conflict, to pause and reflect upon the astonishing speed with which aviation developed in those four decades from the lead which the Wright brothers gave it in North Carolina at the start of the century to the sounding of the first air-raid siren in Britain on 3 September 1939.

It was an exhilarating yet tortuous era, stimulated by the impulse which the exchanges between the Royal Flying Corps and the old Royal Naval Air Service on the one hand and the Kaiser's forces on the other gave it during the appalling holocaust of 1914–18. Nothing promotes aviation like war, and the twenty years that followed the armistice of 1918 bear testimony to the fact.

The stirring flights of the great pioneers – Alan Cobham, Charles Kingsford-Smith, Jimmy Doolittle, Amy and Jim Mollison, Jean Batten, Charles Lindbergh and the rest, who, between them traversed the globe.... The opening up of the air routes to Egypt, the Persian Gulf, India and the Far East, and down the length of the African continent to the south.... The crossing of the United States in a day.... And then, finally, the hideous upsurge of military air power in the 1930s as Europe stumbled on fatally towards war – such was the sequence of an age in which a select but growing number of young men and women began to savour what Douglas Bader was always so fond of calling the *fun* of flying.

The Luftwaffe in Germany, the Regia Aeronautica in Italy, the Armée de l'Air in France, the Royal Air Force in Britain, supported by its volunteer components, the Auxiliary Air Force, the Volunteer Reserve and the University Air Squadrons all began to 'muscle up' for the clash which was so soon to follow as time and the decade of the thirties ran out.

There was magic in the names of Richthofen and Bölke, of Bishop, Mannock, McCudden and Ball. It had kindled the imagination of their young successors. The lessons which had been learnt – the hard way – in the skies over France and Flanders twenty years before, had to be rethought, absorbed and memorized as the massive Teuton war-machine rolled westwards across Europe to the sea.

The Germans and, to a much lesser extent, the Italians enjoyed a significant initial advantage over Britain and her European allies. The new Luftwaffe and its Condor Legion had been exposed during the second half of the thirties to the modern face of military air power in the Iberian Peninsula.

The fighting in the Spanish Civil War, and the experience it bequeathed, gave Göring's forces a head start on the rest.

But there were also lessons that Göring and the Luftwaffe General Staff, to their cost, failed to learn. One was that blitzkrieg alone was not enough. Strategic air power was another game. By the time they had learnt its rules, the initiative had passed to the other side.

Foresight is an indispensable asset of war.

Kittyhawk, North Carolina, 17 December 1903

Douglas Bader, *The Twelfth Man* (Cassell Collier Macmillan, 1971). Commissioned by the Lord's Taverners in honour of HRH Prince Philip, Duke of Edinburgh, on his fiftieth birthday

Greek mythology gave us the original aviators, Daedalus and Icarus. Father Daedalus – obviously a founder member of the Royal Aeronautical Society – fixed wings on himself and son, Icarus, and off they both took. Regrettably, Daedalus had done his sums wrong. As a result, so the story goes, the young and press-on Icarus flew too near the sun and suffered wax fatigue of his shoulder harness. Off came his wings and down he fell into the sea. Thereafter this was named the Icarian Sea,* and is to this day witness if I lie. It was a pity Icarus did not last. He had the makings of a fighter pilot.

Down the ages in tale and fable flying has always fascinated man. Yet the great moment did not arrive until 17 December 1903 at Kittyhawk, North Carolina, when Orville and Wilbur Wright achieved the first powered controlled flight. It is true that Montgolfier and others had gone aloft in hot-air balloons or similar devices, but this was the real thing. In the early morning of 17 December Orville actually left the ground and stayed airborne for a distance of 120 feet (40 yards), at a height of probably no more than 6 feet. What a tremendous thrill it must have been. Possibly a greater surge of elation for Wilbur as he stood and watched it happen than for Orville who might not even have been sure he was off the ground until his brother confirmed it. Kittyhawk has not changed. I was there some years ago. It is reminiscent of Rye in Sussex. Flat grassland by the seashore with the sandhills of the golf links behind. In the shelter of a sand dune stands a black wooden hut with a single door in the middle of the front wall, flanked by square glass windows. The door is locked. Through the left window you can see the sleeping room with its two wooden bunks, through the right the living room with a table, chairs and some plates and dishes beside the stove in the back of the room....

As always when one stands on ground hallowed by history, the years roll back and the very scene unfolds before you. The pale morning light, no wind, the sea like glass, the temperature cool. The door of the hut opens and there stands Orville, savouring the full beauty of the dawn with those fresh smells which only the early morning brings. His gaze lingers on the biplane

*Now called the Aegean Sea.

contraption of wood and fabric on which he and his brother have been working for so long.

He turns and speaks into the hut: 'Come on, Wilbur. It's perfect.' Out they go and line up the aeroplane in the most favourable direction. Orville climbs into the seat, the heavy, unreliable engine is started. He opens the throttle and the machine slowly trundles over the flat sage grass. At perhaps 30 m.p.h. it comes off the ground.

As you walk away from the hut by the sea you notice white marker posts like old English milestones; you look down at the first and read on it 'Orville Wright December 17 1903 120 feet'. A little farther on the second post tells of 'Wilbur Wright', same date, a few more feet. You continue a few more yards to the third post which says 'Orville Wright', same date, yet more feet. You walk quite a distance to the fourth and final post and there it is: 'Wilbur Wright December 17 1903 852 feet' – the last and longest of the four flights made by those brothers between dawn and midday. As you slowly turn away, you reflect: 'Two hundred and eighty-four yards – a long hitter can drive a golf ball farther.'

Orville Wright made the first flight of 40 yards at a height of a few feet. Sixty-five years and two world wars later another American stepped on the Moon. There are people alive today who can recall both historic events....

To my mind, one contribution of the Wrights to succeeding generations has been to provide them with the *fun* of flying. In this day and age most people fly: in fact they travel in comfort in a pressurized tube with upwards of a hundred others. That is not flying. In the same way that the sailing enthusiast enjoys measuring his skill against wind and tide in the small boat, the flying enthusiast likes the little aeroplane for the same reasons.

I learned to fly in 1928 when cockpits were open, undercarriages were fixed (as was the propeller pitch), and the instruments consisted of a rev counter, airspeed indicator, altimeter, an oil pressure gauge and a spirit level set horizontally across the dashboard. When you did a correctly banked turn the bubble stayed in the middle. When you hurtled out of a cloud out of control but in a perfectly banked steep spiral the bubble was in the middle also. It was all good fun.

Due to an error of judgement in December 1931, Isaac Newton's law of gravity caught up with me. I left the Royal Air Force in May and joined the great Shell Company in June 1933....

Before I was Four

Air Vice-Marshal D.C.T. Bennett, *Pathfinder* (Frederick Muller, 1958)

...I often wonder whether I was influenced by the great pilots I had seen. Before I was four years of age – and I can remember it – I saw the Wright brothers give a flying display at Toowoomba, South Central Queensland. Later I saw Hinkler arrive on his flight from England, and also Amy Johnson, the first woman to do that flight (she overshot on landing at Eagle Farm and ran through into some corn, where she turned over). I also saw the arrival of Kingsford-Smith at the end of what I consider one of the greatest flights in the history of aviation – across the Pacific....

Eagle in the Sun

Group Captain W.G.G. Duncan Smith, *Spitfire into Battle* (John Murray, 1981; Hamlyn Paperbacks, 1983)

The great shadow sailed across the hillside, floated over my prone body, unsighting me. Startled, I rolled onto my side, eyes sweeping the sky, trying to focus against the deep blue. I let go my ·22 rifle and, shading my eyes, squinted into the bright light. At last I saw him, a golden eagle, wheeling, his head tucked into his body, soaring with effortless ease on widespread wings, bright spots of light glinting on the feathers.

Quickly I resighted the rifle on the hare, 30 yards in front and slightly below my hidden position. It was going to be a race between us, for the eagle had seen the hare too and was now getting into position for his attack. I wanted the hare as much as he did.

The hare wriggled deeper into the grass, ears flat along his back, trying his best to melt into the colours of the hillside. He had not seen me, but he had seen the winged enemy wheeling above and knew what was in store if he so much as moved a whisker. His only chance of survival was absolute stillness.

The eagle marked the spot and, half closing his wings, dived. I held the sights of the ·22 steady on the hare's back, raised them a fraction to allow for bullet-drop and squeezed the trigger. There was a spurt of dust short of the hare. Realizing I had missed, I frantically tried to reload, but it was too late. The smack from the bullet had made the hare jump up and at full speed he was off down the hill, jinking as he ran.

Fascinated, I watched the race for survival. The golden eagle seemed to spin round in his own length. Still diving at incredible speed, his legs dropped rigid, the huge talons wide stretched on his feet. He snapped up the hare with almost casual ease, talons digging into the flesh, and with one swift movement zoomed back into the blue sky, the struggling hare dangling, kicking for dear life. . . .

The eagle's flat, powerful head rose upwards with his razor-sharp beak poised; then with a vicious stroke he hammered into the softness of the hare's neck. He made the kill as surely and painlessly as any well-aimed bullet would have done.

Carrying the limp body the more easily now, the eagle swung in a great arc, climbing for the upper heights of the crags which crowned the steep hillside. I watched motionless as he disappeared, every detail of the scene etched on my memory by the professionalism of that technique. . . .

I realized suddenly, that to soar, zoom and strike, like a golden eagle, would be wonderful. I promised myself that, though I was still a prep-school brat, come what may I would learn to fly.

The Eagle

He clasps the crag with crooked hands;
Close to the sun in lonely lands,
Ring'd with the azure world, he stands.

The wrinkled sea beneath him crawls;
He watches from his mountain walls,
And like a thunderbolt he falls.

Alfred, Lord Tennyson (1805–92)

'Ever the Pioneer'

Antoine de Saint-Exupéry, *Wind, Sand and Stars* (William Heinemann, 1939). Translated from the French by Lewis Galantière. Reproduced by kind permission of the British Publishers' Guild. Chosen by Air Vice-Marshal J.E. 'Johnnie' Johnson

Mermoz is one airline pilot, and Guillaumet another, of whom I shall write briefly in order that you may see clearly what I mean when I say that in the mould of this new profession a new breed of men has been cast.

A handful of pilots, of whom Mermoz was one, surveyed the Casablanca–Dakar line across the territory inhabited by the refractory tribes of the Sahara. Motors in those days being what they were, Mermoz was taken prisoner one day by the Moors. The tribesmen were unable to make up their minds to kill him, kept him a captive a fortnight, and he was eventually ransomed. Whereupon he continued to fly over the same territory.

When the South American line was opened up Mermoz, ever the pioneer, was given the job of surveying the division between Buenos Aires and Santiago de Chile. He who had flung a bridge over the Sahara was now to do the same over the Andes. They had given him a plane whose absolute ceiling was 16,000 feet and had asked him to fly it over a mountain range that rose more than 20,000 feet into the air. His job was to search for gaps in the Cordilleras. He who had studied the face of the sands was now to learn the contours of the peaks, those crags whose scarfs of snow flutter restlessly in the winds, whose surfaces were bleached white in the storms, whose blustering gusts sweep through the narrow walls of their rocky corridors and force the pilot to a sort of hand-to-hand combat. Mermoz enrolled in this war in complete ignorance of his adversary, with no notion at all of the chances of coming forth alive from battle with his enemy. His job was to 'try out' for we others. And 'trying out' one day, he found himself prisoner of the Andes.

Mermoz and his mechanic had been forced down at an altitude of 12,000 feet on a tableland at whose edges the mountain dropped sheer on all sides. For two mortal days they hunted a way off this plateau. But they were trapped. Everywhere the same sheer drop. And so they played their last card.

Themselves still in it, they sent the plane rolling and bouncing down an incline over the rocky ground until it reached the precipice, went off into the air, and dropped. In falling, the plane picked up enough speed to respond to the controls. Mermoz was able to tilt its nose in the direction of a peak, sweep over the peak, and, while the water spurted through all the pipes burst by the night frost, the ship already disabled after only seven minutes of flight, he saw beneath him like a promised land the Chilean plain.

And next day he was at it again.

When the Andes had been thoroughly explored and the technique of the crossings perfected, Mermoz turned over this section of the line to his friend Guillaumet and set out to explore the night. The lighting of our airports had not yet been worked out. Hovering in the pitch-black night Mermoz would land by the faint glimmer of three petrol flares lined up at one end of the field. This trick, too, he taught us, and then, having tamed the night, he tried the ocean. He was the first, in 1931, to carry the mails in four days from Toulouse to Buenos Aires. On his way home he had engine trouble over a stormy sea in mid-Atlantic. A passing steamer picked him up with his mails and his crew.

Pioneering thus, Mermoz had cleared the desert, the mountains, the night, and the sea. He had been forced down more than once in desert, in mountain, in night and in sea. And each time that he got safely home, it was but to start out again. Finally, after a dozen years of service, having taken off from Dakar bound for Natal, he radioed briefly that he was cutting off his rear right-hand engine. Then silence.

There was nothing particularly disturbing in this news. Nevertheless, when ten minutes had gone by without report there began for every radio station on the South Atlantic line, from Paris to Buenos Aires, a period of anxious vigil. It would be ridiculous to worry over someone ten minutes late in our day-to-day existence, but in the airmail service ten minutes can be pregnant with meaning. At the heart of this dead slice of time an unknown event is locked up. Insignificant, it may be; a mishap, possibly: whatever it is, the event has taken place. Fate has pronounced a decision from which there is no appeal. An iron hand has guided a crew to a sea landing that may have been safe and may have been disastrous. And long hours must go by before the decision of the gods is made known to those who wait.

We waited. We hoped. Like all men at some time in their lives we lived through that inordinate expectancy which like a fatal malady grows from minute to minute harder to bear. Even before the hour sounded, in our hearts many among us were already sitting up with the dead. All of us had the same vision before our eyes. It was a vision of a cockpit still inhabited by living men; but the pilot's hands were telling him very little now, and the world in which he groped and fumbled was a world he did not recognize. Behind him, in the glimmer of the cabin light, a shapeless uneasiness floated. The crew moved to and fro, discussed their plight, feigned sleep. A restless slumber it was, like the stirring of drowned men. The only element of sanity, of intelligibility, was the whirring of the three engines with its reassuring evidence that time still existed for them.

We were haunted for hours by this vision of a plane in distress. But the hands of the clock were going round and little by little it began to grow late. Slowly the truth was borne in upon us that our comrades would never return, that they were sleeping in the South Atlantic whose skies they had so often ploughed. Mermoz had done his job and slipped away to rest, like a gleaner who, having carefully bound his sheaf, lies down in the field to sleep....

Pupil Pilot

Wing Commander H.R. 'Dizzy' Allen, *Fighter Squadron* (William Kimber, 1979)

On all accounts it was essential to avoid being given a rating as an 'exceptional' pilot at the end of the course, because the tendency then was to take exceptional *ab initio* pilots and groom them to be flying instructors in their own right. The last thing I wanted was to be selected for Flying Training Command.

That Carr was an exceptional pilot was in no doubt, not even in my innocent eyes. He taught me aerobatics – or as we called them in those days acrobatics – but I made sure I was never perfect in loops, slow rolls, flick rolls, although I was always in a hurry to get out of a spin, because an aircraft in a spin is literally out of control and I never enjoyed not having full command over my various aeroplanes. . . .

When I had achieved 5 hours flying Carr decided that I should fly solo. . . . He watched me strap myself in, muttered 'Good luck' and wandered back to his hut, lighting his pipe as he did so. Flying instructors always try to appear unconcerned when they send their trainees off on their own, but in fact they tremble like they've got St Vitus dance until he lands safely. Carr didn't fool me! Five hours dual was pretty fast to be sent solo; some people take 8 or even the top limit, 10 hours dual, before they are sent off. Some, of course, are never sent solo, I suppose 25 per cent of my course failed to get into the air on their own.

I wasn't afraid to go solo because I thought I would make it. But it was an odd experience when one turned one's head and noted that Carr was not sitting in the back seat. I also observed when I took off that the trim of the aeroplane was slightly affected since there was not a 12-stone man sitting behind. But as I climbed I felt free as air for the first time – with the blue above and the green below. There was no Gosport tube and there was no Carr bawling down it. If I felt like diving into the ground or colliding with a big Scottish mountain, no one could say nay to me. I climbed to about 5000 feet stuffed the nose down and performed a rather nasty loop. This was strictly unauthorized, but no one could see me for there was a layer of thin cloud keeping me invisible from Carr's searching eyes. Then I practised a few sideslips, high above the ground. I loved sideslipping; I even sideslipped the big jet fighters I went on to fly years later. But in the Tiger Moth with its open cockpit, one could feel the gust of air on one's face as one proceeded along sideways, a most exhilarating sensation. . . .

When I had been in the air for my allotted time, I decided to come in too high on my approach for landing just to give Carr a fright. At about 500 feet I sideslipped to the right and then to the left, straightened her out and touched down on the grass just alongside Carr's office. It wasn't a bad landing. I taxied her in and Carr came leaping along like an injured rhino.

'What the bloody hell do you think you're doing?' he shrieked.

'Sorry sir,' I replied, 'I made an error of judgement and had to slip off a bit of height.'

'Slip off a bit!' he screamed. 'You slipped off five hundred feet of bloody height!'

'Well I got down in one piece, didn't I?' I remarked mildly.
He grunted.
'Come along to the mess,' he said. 'You didn't do my ulcer much good.'
In about 1941 a disillusioned flying instructor, who was also no mean poet, wrote a poem for *Punch*. An extract reads:

'What did you do in the war, Daddy,
How did you help us to win?'
'Circuits and bumps by the score, Laddy,
And how to get out of a spin.'

Carr did not give me an exceptional assessment, neither did I deserve one. I just wanted to be a fighter pilot.

Awakening

Group Captain Leonard Cheshire VC, *Bomber Pilot* (Hutchinson, 1943; White Lion Publishers, 1973)

... As an illusion it did not last long, no longer than it took me to realize that bombing is technical, a matter of knowledge and experience, not of setting your jaw and rushing in. And when you have the knowledge and the experience, the crux of the issue is crew cooperation. To achieve it you have to lay the foundation of confidence: confidence in the crew partly, and partly confidence in yourself, but more than anything confidence in the captain. Personally, I would prefer to be a lone wolf, but since this is impossible I pray God I may see cooperation at its best. To see it there is only one way. You have got to be good, and they have got to know it, and they will only know it by results: for once you cannot bluff. It is not difficult, anyone can do it; merely a matter of working hard enough, but, as I said, you cannot bluff. I trust Lofty implicitly; I know what that trust is worth. It is worth working for.

First and foremost I had to learn to fly; learn, and then cast the thought of flying away into the background. Flying in itself is wholly unpredominant: to have a perfect pair of hands is important, but it is only a question of degree, not the end-all and be-all. Smooth landings do not affect the success of an operation; it is finding the right way to the right place that matters. In other words, flying must be subconscious. A soldier does not learn to walk or he would not be a solider, but he learns to walk far and in the right direction. Lofty let me fly all I wanted, at first over the sea and over England, then, as I grew more experienced, over Germany, and once over the target itself. When I was not with Lofty I took an aircraft and flew myself, by day and sometimes by night as well. I practised flying relaxed till I found it no more tiring than sitting down. I learned to fly on instruments till it was no more tiring than reading a book. At the same time I settled down to learn the weapons I handled, what there was and where it was. I blindfolded myself and moved round the aeroplane until I could lay my hands on everything without the use of eyesight. I learned to load the guns, to clear stoppages, to set the bombsight, and what to do with the selector switches. I learned about engines, too: how to take care of them: their limitations and their possibilities. Once we had engine trouble over the North Sea. The oil temperature rose and the engine revolutions flickered. Lofty looked worried and asked me whether we ought to turn back. I realized then that this was a situation I could not meet. To turn back for no good reason is a crime, and anyway you will never persuade people you were not yellow; to carry on with no chance of return – suicidal: just as well to know, but I knew nothing about engines. For this I turned to the ground crews, and their anxiety to teach me was the greatest stimulus I could ever want....

Oxford

Air Chief Marshal Sir Christopher Foxley-Norris, *A Lighter Shade of Blue* (Ian Allen, 1978)

Lifelong friendships. But how sadly in so many cases the life was not long. Alex Obolensky; Mike Marshal; Richard Hillary; Melvin 'Dinghy' Young; Noel Agazarian. So many others who went so quickly and so prematurely. Of all my friends, perhaps the most improbable survivor because the most harebrained risk-taker was Leonard Cheshire, who proved even more improbably in later life to be consecutively the greatest bomber pilot of the war and perhaps the best candidate for canonization of our generation. If someone had told me then that forty years later I should find myself chairman of a global charitable foundation, maintaining one hundred homes for the sick and the relief of suffering and founded by the selfsame Cheshire, I doubt if I would have given him much credence.

In 1939 one's days, however, could still be carefree and it was with real regret that I finally left the University and its air squadron. Having won a Harmsworth Scholarship to the Middle Temple I had stayed up at Oxford an extra year to read for the Bar; and thereafter I moved to London to read for my finals. I also planned to join No. 615 (County of Surrey) Auxiliary Squadron to continue my flying, if they would have me, for the Auxiliary Air Force was a *corps d'élite*, and competition to join was intense.

But greater men were making larger plans in wider fields, and mine duly went the way of those of mice and men.

Golden Age of Flying

Air Vice-Marshal J.E. 'Johnnie' Johnson, *Full Circle* (Chatto and Windus, 1964)

...For the pilots this was the golden age of flying, when their bright, stubby biplanes, only a few miles faster than the SE5, handled beautifully and, in the rare event of an engine failure, could be safely put down on a soccer field. One young pilot, on being admonished by his flight commander for a poor landing, replied, 'Well, sir, I was landing with my left hand in case my right ever gets shot off!' Aerobatics gave ample scope for brain and artistry, and there were opportunities for skilled pilots to excel, for Harry Broadhurst and Teddy Donaldson to fight out the gunnery contest and for Dick Atcherley and Dermot Boyle to perfect the art of pure flying.

The annual display at Hendon expressed the junior Service's lighthearted approach to life. The Prince of Wales arrived in his private aeroplane to join the vast crowd that watched air refuelling, autogiros for army cooperation, bombers from some of the Auxiliary squadrons trailing smoke across the sky, teams of dumpy, chequered fighters, tied together with coloured rope, looping and stunting over the airfield and landing in immaculate formation with the rope unbroken. . . . Imperialism was unfortunately portrayed when the fighter boys beat up with flour bags a native fort whose coloured occupants were about to commit some dastardly crime. After the natives had abandoned the fort it was destroyed by means of a small quantity of explosive which went off with a large bang. Unfortunately, on one occasion the timing of the various events went astray and the fort was demolished just as Dick Atcherley was flying upside down a few feet overhead. The blast blew him the right way up and Atcherley and the display proceeded. . . .

When the Luftwaffe was rapidly growing Britain rated, thanks to our peace-at-any-price policy, as the world's fifth air power and when, in the mid-thirties, the RAF was expanded to meet the German threat our plans not only included the regular forces but fortunately provided for a cheerful and growing band of weekend flyers. The bomber squadrons of the superb and exclusive Auxiliary Air Force were converted to fighters, so that by the outbreak of war no less than one-third of our fighter squadrons were Auxiliaries, and this *corps d'élite* was reinforced as the expansion proceeded by another enthusiastic community of amateur pilots, the Royal Air Force Volunteer Reserve. . . .

'The Weekend Flyers'

Group Captain Sir Max Aitken, 'The Aitken Papers', London, 1982

. . . I was introduced to aerial warfare at the age of five – or maybe six. One night in Cherkley, our house at Leatherhead, I was roused by the insistent pealing of the firebell. My father made me get up from bed and dress. Then he took me up to the top of the downs nearby. And there it was! In the sky, just overhead, lit up by the beams of the searchlights was a long, silvery cylindrical shape. I heard the drumming of its engines. It was a zeppelin on its way to bomb London. I was very frightened, as I remember, although, in fact, the airship was not likely to waste a bomb on the empty darkness of the Surrey hills. But at that age I was so impressed by the menace overhead, the killer in the sky, so large, so invulnerable, that I was in no state to work out the probabilities. My next meeting with air warfare was different; I was the menace in the sky. But that was thirty years later.

One day in 1934 a friend of mine, Roger Bushell, said to me 'Why don't you join the Auxiliary Air Force, Max?' Bushell was one of a group of excellent skiers of my acquaintance, an adventurous, hell-raising collection of men who, at the right time of year, would cross the Channel with their cars and drive fast down the highway to St Anton, where I had a house. Bushell also flew. The idea of flying an aircraft attracted me very much and when Bushell said, 'I'll arrange for you to meet Philip Sassoon at lunch,' I at once agreed. Sir Philip Sassoon at the time was Under-Secretary of State for Air, with a special interest in the Auxiliary air squadrons. I met him, liked him very much and, in consequence, I found myself posted to 601 Fighter Squadron with its headquarters at Hendon. My companions there were, as you would expect, a pretty wild and high-spirited gathering many of whom I already knew from skiing – and after-skiing – parties at St Anton. They were the sort of young men who had not quite been expelled from their schools, whom mothers warned their daughters against – in vain – who stayed up far too late at parties and then, when everyone else was half dead with fatigue, went on to other parties. Does that sort of young man still exist? I do not know. But in those days they were quite common. And they clustered in unusual density at the headquarters of 601 Squadron. . . .

No. 603 City of Edinburgh (Fighter) Squadron, Royal Auxiliary Air Force

Wing Commander J.L. Jack, A Record of the Service and Achievements of No. 603 (City of Edinburgh) Squadron, Royal Auxiliary Air Force, 1925–57. 1960

Nominal Roll of Officers as at 3 September 1939

Rank Sept. 1939	*Name*	*Rank* Dec. 1942	*Civil Occupation*
S/Leader	E.H. Stevens	G/Capt.	Writer to the *Signet*
S/Leader	Rev. J. Rossie Brown	W/Com.	Chaplain
F/Lieut.	J.L. Jack	W/Com.	Bank Agent
F/Lieut.	I. Kirkpatrick	W/Com.	Writer to the *Signet*
F/Lieut.	P. Gifford	Killed	Solicitor
F/Lieut.	G.L. Denholm	W/Com.	Timber Merchant
F/Lieut.	I.A.G.L. Dick	W/Com.	Surgeon
F/Lieut.	T.O. Garden	S/Leader	Chartered Accountant
F/Officer	F.W. Rushmer	Killed	Electrical Engineer
F/Officer	H.K. Macdonald	Killed	Writer to the *Signet*
F/Officer	J.G.L. Cunningham	Killed	Grain Merchant
F/Officer	G.T. Wynne-Powell	S/Leader	Apprentice Chartered Accountant
F/Officer	R. McG. Waterston	Killed	Engineering Student
F/Officer	J.G.E. Haig	S/Leader	Paper Maker
F/Officer	I.S. Ritchie	S/Leader	Writer to the *Signet*
F/Officer	J.A.B. Somerville	Killed	Bank Clerk
F/Officer	C.D. Peel	Killed	Apprentice Chartered Accountant
P/Officer	G.C. Hunter	S/Leader	Apprentice Chartered Accountant
P/Officer	G.K. Gilroy	W/Com.	Sheep Farmer
P/Officer	J.S. Morton	S/Leader	Mining Engineer
P/Officer	D.K.A. Mackenzie	Killed	Apprentice Chartered Accountant
P/Officer	W.A. Douglas	F/Lieut.	Student
A/P/O	C.E. Hamilton	Killed	Student
Pupil Pilot	R. Mackay	F/Lieut.	Travel Agent

Lord Trenchard's Gospel

Air Chief Marshal Sir Wallace Kyle in a Foreword to *The Golden Eagles* by Peter Firkins (St George Books, Western Australia, 1980)

Without exception these Australians were young men of immense resource and true courage. They responded to the opportunities created by the Imperial authorities when air warfare was developing during World War I and later when independent air forces became part of the defence organization of Britain and Australia.

In Britain, Lord Trenchard preached the gospel of a separate and autonomous third service and the indivisibility of air power. The creation of the Royal Air Force was the direct result of this doctrine. Australia, through Sir Richard Williams, was an ardent supporter of that theory. Thus a close association between the Royal Air Force and the Royal Australian Air Force was automatic and to their mutual benefit.

Methods of flying training, in air tactics and strategy, developed along the same lines and the stage was set for Australian and other young men of the Commonwealth who wanted to join one of the new flying services and see the world.

And so it was that more than four hundred Australian pilots were in the Royal Air Force at the outbreak of World War II and the machinery existed for others, including complete squadrons, to join in the early battles.

I think it is and will remain a historical fact that the period when adventurous men had the opportunity to display a combination of personal bravery, skill and special prowess in air operations, which made them stand out as individuals, was a brief one. . . .

Those of us who were privileged to serve our country in those thrilling days will regret the passing of that era. Perhaps we should take comfort in the thought that world wars are less likely because the power and accuracy of modern weapons will serve as deterrents to any nation risking another big war. . . .

Government House, Perth
Western Australia

The Rise of the German Air Force

With the accession of Hitler and the Nazi Party to supreme power in Germany there began, in the fateful 1930s, the build-up of the new Luftwaffe under Göring. Initially, the process was wrapped in secrecy. The story of the remarkable development of the Fighter Arm was told, during prolonged interrogation in 1945, by General Adolf Galland, 'the man who, more than any other, shaped its destiny and led it'.

Here is a condensed version of what those who questioned him after the war wrote about the critical formative stages. The preamble is set in generous terms which pay tribute to the achievements of this impressive officer.

Starting as a glider enthusiast under the Deutsche Luftsport Verband, becoming a civilian Lufthansa pilot, and finally entering the new Luftwaffe under circumstances of comic-opera secrecy, Galland flew in the Legion Kondor in the Spanish Civil War. In the first two years of World War II he became one of Germany's greatest fighter aces, with claim to over 100 victories. In December 1941, he was promoted to General der Jagdflieger (AOC Fighters) where he had a great influence on operations in his function as arbiter of tactics, training and equipment for the fighter force.

Throughout the course of the interrogations Generalleutnant Galland has demonstrated a passion for objectivity, a respect for truth, and a searching self-criticism which lend his statements an added credibility.

Family Background

The circumstances of the General's family were briefly described:

On the paternal side, Galland's forbears have, for almost two hundred years been estate directors for a titled family, the Westerholt-Giesenberg line, at Westerholt in the Ruhr. Oddly enough, the first Galland came to Germany as a Huguenot refugee in the seventeenth century, so that in terms of discredited racial science, the present Gallands are more or less non-Nordic. The mother of our subject came from a fairly well-to-do business family in Bochum. Both sides of Galland's family are still devout Roman Catholics, but he himself takes no notice of religion at all. Adolf had three brothers; all became fighter pilots. The two younger ones were killed in action and the eldest ended the war as a reconnaissance pilot in Italy.

CIVILIAN FLYING AND SECRET MILITARY DAYS

Training for the Lufthansa

The sequence of events leading up to the opening of the Spanish Civil War was varied and progressive.

The best opportunity for learning powered flying seemed to be to enter the Deutsche Verkehrsflieger Schule (German Airline Pilot School), which had branches at Braunschweig/Broitzem, Warnemunde, Schleissheim and List on Sylt. Competition for entrance was stiff, and 100 men were chosen from among 20,000 volunteers for the examination.

The training course at Broitzem for beginners was very strict. All wore a uniform blue civilian dress, lived in fine clean barracks and had high morale. They were continually threatened with expulsion if they relaxed for one minute.

The school had an enrolment of about fifty beginners, a number of advanced pupils flying heavy aircraft, and a special secret course for 'discharged' Reichswehr officers also on heavy aircraft. Relations between these officers and the bona-fide civilians were not good. By the end of 1932 50 per cent of the beginners had failed the course.

First Military Contacts

Galland's first contact with military flying was begun with an unexplained order for himself and four other trainees to report to the Zentrale der Verkehrsflieger Schule in Berlin. Upon arrival, they were asked if they would care to participate in a secret training course for military aviators. All accepted and they were sent in civilian clothes to Schleissheim, near Munich.

The training at Schleissheim was all under the auspices of the Verkehrsflieger Schule and was supervised by a Major Beyer, who had himself been trained in a clandestine programme in Russia. (According to other German Air Force generals, such a programme took place at Lipetsch, Russia from 1926 to 1932. German civilians and 'furloughed' Reichswehr officers were trained in three groups by Russian and German fliers.)

In May 1933, after Galland finished the short course at Schleissheim, he attended a most secret meeting of seventy-odd young fliers, all from the Verkehrsflieger Schule or other illicit training courses, at the Behrendstrasse, Berlin. Most of the men were officers or NCOs lately with the Reichswehr. The main attraction of the meeting was a speech by Göring inviting the assembled fledglings to visit Italy for a 'marvellous' course of fighter training to be given by the Italian Air Force. The journey was to be accomplished in civilian clothes although all but twelve, one of whom was Galland, were either technically or actually soldiers.

Upon arrival the Germans were rushed to the airfield and locked up in barracks, where they were given worn-out uniforms of Italian Air Force candidates. For two months they remained on the field and were under constant observation.

In September 1933 they all returned to Germany, disillusioned and sunburnt. Galland believes that the Italians were really good fliers and had good fighter units, but he does not believe that the German Air Force benefited by Italian experience at all.

Airline Flying

Back at the Verkehrsflieger Schule in Breitzen, Galland got his first experience in multi-engined transport machines. In October 1933 he began flying on a regular Lufthansa passenger route as second pilot. His usual route was Stuttgart, Geneva, Marseilles, Barcelona. This routine work did not hold him for long, for in December 1933 he was called to school headquarters in Berlin and asked if he would care to become a soldier and eventually enter the still unborn air force. To the delight of his father, he accepted and awaited the call-up.

THE NEW LUFTWAFFE

Preliminary Training

In February 1934 Galland reported at Inf. Reg. 10 at Dresden. In that unit were fliers of all types, including airline pilots, weather fliers, club fliers, and some veterans of the Reichswehr. The oldest was thirty-eight and Galland, at twenty-one, was the youngest. After three months of basic infantry training most were promoted to Fahnejunker Unteroffizier (officer candidate) and sent to the Kriegsschule (officers' school) at Dresden, where they received the usual officer training with an air force flavour but all disguised as 'sports training'.

The entire school was under the control of the army and hence the role of the future independent air force was overshadowed by the importance of army cooperation.

In the autumn of 1934 the officer training came to an end and the men were discharged from the army to be absorbed into the still secret air force. Ranks from Leutnant to Hauptmann were granted immediately, depending upon experience and age.

The Luftwaffe had already set up several flying schools, and Galland was posted to one at Schleissheim. An intensive training programme was in progress.

The Unveiling and Growth of the Fighter Arm

In February 1933 Göring put in an appearance at Schleissheim, and, to everyone's delight showed off the first Luftwaffe uniform, essentially like the later ones. He announced that the whole Luftwaffe would soon be brought out into the open and, indeed, in March all restrictions on publicity of a favourable kind were lifted.

Spain

With the Spanish Civil War now well under way, new opportunities were opening for enthusiastic and ambitious young pilots.

Entranced by stories of flying and good pay in Spain, Galland volunteered for the Legion Kondor and was ordered to Doberitz, where a troop of three hundred and seventy Luftwaffe men had been assembled to make the trip. Dressed in civilian clothes and equipped with civilian papers, they went to Hamburg by rail and embarked on a tramp steamer for an unknown port in Spain. On the twelfth day they disembarked at El Ferrol.

Ground Attack Flying

At Vittoria, Galland was given command of the Stabs-Kompagnie of the German Fighter Gruppe J/88 and spent a few weeks supervising the flying in of new aircraft, setting up repair stations, and organizing motor transport. He was then given command of the 3rd Staffel of the Gruppe, equipped with the Heinkel 51 biplane. The other two Staffel used the new Me 109.

The role of 3/J/88 was not strictly that of a fighter unit; it was actually the first Schlachter or ground attack unit of the modern German Air Force.

Throughout the time Galland was in Spain, he experienced little fighter combat. Republican fighters rarely crossed the front looking for trouble and 3/J/88 was rarely molested. Whilst Galland and his He 51s had been mainly concerned with flying in direct support of ground operations, the other Staffeln of J/88, equipped with Me 109s, had been flying some pure fighter sweeps and escort missions. On some occasions sweeps were flown to lure the Republican fighters into the air and when the Russian and Republican fliers rose to the bait, large air battles resulted. The Russians flew well and showed thorough training.

It can definitely be said that the Luftwaffe utilized the Spanish Civil War as a proving ground for a great number of new ideas and drew much valuable information from the study of results.

Tactical Conclusions

Galland believes that the Spanish Civil War clearly showed the course of aerial warfare as it came to pass in World War II, but the lessons learned by the operational fliers were not heeded by the Luftwaffe General Staff, which continued to dream of unescorted and unopposed clouds of bombers laying waste any and every country which might engage the German colossus. Unfortunately for Germany, the weak air opposition later experienced in Poland in 1939 and in France in 1940 tended to confirm those misconceptions, whereas experience in Spain had actually proved the folly of unescorted bomber missions in daylight. It so happened that the He 111 bomber was fast enough to outdistance most of the fighter aircraft which appeared in Spain, and the Luftwaffe thought that this superiority would forever be a decisive factor in the employment of bombers. Later, over England, they learned how wrong they were.

As a result of experience in Spain the German Fighter Arm stopped flying in close formation made up of elements of three. The great speed of modern fighters made World War I close formations impractical.

The best organization for air combat at the time was judged to be the Staffel of twelve aircraft, made up of three Schwarm (elements of four).

Galland asserts that the RAF did not adopt the loose formation until well into the Battle of Britain and that the Italians never gave up their old close formation.

The principle of loose formations based on the element of two, the Rotte,* and the element of four, the Schwarm,† was incorporated by Mölders, the great German ace who relieved Galland in Spain as chief of 3/J/88, in a report on fighter experience in the Peninsula. When Mölders returned to Germany at the close of the Spanish War in 1939 he was posted to LI3, the Inspectorate of the Fighter Arm, and commissioned to analyse the experience gained in the Luftwaffe's baptism. Mölders included in his report a recommendation that all fighter formations should fly with their elements at varying heights for mutual cover and mutual visibility as well as to render the formation less conspicuous. This principle was also adopted.

Mölders formulated at the same time a conception of the ideal fighter aircraft, which was much talked of among fighter pilots at the time. The various features are recorded below in order of precedence:

(a) Speed and climbing. All agreed that these were the most vital considerations.
(b) Manoeuvrability. Mölders deemed this quality more vital than firepower, which follows, but Galland disagreed and now claims that the later increase in armament on Allied bombers bore him out.
(c) Firepower. Despite relegation of this feature to a lower position, German fighter armament grew steadily and always tended to use larger calibres.
(d) Endurance. This quality was given little attention. Germany's central position in Europe and the general undervaluation of the importance of fighters caused the Luftwaffe to neglect the development of long-range fighters. Galland admits that no one saw the need before it was too late.

*Rotte: A pair of fighters flying in line abreast, 200 yards apart.
†Schwarm: A section of four fighters, flying in line abreast, 200 yards apart i.e. two pairs (Rotten) flying together. This became the basis of all fighter operations in the German and Allied air forces. (Ed.)

'Munich'

On 1 October 1938, Neville Chamberlain, the British prime minister, returned to London by air from Munich after his meeting with Adolf Hitler, the Nazi leader. On the tarmac at Heston Airport he claimed that, by his deal with the German Chancellor over terms for Czechoslovakia, he had gained 'peace for our time'. When he got back to Downing Street he added that he had brought 'peace with honour'.

Four days later on 5 October, in the debate in the House of Commons on the Munich Agreement, Winston Churchill delivered a searing onslaught on the Government. In the euphoria which had been created by Chamberlain's announcement, he ended his speech with a warning and a call.

I do not grudge our loyal, brave people ... the natural, spontaneous outburst of joy and relief ... but they should know the truth.

They should know that there has been gross neglect and deficiency in our defences; they should know that we have sustained a defeat without a war, the consequences of which will travel far with us along our road; they should know that we have passed an awful milestone in our history, when the whole equilibrium of Europe has been deranged, and that the terrible words have for the time being been pronounced against the Western democracies: 'Thou art weighed in the balance and found wanting.' And do not suppose that this is the end. This is only the beginning of the reckoning. This is only the first sip, the first foretaste of a bitter cup which will be proffered to us year by year unless by a supreme recovery of moral health and martial vigour, we arise again and take our stand for freedom as in the olden time.

Winston Churchill, Hansard, 5 October 1938

PART TWO

Tumult in the Clouds 1939–40

The invasion of Poland on the first day of September 1939 and the German blitzkrieg which sustained it, brought the Polish people under the heel of the jackboot and touched the spark that lit World War II.

There followed a winter of unnerving quiet. Fantasy accompanied inaction. There was wistful talk of a 'negotiated peace', of the war being 'over by Christmas' – without being too sure which Christmas. Rumours of 'overtures' abounded; happy, pre-war days would soon return. Unreality was everywhere.

The less gullible were not so sure. The rugged Czechs and Poles, having seen the rules of the new game, concluded that they wouldn't be playing to the dictator's whistle. Many a young air force officer left his home and his country to take the fight to the enemy elsewhere.

Then, in May 1940, Western Europe erupted to the crack of thunder as German panzers, supported by a well-drilled air force, drove on westwards towards the sea, cutting multiple tracks through the opposing forces, soon to reach the Channel and, later, the Atlantic coast. Blitzkrieg had scored again.

In Denmark and Norway, enemy troops, with massive air backing, had already fought their way relentlessly northward. What Churchill had warned against for much of the thirties had now come to pass. By the end of June 1940, from the North Cape to the Pyrenees, the Wehrmacht possessed the field – and Göring's Luftwaffe the air.

The air fighting in France and Norway had revealed manifest Allied weaknesses – ill planning, poor or non-existent communications, lack of coordination, an absence of effective radar and ground control, obsolete aircraft. Mistakes abounded – but so did courage.

True, the evacuation from Dunkirk had been a God-given deliverance. Like the Falklands campaign forty-two years later, it was hard to comprehend how we had been allowed to get away with it. Even so, we were back on our heels against the ropes. The aggressive contender for the title glowered at us, poised to move in for the kill.

Joe Kennedy, the United States' ambassador to the Court of St James, buddy of Ribbentrop and one of Wall Street's smarter slickers, wrote Britain off and advised his president that John Bull was through. He wasn't alone in the error. Independent advice was soon being heard. Sue for peace, it said; cut your losses and make a deal. It's pointless to slug it out.

But that wasn't the way Churchill and his ministers in the new Coalition saw it. Nor was it at all how the Royal Air Force, soon to be strengthened by the arrival of Allied and Commonwealth pilots and, perhaps most of all, by

Beaverbrook's astonishing flow of aircraft from the factories, viewed it. Dowding at Fighter Command, with undramatic courage, had stuck fast against pressures to release more squadrons to prop up the faltering French. Defeat was never in the heads of his pilots as they braced themselves for the attacks which must come. In Keith Park, the New Zealand commander of 11 Group, who stood eyeball to eyeball with the enemy across the Channel, the C-in-C had the outstanding fighter tactician of the war.

In all the controversies that were to follow victory in the Battle of Britain, few doubted that, while Hugh Dowding had prepared the way, it was Park's genius which won the day.... This, and the German High Command's single, fatal blunder in switching the attack, early in September, away from the airfields of Sussex, Kent and Essex, onto London and the major cities when the Luftwaffe was within – what? – maybe a fortnight of success.

As 1940 closed, the long climb back from near disaster had begun. Already, in the air, we were beginning to look outwards to the importance and needs of other theatres – to the Atlantic, the Mediterranean, the Western Desert, East Africa and beyond. Two years and more, however, were to pass before the bombing offensive against Germany itself could be said to be an effective reality.

We had won a breathing space. But as we waited for the bell for the next round, anxious ringside glances were still being cast towards our corner.

FANTASY ... AND REALITY

'Moving Battalions means War....'

Group Captain Sir Max Aitken, 'The Aitken Papers', London, 1982

... In England there were already rumblings of war. It would be absurd to pretend I had any particular premonition of what was coming. But, on the other hand, what was I doing training in aircraft; why had I been recruited so persuasively by Roger Bushell all those months ago; what was at the back of Philip Sassoon's mind when he worked so hard to build up the Auxiliary Air Force? Somebody else had no doubt at all of what lay ahead. When we, 601 Squadron, were at Biggin Hill, Winston Churchill used to come over sometimes from Chartwell. He was an honorary air commodore and took particular pride in wearing the uniform. We looked on him as a father figure. One day in the mess, I asked him, 'Do you think there is going to be a war?' He took the cigar out of his mouth. 'Yes,' he said. 'Moving battalions means war and that is what the Germans are doing. Moving battalions nearer the French frontier. So be ready, be ready!' I was not taken entirely by surprise when war broke out.

Czechoslovakia

Squadron Leader Marcel Ludikar, Free Czech Air Force Archives, Northwood, 1982

In 1938, the Czechoslovak Army Air Force was one of the most modern air forces of Europe, determined to fight in the threatened conflict for the freedom and integrity of its country. However, the morale of its personnel sunk from the highest peak to the deepest depression when, by the Munich Agreement, the great powers, Great Britain, France, Germany and Italy crippled Czechoslovakia and rendered it defenceless. Six months later, on 15 March 1939, the remainder of the country was occupied by Hitler's forces without a shot being fired.

On the last day of March, former Czechoslovak airmen left their stations and airfields for the last time to disperse to their homes. On one of the fighter stations, Major Alexander Hess, the Commanding Officer, bid his airmen farewell ending '. . . and I trust, as I know you and see you here in front of me, that we shall all meet again soon to fight our enemy as we have sworn on oath. I believe that we, airmen, have not yet lost our fight.' Many unit commanders spoke to their airmen in similar terms.

Fighting On

Frantisek Fajtl, Praha, Czechoslovakia, 1982

Frantisek Fajtl saw it all – or most of it. One of Czechoslovakia's first air force officers to leave his native land in 1939 after the Nazi invasion, he made his way through Poland to France where he fought with the French Air Force in the spring and summer struggle of 1940 until the collapse. Slipping away to North Africa, he then moved on to Britain to play his part in the September battles against the Luftwaffe.

Fajtl was the first Czech to command a British squadron (No. 122). He was shot down over France in 1942, evaded capture and, with the help of the French Underground, found a passage to Spain only to be clapped into a concentration camp. After organizing his release he returned to Britain, ultimately to take over one of the Czech squadrons in the Royal Air Force.

In 1944 he was assigned, with twenty other pilots, to the Soviet Union for the purpose of forming the First Independent Czechoslovak Air Regiment, equipped with Russian aircraft. This he commanded on the Eastern front until the end of the war.

Since those days, Frankie Fajtl has become a successful writer in Czechoslovakia. Here, he remembers the first months of Nazi domination of his country and an incident which contributed to his decision to quit and link up with the Allies.

For me, the war started long before hostilities broke out.

It all began on 15 March 1939 in the small town of Prerov, in Moravia, Czechoslovakia. Mirek Holomucky and I (we were both lieutenants) were sitting with our commanding officer, Major V. Skoba, on Prerov airfield awaiting the arrival of officers of the German Forces of Occupation who were to take over the aerodrome. Eventually, two Luftwaffe colonels, accompanied by Lieutenant Schubert, arrived. They were very polite and said they were sorry to have to perform this duty. With that they left, leaving Schubert to finish the job of possessing the airfield.

Like the whole Czechoslovak nation we were sad and silent. 'Frankie and

Mirek,' said Major Skoba, 'come to the café tonight.' We both agreed.

Major Skoba was in mufti and Holomucky and I were in uniform, as we sat that evening drinking wine at our favourite table in the café.

About half an hour later three Germans, a sergeant major, a corporal and a private soldier marched in and sat at a table close to ours. The sergeant major undid his belt with a revolver holster attached to it, and threw it noisily on the floor. His companions did the same. Although already drunk, they ordered six cognacs. Meanwhile a small group of people came into the café and joined our table. There was a girl among them with whom we danced. Suddenly the sergeant major got up (it was a wonder he did not fall) and asked the girl to dance, but she refused. Crimson in the face, he made his way back to his table where he ordered more cognac. Noisily, they then left the café.

At about half past ten Holomucky and I took leave of our friends. As soon as we were in the street we heard a hoarse voice roaring at us through the darkness: 'Halt. Hände hoch!'

'What's that?' asked Holomucky.

'Hands up,' I explained.

'Hands up, you swine,' yelled the sergeant major who was with his gang.

'Mirek, do as I do,' I whispered, putting my hands up. 'What is the meaning of all this?' I asked the Germans directly.

'Shut up, and take two steps back to the wall,' came the answer.

The leader followed us, waving a revolver like a madman; the other two Germans also aimed revolvers at us.

'This won't do, Frankie,' whispered Holomucky. A moment later he almost lost control of himself, saying to me: 'I am going for them.'

'Stop, Mirek,' I said, 'and listen to me. They are drunk. We must be careful.'

'What are you blathering about?' raged the sergeant major.

'We don't know what we have done,' I replied. 'I insist that you let us go immediately, or we shall report you to our commander. We are Czech officers and will not tolerate insults.'

'Shut your mouth. I am going to shoot you.' The drunkard's finger was on the trigger.

'Who is that civilian,' he asked, pointing at Major Skoba, 'is he a Communist? He has a red star on his coat.'

'He is not a Communist,' I retorted sharply almost losing my temper. 'He is a major and the medal on his coat is a Rumanian decoration which may be worn in mufti.'

The sergeant major was taken aback, but after a while, in a fit of fury, he jumped up again.

'Up with your hands,' he shouted, 'I am going to shoot you because you were speaking Russian and singing Russian songs.'

'That's not true,' I protested.

Almost half an hour had passed since they had started their 'little game'. Then a number of German soldiers – guards from the main garrison and perfectly sober, appeared at the window of a house across the street. They aimed their rifles at us, making bets about who would hit the target first.

The situation was getting more serious every minute. Seeing this I used my most persuasive manner to argue with the drunkard.

'I insist that the commander of the main guard or Lieutenant Schubert who took over the airfield from me be sent for. If you fire on us you will be for it because your own command have guaranteed the security of all peaceful citizens. We have done nothing wrong and I must ask once more that you leave us alone.'

I deliberately raised my voice so that the guards and their sergeant should hear me. This seemed to work, for the Germans then began to argue among themselves.

At that moment Lieutenant Schubert arrived demanding an explanation, first from the sergeant major and then from me. I told him what had happened and demanded full satisfaction.

'I am sorry, Lieutenant,' said Schubert, 'they are drunk and will be duly punished.'

The following day Mirek and I, together with Major Skoba and the commander of the Czech garrison, handed in our evidence in writing to the German garrison commander. The colonel in command apologized with almost embarrassing politeness.

'I can hardly believe it, gentlemen,' he said. 'The sergeant major is an excellent instructor and one of my best soldiers. Nevertheless discipline comes first. He will be punished. I am extremely sorry, gentlemen.' Then the colonel gave us a slight bow. He had nothing more to say.

We were surprised to come across that vile trio again the following afternoon. There was a cynical grin on their faces as they passed us. These representatives of the Herrenvolk had clearly not been punished, but probably praised for their outrageous behaviour towards men of 'inferior race'.

This was one of many incidents which decided me against serving under Nazi rule. Holomucky, my dearest friend, had the same feeling. He could not leave the country because he was married and his wife was expecting a baby. Instead he joined the anti-Nazi underground movement and started his war under the nose of the occupants.

Sadly, he was very soon captured and sent to a concentration camp. One day he felt sick and asked for treatment. The leading camp doctor, Krepsbach, instead of prescribing a remedy gave him a killing injection. It happened on 22 December 1941 in Mauthausen concentration camp.

Mirek Holomucky died without having seen his little daughter. She was born shortly after his death.

I fled my country on 11 June 1939 to Poland with four of my friends, all pilots and all lieutenants. O. Korec was eventually killed in action in France on 5 June 1940. B. Dvořák was shot down on 3 June 1942. He succeeded in escaping from the prison camp at Sagan, got safely to his country, Bohemia, where he was then captured and sentenced to death. Fortunately, Germany lost the war so that the execution could not be carried out. Of the remaining two, R. Fiala was shot down over Paris on 25 May 1940, losing his right eye. Although seriously wounded, he hid in France and eventually reached England in 1942 travelling via Spain and Portugal on a forged French passport. B. Kimlička became a squadron commander, married an English girl and has lived in England ever since.

Blitzkrieg...

Generalleutnant Adolf Galland, *The First and the Last* (Methuen, London, 1955; Bantam Books, New York, 1978)

... The corn was ripening in the fields, and the peasants would soon start to bring in the harvest. Already once before a world war had broken out in Europe during the sultry late summer months. The wildest rumours were afoot. One of them maintained that von Ribbentrop had flown to Moscow to negotiate with Stalin, and this was soon confirmed officially. The German–Soviet Non-aggression Pact was born, and with its signature on 23 August 1939 Hitler had removed the danger to our rear should it come to war with the Western powers over the Polish conflict. This was the general reaction in Germany to this astonishing pact which could not be reconciled with the official ideology in any other way. From a military point of view, it appeared to be an ideal solution, and it coincided with the ideas of the German General Staff, which considered that it was absolutely essential to avoid a war on two fronts.

Great Britain reacted by giving a guarantee to Poland, and Hitler started negotiations with Poland about Danzig and the corridor. Initially the talks went well, but they finally collapsed. This was the external cause of the Second World War.

Four German armies were already drawn up along the German–Polish border. Under the command of Richthofen, we were attached to General von Reichenau's army in the Silesian sector....

It was still dark on the morning of 1 September 1939 as we climbed into our cockpits. Blue flames spurted from the exhausts of our engines as they warmed up, and at the first signs of dawn the fireworks started. The targets for our unit were Polish Staff HQ and barracks. The operations had been planned and prepared with the greatest care, and the spine of the Polish army was to all intents and purposes broken on this first day of the campaign. After a few heavy strategic blows by the Luftwaffe which gave the Germans absolute air supremacy, the splendid coordination and cooperation between the fast-moving mechanized army and the Luftwaffe began.

It was something completely new in modern warfare. Everything went off with the precision of clockwork. Five Polish armies, standing in readiness opposite the four German armies, were annihilated in a flash by the revolutionary German strategy, combined with the efficiency of the German General Staff, the modern armament and equipment and, finally, the incomparable fighting spirit of the German soldier. Thus occurred the miracle of a German blitz victory in eighteen days. The Polish Air Force was destroyed mainly on the ground during the first days of the campaign. Richthofen* had ample opportunity to exploit his Spanish experience in the art of giving the most precise tactical air support to the army. As we hardly met any Polish fighter planes, our own fighter force was used predominantly for army support. One sortie followed the other, and our base was constantly moving forward. We advanced at a breathtaking speed.

*Wolfram von Richthofen, cousin of Baron Manfred von Richthofen, Germany's great World War I 'ace'.

. . .After the blitzkrieg in the east followed the sit-down war in the West. It was a terrific nervous strain for all concerned. I commanded all three wings of our group in turn for a fortnight while the respective commanders were on leave. There were continuous take-off alarms – false ones, of course, because the ominous siren wail or control-tower orders which sent us zooming into the air, consuming considerable amounts of material and fuel, were usually based on errors or illusions. One radar report of a mass approach of enemy aircraft, for example, turned out to be a flight of birds. One day, however, in the Lower Rhine area somebody was really shot down during one of these alarms. It was one of our planes, a FW 58 'Weihe', piloted by the regional fighter commander. What irony! It was as well that nothing else happened. . . .

The postponement some dozen times between November 1939 and May 1940, of the German offensive on the West front was due not only to weather considerations, politics, personnel, armaments, etc., but also to chance. This could be qualified as lucky or unlucky according to the time and the point of view under consideration. On 10 January a Luftwaffe major flew from Münster to Bonn, carrying with him the entire plan of operations for the offensive in the West. In bad weather he was blown off his course and made a forced landing on Belgian soil. He did not manage to destroy the documents in time, and they fell into the hands of the Allies. A new plan had to be worked out, which, after a further delay caused by bad weather, was finally put into execution on 10 May 1940. . . .

Borkum, 27 November 1939

Group Captain Sir Max Aitken, 'The Aitken Papers', London, 1982

On the first day of the war I was flying over the English Channel. I went as far as Le Touquet on a mission for which there was no real necessity. The squadron's first experience of war came on 27 November when we raided the German seaplane station on the island of Borkum. In addition to six aircraft from 601 Squadron, there were six from 23 Squadron. We were not allowed to carry any bombs; we flew from Tangmere to Northolt where we had a conference with Orlebar who told us, 'You are lucky to have a first crack at the enemy.' I am not sure whether we saw it quite in this light. We stopped for lunch at Bircham Newton but, curiously enough, none of us was inclined to eat. After flying over the North Sea for half an hour, we were fired at by one of our convoys, but continued flying along the Dutch coast to Borkum, which we approached very low. It was my job to deal with the German fighters. After that we turned towards the seaplane base. I have a vivid recollection of the attack. I saw a man standing on a ladder painting one of the hangars. He fell off the ladder. His hat fell off with him. So far as I know that was all the damage we did at Borkum that day.

Flying home again, we were once more fired on by our own convoy. By the time we reached the English coast it was dark. We had no wireless. The balloon barrage which might have made things awkward was up, but the Station Commander at Debden had the good sense to turn on a searchlight which he pointed straight up in the air as a beacon. All the aircraft made for it as a result of which there were a great many near misses. I was the last to land, very short of petrol. We had encountered a great deal of anti-aircraft fire but fortunately no German fighters. On the following day we were back at our home at Biggin Hill, and to celebrate our safe return the Station Commander renamed his house Borkum.

Brilliant Generalship

The post-war questioning of General Galland revealed the respect in which Werner Mölders was held by the German Air Force.

'From the outbreak of war in September 1939 until the beginning of the offensive in the West in May 1940, the Luftwaffe hoarded its strength and allowed only a few units to participate in aerial combat. For most of this period it was strictly forbidden to cross the front, but since the line was not recognizable from a height of 15,000 feet, air battles between German and French formations often took place.

Mölders, a Gruppenkommandeur in JG 53 at Wiesbaden, was considered at this time to be the most experienced German fighter pilot.... As a formation leader he showed brilliant generalship in the air, a great talent for teaching and outstanding ability as a pilot. German formations over the Western front already flew in the open formation prescribed by him and enjoyed thereby a considerable advantage over the French who still flew closed up.

At this time Mölders and his unit achieved a large number of victories against the French formations which were not as well disciplined as they should have been. Both sides regarded the fighting as chivalrous jousting, and there was little of the frenzied and furious clashing seen later in the May blitzkrieg or in the Battle of Britain.

During the period of waiting and building up, the Germans had their share of false alarms and false starts, and it became clear that an improved reporting service was needed. Little was done, however, until much later, as at that time the Luftwaffe rightly considered Germany safe from air attack.'

Well-liked German

In the eerie quiet of the Western front during the first winter of war, C.G. Grey, the outstanding aviation commentator of his time, caught the strange mood in a short vignette he wrote in the Aeroplane *of 1 March 1940, on Wolfgang Falck, a Luftwaffe officer of unusual charm and ability. Falck, who was to enjoy considerable success against the Allies during the next five years, had visited England shortly before the war.*

Those who were at the opening of the York Aerodrome and the festivities associated therewith will no doubt have recognized Captain Falck as the right-hand figure in the photograph on page 246, 1 March 1940 issue, of a cheerful group of German pilots with Lieut.-Col. Schumacher and Dr Dietrich.

I met him first when he was an Oberleutnant in a Staffel of the Richthofen Geschwader at Damm, near Jüterbog, where there is a monument to record that the great Immelmann learned to fly there.

Afterwards various of our English sport flyers and officers of the RAF met Falck at Frankfurt meetings and at York, to which meeting he flew with about half a dozen young German pilots in sporting aeroplanes. He was particularly well liked.

His name fits him well for he looks like a falcon. He is not big enough to be an eagle. But he has the aquiline features which artists and novelists love to ascribe to heroic birdmen.

He speaks excellent English and is a charming companion. He reminds me much of our great air fighter of the last war, Jimmy McCudden, VC, DSO, MC, etc. He has the same boyish manner, the same straightforward look, the same attitude towards killing his official enemies, who might be his personal friends.

McCudden always described himself as a hired assassin, and never had any animosity against anybody. That sort is always the most dangerous fighter, for the coolness of his head is never disturbed by the heat of his emotions.

The day on which the great von Richthofen was killed McCudden came into my office really moved and said mournfully – 'And I did so want to talk it all over with him after the war.' I can only hope that Falck's English friends will be more fortunate.

Two months later, all was changed.

An Irish Airman Foresees his Death

I know that I shall meet my fate
Somewhere among the clouds above;
Those that I fight I do not hate,
Those that I guard I do not love;
My country is Kiltartan Cross,
My countrymen Kiltartan's poor;
No likely end could bring them loss
Or leave them happier than before.
Nor law, nor duty bade me fight,
Nor public men, nor cheering crowds,
A lonely impulse of delight
Drove to this tumult in the clouds;
I balanced all, brought all to mind,
The years to come seemed waste of breath,
A waste of breath the years behind,
In balance with this life, this death.

W.B. Yeats

'After Today Nothing Would Ever Be the Same....'

Group Captain M.M. Stephens, Roquefort Les Pins, France, 1982

... 10 May 1940, dawned bright and clear with patches of mist filling the valleys around Kenley. It was the start of a glorious spring day. But this was not to be a day like any other. After today, nothing would ever be the same – not in Europe, not in the world. We could not know it; but already the Belgians and Dutch were reeling under the first blows. The 'Phoney War' was over.

We were sitting on our beds in the dispersal hut, on normal 'dawn readiness', when the telephone rang for No. 3 Squadron. 'Scramble Blue section.' We were out of the hut and climbing into our Hurricanes almost before the airman's words were out of his mouth.

In a moment we were airborne, climbing at full speed over the bright, sleeping countryside. 'One bandit, angels twenty, vector one-zero-zero.' It was the voice of the Controller. We acknowledged briefly and turned onto course. In a few minutes the Controller spoke again: 'He's turning back. We think it's a Junkers 88. He's straight ahead of you, same height – can you see him?' The enemy was too far away for us to see him. After chasing the fleeing Ju 88 out over the coast of England and over the Channel, the Controller came through again. 'He's got his nose down,' he said, 'and is drawing away. You won't catch him, now. Return to base and "pancake".'

As we turned in for landing, wheels and flaps down, we could see at once that something was afoot. The station was a hive of activity, whereas normally, apart from the 'readiness' aircraft it would only just be coming to life. I taxied back to my dispersal point. The refuelling tender was already there, and almost before I had rolled to a stop the mechanics were climbing onto the wings, opening the filler caps and filling the tanks. My rigger jumped up and took the straps of my harness. 'Well, sir,' he said, 'you've done your last trip from Kenley.' 'What do you mean?' I asked. 'We're off to France in a couple of hours,' he replied. 'The Jerries have invaded Belgium and Holland. All Hell's let loose!'

Shortly after midday we were on our way to France. There were not sufficient maps for everyone; I, as the most junior member of the squadron, was not one of the favoured few. We had, therefore, to content ourselves with following our leaders.

Soon we received the signal to 'break' for landing. We touched down on a grass field with three Nissen huts on either side. It was Merville. On the far side of the airfield another Hurricane squadron had just arrived. There was feverish activity as pilots and ground crew sorted out the mass of equipment which had been hastily unloaded from the transport aircraft. It took off again immediately the last item had been unloaded.

We snatched a hasty lunch of bully-beef and biscuits, with the inevitable mug of strong over-sweet tea, and over the field telephone came an order for one flight [six aircraft] to patrol between Maastricht and Bree. Faithfully, we followed the flight commander, who was the only one with a map! We saw nothing on that first patrol except roads packed solid with the pathetic stream of refugees. It was to become a depressingly familiar sight.

Back at the airfield, we had just finished refuelling when, without warning, a formation of Heinkel 111s flew across the airfield at about 8000 feet, dropping their bombs with accuracy. We took off in whatever direction we happened to be pointing, hoping to catch the Heinkels. It was hopeless. There was no radar, no fighter control at all! We were just wasting effort and hazarding aircraft in the hopes of finding our quarry in the gathering darkness. I still had no map but seeing a Hurricane silhouetted against the western sky I decided I would join up with him. It was Walter Churchill, the Flight Commander of 'A' Flight. As I formated on him he led me back to our field, where the ground crew had hastily organized an improvised flarepath with the headlights of available trucks and cars. In the circumstances, we were lucky to lose only one aircraft. Its pilot, having no map, had decided to head west in the hope of getting into one of the airfields in the southeast of England. Unfortunately, he ran out of fuel just as he reached the coast. He baled out over St Margaret's Bay. The next day he rejoined the squadron with a replacement Hurricane. . . .

Already rumours were running thick and fast. The 'Fifth Column' – two German agents disguised as 'Sisters of Mercy' – had been caught shaving in the Forêt de Nieppe, where they were operating a radio transmitter. The telephones were being operated by agents dropped by the Germans; false orders were being issued, no one was to be trusted over the telephone unless positively identified. Convoys were being misrouted by fifth columnists disguised in British/French/Belgian uniforms. The Germans were using captured British/French/Belgian aircraft. I do not know to what extent agents and fifth columnists were being used. It is certain, however, that the reports of their activities, greatly exaggerated though they may have been, caused considerable confusion and mistrust.

On the following day a party of our airmen was arrested as 'fifth columnists' because they were carrying rifles which were not of the usual British type, and they were wearing 'jackboots'. The rifles were, in fact, Canadian Springfields, whereas the British rifle was the Lee-Enfield. The 'jackboots' were ordinary rubber Wellington boots. In the prevailing atmosphere it was fortunate that they were soon cleared. There was a real danger of their being shot summarily as enemy agents. The incident emphasizes the state of alarm and confusion provoked by the reports of fifth column activity.

We slept at the airfield, huddled together in the few Nissen huts, and were awake and ready to go before dawn. Off we went throughout the day, flying patrols of six aircraft. My first encounter with the enemy came early the next morning, when we were sent to patrol the area between St Trond and Diest. As usual, there was no ground control, no radar. But the scale of enemy air activity was so great that the odds were very much in favour of making contact.

Suddenly we spotted about sixty tiny black dots in the sky, flying west like a swarm of midges. The next moment we were among them – Stukas, with an escort of about twenty Me 109s. I got one lined up in my gunsight, and opened fire from about 50 yards. After a short burst he blew up in an orange ball of flame, followed by a terrifying clatter as my Hurricane flew through the debris. My speed was reduced to that of the swarm of Stukas, and I realized that I was in the middle of the formation, with Stuka rear gunners firing at me from all sides. I disengaged hastily, setting course in what I hoped was the right direction for base. Just then, from out of the cloud a few hundred yards away, emerged a Dornier 17. I gave him a short burst from short range, hitting his starboard engine which started smoking. I had the satisfaction of seeing the pilot belly land the aircraft in a ploughed field. I then continued again on my original course. After flying for about 20 minutes, and with my fuel beginning to run ominously low, I found beneath me an airfield under construction. It was Grevillers. I put my aircraft down, and was collected almost immediately by a major of the Royal Artillery, who whisked me off in a staff car to the nearby British Corps Headquarters. There I was given an excellent breakfast before being taken back to my aircraft, for which some petrol had been found. I refuelled my Hurricane, persuaded one of the Royal Engineers from the airfield to wind the starting handle, and was soon on my way, following the roads back to my base at Merville.

The rumours about the Germans flying captured British aircraft were obviously having some effect. I remember one morning coming back alone from a fight over Belgium – as usual without a map – and with only the vaguest idea of where I was. I spotted a Lysander, and decided that, if I were to formate on him he might lead me back to his field. But he wasn't having any of it! Whenever I got anywhere near him he began the most violent evasive action, convinced that I was about to try to shoot him down. I put my wheels and flaps down, reduced my speed to the minimum and opened my canopy. Eventually he seemed persuaded that my intentions were not evil, and allowed me to formate on him. Then began quite a pantomime. I tried, by mime, to indicate to him that I was lost. It seemed impossible to get him to understand. Finally I reached down into the cockpit and produced a map of the southern sector of England. I held it up for him to see and then flung it over the side. He got the message! He led me straight away to the airfield at Wevelegem, near Courtrai. On landing he taxied immediately into a bomb-battered hangar, while an officer climbed on my wing and urged me to do the same. 'We've just had Hell bombed out of us', he said. 'If they see your aircraft here we'll catch it again. Please get your aircraft under cover.' At this, I taxied into the hangar. Leaving my engine running, I climbed out and asked for directions. I was given the most marvellous present of a real map! I took off immediately for base, guarding my newly-acquired map as if it had been the title deeds to a gold mine!

And so the battle went on. With each mission the patrol area drew inexorably closer to our base. By now many of our combats were over and around Lille. On 14 May we were suddenly directed to fly down to Berry-au-Bac, near Reims, to take part in an all-out air effort to stop the Germans who were breaking through at Sedan. It turned out to be a very hectic evening, with the sky thick with German aircraft – Stukas, Me 109's, Me 110's and Ju 88's. We destroyed 17 for the loss of 5. Every day, with one exception, we lost aircraft. Sometimes we got the pilots back, weary and often foot-sore, having walked back through the endless stream of refugees, or having 'borrowed' transport. Each day more of our aircraft were damaged beyond repair and each day fresh pilots and Hurricanes arrived from England to make good our losses. The squadrons at home were being depleted to keep us up to strength. On 14 May, four pilots flew in from 32 Squadron, two from 56 Squadron and three from 253 Squadron. On 17 May a further six pilots and twenty-eight ground crew joined us. Mostly the new arrivals were in action within an hour or so, some to disappear on their first mission.

By 20 May, the Germans had already passed us to the north, heading towards St Omer, while we were strafing German motorized columns on the outskirts of Arras, some 20 miles to our south.

Our field was now in imminent danger of being overrun, and the order was given to evacuate. The transport aircraft came in – not enough of them, but all that could be made available. Of these several were shot up on the ground and were soon burning hulks. Our ground crews piled hastily into trucks and headed off for Boulogne, before the road was cut. Having burned our unserviceable Hurricanes, we made a last strafing attack on the Germans outside Arras, and then flew from this mission straight back to Kenley.

In our ten days in France, 3 Squadron had destroyed 74 German aircraft for the loss of 21 Hurricanes. Nine pilots were killed and others missing. For us the Battle of France was over.

The impression I was left with was of chaotic conditions, a fight against overwhelming odds and appalling communications – or none at all. One's memory is of the desperate, untiring efforts of our gallant ground crews to keep the aircraft airworthy – cannibalizing damaged aircraft, improvising, and working all hours of the day and night.

People ask how much more effective would our effort have been had there been a reasonable early-warning system and control organization – even a reliable telephone network? Could this have affected the outcome of the battles of 1940? It is impossible to say, although the answer is probably not. We should certainly, however, have extracted a higher price from the Germans for their temporary victory.

Four years later, when we advanced northeastwards across the same countryside, the lessons of those early and traumatic days in France had been fully absorbed.

Blenheims Attack Maastricht, 12 May 1940

James Sanders, *Venturer Courageous* (Hutchinson, New Zealand, 1983)

By noon on 10 May . . . the invaders were pressing on into Belgium over undestroyed bridges near Maastricht. . . .

When, on 12 May, it was learned that two bridges across the Albert Canal had still not been destroyed – and that the enemy was pouring across them – 15 Squadron was detailed to bomb the crossings.

On that Sunday in spring, while the good villagers around Wyton, Norfolk, were going to their churches, twelve Blenheims took off for

Belgium and the bridges over the Albert Canal. Trent [New Zealand's Group Captain L.H. Trent, VC] piloted one of the six aircraft in Squadron Leader Glenn's flight.

They approached their target at 5000 feet, vigilantly on the alert for enemy fighters but thankful that, so far, none had appeared. And then came the anti-aircraft fire with such intensity that Trent felt none could survive. Shrapnel was making jagged rents in the wings and fuselage of his Blenheim when Glenn waved his flight into individual dive-bombing attack. It was an awful prospect that presented itself to the crews – a situation from which, all were sure, none could escape alive. The air was thick with flak and the stuff was being hosed up at them in terrifying streams.

Trent gave his machine its injection of plus-nine boost and pushed the column forward, keeping a broad aim on his target bridge while he pushed hard on each rudder pedal, skidding the aircraft away from any steady bead on the ground ack-ack batteries might hope to fasten on him.

Bombs gone! And, screaming down towards the deck – with his engines still roaring with plus-nine power – he hedge-hopped, weaving and jinking, away from the target area. A line of German soldiers, stringing across a field, crossed the sights of his wing-mounted ·303 gun. He thumbed the firing tit on his control column and saw the tracer snaking out, guiding his aim. 'I'll get a few of them,' he promised himself, dropping the nose of the Blenheim still lower.

But, in his concentration on the now wildly scattering infantrymen, he had not seen a line of tall poplars ahead. In a heart-stopping moment he had hauled back on the stick, clipped the tender spring leaves of one of the trees and was pushing his aircraft down again to ground level. The engines hiccupped characteristically and, again in rapport with gravity, they took up the strain of headlong flight.

Everywhere were the red flames of invasion. Trent and his crew flew past a Belgian airfield with hangars and dispersed aircraft blazing furiously; and the enraged defenders of both aerodrome and a homeland now doomed, opened fire with rifles and machine guns, believing the Blenheim to be a hated tool of Hitler.

As if the battlefield in Belgium was not shocking enough on that Sunday of 12 May, an incident on the circuit at Wyton, as Trent joined the landing pattern, gave the young officer and his crew an unnerving welcome home. Trent tells it as he remembers it.

'I was horrified to see a Blenheim, turning on to its final approach with a dead port motor, overbank and stall. It dropped out of control to crash in flames ahead of us.

'Our knowledge and experience of asymmetric [want of balanced power] flight was practically nil; but in those days, I seem to remember, we were told never to turn towards the dead engine. . . .'

After Len Trent and his crewmates had landed they heard that, of the 12 squadron aircraft that had set off on the Maastricht mission, 6 had been lost – a 50 per cent casualty rate that did not make an encouraging start for 15 Squadron in the first phases of the fighting war.

'Casualties . . . Sometimes Reaching 100 Per Cent'

Air Chief Marshal Sir Christopher Foxley-Norris, *A Lighter Shade of Blue* (Ian Allan, 1978)

. . . The RAF in France was soon as tightly and agonizingly racked as its comrades elsewhere. The Lysander proved a robust and reliable fighting aeroplane but totally inadequate in performance and invariably heavily outnumbered: also the tactics and practices we had learnt at Old Sarum proved murderously unrealistic; enemy anti-aircraft fire and fighters demonstrated this, but at a dreadful cost. Even so our losses were less dramatically crippling than those of our fellow squadrons equipped with the Fairey Battle light bomber. At the time of its design in the early thirties it was one of the earlier monoplanes, a precursor of the Hurricane and Spitfire; it was widely acclaimed and put into extensive squadron service with the RAF (and also the Belgian Air Force). But by the time it was called on for use it was sadly outmoded, too slow and clumsy to evade enemy anti-aircraft fire and fighters, and too lightly armed to have any hope of beating off the latter. These defects, together with the deterioration of the allied ground situation which demanded its use in what amounted to suicide missions, led to shocking casualties, sometimes reaching 100 per cent of the aircraft launched against a particular target. The courage of the crews was superb but it was the courage of the doomed and the desperate.

This was brought home to me very sharply one day. I had landed at a Battle Wing airfield south of Rheims. During my short stay there, a new Battle flew in from England to replace one of yesterday's casualties. The pilot joined the circuit and made his approach but failed to lower his wheels. The aircraft crashed on its belly in a cloud of dust and the chastened pilot stepped out unhurt but due for at least a severe official rebuke from the Wing Commander for his carelessness. No doubt he received one; but to my surprise he was more or less feted by the other aircrew. I soon realized why; his accidental destruction of the aircraft had prevented the inevitable loss to enemy action of the crew that would have flown it the next day.

The twin-engined Blenheim bombers fared little better. They were faster and somewhat better armed than the Battle, but were called on to penetrate far deeper into enemy-dominated airspace; and their fate was inexorably the same. So too for our Lysanders. After quite a short period of fighting, my own squadron had lost all its serviceable aircraft; and those of us who survived joined the pathetic rabble of refugees fleeing westwards across France before the German armoured thrusts, constantly harassed and lacerated by enemy air attacks. These latter seemed to us to be almost entirely unopposed, although here we made the same mistake as our soldiers later made at Dunkirk. Our fighters were indeed meeting the enemy and although heavily outnumbered, were blunting many of his attacks. But the air fighting was rightly conducted as far forward (that is eastward) as possible, to intercept the German aircraft before they could reach their targets. So we witnessed very little air fighting and wrongly deduced that there was none. We were tired and frightened, we misunderstood the realities and became embittered; so too at Dunkirk our men were tired, frightened, misunderstanding and embittered.

'Where was the RAF at Dunkirk?...'

Winston Churchill gave his answer to the question 'Where was the RAF at Dunkirk?' in the House of Commons on 4 June 1940:

'... We must be very careful not to assign to this deliverance the attributes of a victory. Wars are not won by evacuations. But there was a victory inside this deliverance, which should be noted. It was gained by the air force....'

Ducal Mission

At this bleak juncture, as the French armies were falling back in the face of the German onslaught, Hugh Dowding (Air Chief Marshal Lord Dowding), Air Officer Commanding-in-Chief, Fighter Command, sent a personal emissary to France to report directly on the worsening situation.

It was an unusual arrangement. The officer assigned to the task was Wing Commander The Duke of Hamilton, the first man to fly over Everest and one of four remarkable air force brothers, each of whom, at one time, became a squadron commander. Hamilton flew this three-day mission (17–20 May) to the battle area in a tiny Miles Magister, a light, single-engined aircraft with a top speed which would have compared unfavourably with a modern small motor car. He was then serving as a controller under Keith Park (Air Chief Marshal Sir Keith Park) at Headquarters, 11 Group.

For the historical record, it is worth reproducing the relevant entries in the pilot's flying log book (see illustration between pp. 64 and 65).

The report which Hamilton brought back convinced the C-in-C that, with the French crumbling, no further fighter squadrons should in any circumstances be sent to France. From this stance, Dowding fought his corner with great courage. Meanwhile 'Douglo' Hamilton received a mention in despatches for his exploit.

Intervention

Harold Balfour, *Wings Over Westminster* (Hutchinson, 1973)

I remember at the time that after Dowding had insisted on personal intervention with the War Cabinet to stop any further of his meagre force leaving these shores to prop up the failing French, he came to my room to appeal as an old friend that I should use any influence I had with my Air Council colleagues to this end. Within my limited capacity I did my best for, at the finish of the Dunkirk evacuation, he had but three fighter squadrons left that had not been mauled in the continental fighting.

Dunkirk

Group Captain Sir Max Aitken, 'The Aitken Papers', London, 1982

... I found my way to Hendon and from there back to the squadron at

Tangmere. A few hours later I was involved in an air battle over Dunkirk in which my Hurricane was hit and was shot down in the sea. Next day – on the same circuit, Tangmere to Dunkirk and back – my Hurricane was hit in the engine but I got home. On the third day we were involved in a dogfight with Messerschmitt 109s. I had been 5 hours in the Hurricane before lunch that day when I was visited by the Under-Secretary of State for Air, Captain Harold Balfour ... who came out and stood on the wing of my aircraft. He said, 'Are you tired, Max?' Tired! It was a period of my life which I look back on with something like incredulity.

... Life was to become even more exciting before long. In the first week of June, 601 Squadron was visited by Air Vice-Marshal Park, who commanded 11 Group. He was an impressive New Zealander, a regular air force officer of great courage and determination. He had insisted on flying over Dunkirk to see for himself what it was like. He used to fly himself everywhere. Now he made the officers sit round him on the ground. He said: 'War isn't much fun, you'll find out.' We agreed....

Dunkirk was a fantastic sight with a pall of smoke drifting over a hundred miles from the burning oil tanks. We flew on patrol as closely as we could to the smoke in order to hold off the Germans who otherwise would have made a massacre of the soldiers whom we could see in long lines on the beaches waiting to be evacuated. The Germans could not break through the screen of fighters which was kept up 20 miles inland from Dunkirk. So we had already had some experience of what war was like.

Quite suddenly, in the middle of his talk, Park stopped. 'Max Aitken,' he said, 'I am going to promote you to command the squadron.'

This was, to me, a complete surprise and one which naturally pleased me very much indeed. I was less pleased, however, by a turn in events which followed not many days later. First of all, came a signal ordering us to give up our offensive sweeps over France and devote ourselves to training. This was exactly what was wanted, as a preparation for the battle which we knew was coming. But a signal which followed almost immediately, reversed the order given in the first. I read it with horror.

Here we were on the eve of a great defensive battle for which every machine should be reserved and every pilot at the peak of his freshness, fitness and efficiency. And we were being ordered to fritter away our efforts in a series of operations in France where the land battle was already lost! I decided to take action which was certainly bold and unusual in the commander of an Auxiliary squadron. It obviously involved a risk of serious disciplinary penalties. I had no idea what the situation was in the higher reaches of politics and strategy. But the fighter squadrons were faced with madness, as I saw it. I decided to use what influence I could to prevent it.

At once I flew to Reading. There I got into a train and went to London. I went straight to see my father [Lord Beaverbrook: appointed Minister of Aircraft Production, 14 May 1940] at Stornoway House. I told him what the situation was as I saw it. I said that if we went back to carrying out offensive sorties over France it would be a disaster. He heard me but said nothing. I returned at once to the squadron. Next day the order was rescinded.

New Zealand Blooding

Group Captain C.F. Gray, Waikanae, New Zealand, 1982

Colin Gray from Christchurch, New Zealand, was to become one of the Royal Air Force's most successful fighter leaders of World War II, his 28 victories taking him high up in the Allied charts and top of the New Zealand pops. As a 25-year-old pilot officer of 54 Squadron, he had his first taste of real fighting over Dunkirk. Like all such initial experiences, it became fastened upon his mind.

The evacuation from Dunkirk, codenamed 'Operation Dynamo', started on 26/27 May 1940 and was called off nine days later, on 4 June. Fighter Command's involvement had actually begun ten days or so earlier when Spitfire and Hurricane squadrons of 11 Group started to fly patrols over the battle area.

I am well aware of criticism aimed by members of the British Expeditionary Force and the navy at the effectiveness – if not the actual presence – of the Royal Air Force during this critical engagement. Well, we were there, anyway.

My log book records a total of over 30 flying hours in the Calais/Boulogne/Dunkirk area during the twelve days prior to our withdrawal from the front line on 28 May – just when things were, unfortunately, beginning to hot up. Towards the latter part of this period we were flying two and sometimes three sorties a day with an average duration of over 2 hours – and this in aircraft designed for a safe operational endurance of less than half that time. (We had no long-range fuel tanks at that point.)

Furthermore, our hours of 'readiness' covered a period from dawn (say 4 a.m.) to dusk at about 10 p.m. It was quite tiring.

The problem became even more acute during the actual embarkation phase. Demands were made upon the Air Officer Commanding No. 11 Group, Keith Park, to maintain standing patrols over the beach head – some 60 miles from our main bases – during daylight which, at that time, amounted to around 17 hours. Since Park had a total of only sixteen squadrons at his disposal, it is little wonder that the cover was thinly spread.

Dunkirk was altogether a pretty torrid time for us. This was partly because of the long period of readiness and the intensive flying involved during it, but also because for most, if not all, of us this was the first time we had seen shots fired in anger; we had to learn fast.

My squadron, No. 54, flying from Hornchurch in Essex, lost half its aircraft and pilot strength in four and a half days' fighting, from 24 to 28 May. Not surprisingly, we were then withdrawn for a rest to be replaced by another squadron from the north. Fortunately three of our missing pilots were later returned to us from the beach head by courtesy of the Royal Navy.

Squadron operational records show that during the Dunkirk period we claimed to have shot down 31 enemy aircraft and damaged another 16. No doubt, in common with others, these claims were somewhat overoptimistic, but at least they showed that we were there.

My most vivid recollection of this time, which I recall with no pride whatever, occurred on 25 May, when we escorted a squadron of Swordfish – yes, Swordfish! – to dive-bomb Gravelines. Their bombing was fairly ineffectual but they certainly managed to stir up a hornet's nest.

In the subsequent melee, whilst I was minding my own business, some character planted a couple of cannon shells and numerous bullets into my aircraft. It served me right as I was watching an Me 109 pilot bale out at the time – something we had been warned never to do. My opponent must have been a pretty good marksman since it was clear from the angle of the bullets that it was an almost full 90-degree deflection shot from well above – which, I suppose, is why I never saw my attacker. The first cannon shell went through the aft starboard inspection hatch, where it exploded, severing the elevator trimmers and knocking out the hydraulics and the air pressure system. The second shell passed over the top of the cockpit and through the port aileron, neatly removing the pitot head.

Subsequent landing back at Hornchurch was not without its moments. The undercarriage came down with the aid of the emergency CO_2 bottle, but the loss of air pressure meant no flaps or brakes and, with the pitot head shot away, no airspeed indicator.

I made it at the second try. Rugged aeroplane, a Spitfire!

Obituary: Pilot Officer Louis Arbon Strange

St Edward's School Chronicle, July 1966

Incredible stories came out of France as the Germans overran the country and the Royal Air Force squadrons – or what was left of them – hurriedly withdrew. None was more engaging than that of Pilot Officer (formerly Wing Commander) Louis Arbon Strange, distinguished World War I pilot who, at the age of fifty, thereby added to his already sizeable total of decorations.

In the archives of St Edward's School, Oxford, Strange's old school, the tale is recorded in an obituary note.

Louis Strange learnt to fly at Hendon in 1912, and in the First World War commanded No. 23 Squadron and 80 Wing. He left the air force because of ill health after the war with a DSO, an MC and a DFC to his credit. In 1915 when flying a Martinsyde Gnome Scout near Menin he was potted at by the observer of a German aircraft armed with a parabellum pistol. Strange promptly retaliated. Attempting to pull off an empty ammunition drum from his Lewis gun, which was mounted on the Gnome Scout's top wing, Strange's harness slipped and he fell out. The recalcitrant drum saved his life for it bore his weight until, from an upside-down position, he was able to claw himself back into his cockpit feet first and right the aircraft.

He rejoined the RAF in 1940 and was ground control officer at Merville in France when the airfield was hurriedly evacuated. He carried out a hasty and unorthodox repair to a Hurricane and flew it back across the Channel. It was not a type of aircraft he had flown before and was, moreover, unarmed and without several important instruments. Nonetheless, when attacked by eight Messerschmitts, Strange became fairly belligerent and put on such a good show that the Germans, apparently believing that he was trying to bring his non-existent guns to bear, sheered off. For his exploits Strange was awarded a Bar to his DFC.

Naval Grit

Group Captain Sir Max Aitken, 'The Aitken Papers', London, 1982

When the evacuation began, the squadron escorted some naval Swordfish which were attacking German targets north of Dunkirk. The Swordfish, popularly known as Stringbags, were very slow, suitable for operating from carriers but at a great disadvantage when matched against land-based aircraft. We had difficulty in flying slowly enough to protect them.

I remember that the naval pilots all wore white scarves which trailed behind them when they flew; they were extraordinarily brave and suffered heavy casualties that day. We were flying above them at 10,000 feet. I had my first experience of tracer. Looking down, I saw the shell leave the gun; it appeared to rise very slowly but the next thing one knew, it was flying past one's head.

After dropping their bombs, the Swordfish, or what were left of them, flew home. Halfway across the Channel, one of the pilots turned round and headed once more for the French coast. Puzzled to know what was in his mind, I went back with him. What had happened was that his bomb had not been released, owing to some defect in the mechanism and he went back alone and dropped it. I was very much impressed by his bravery....

... Three days after the Swordfish escort battle I received an order which caused me a good deal of surprise. I was to go to London without delay and attend on the Prime Minister at lunch. It was 27 May and Mr Churchill had been at the head of the government for just sixteen days. I was amazed that he could spare the time to entertain a mere flight lieutenant. However, off I went at speed. I flew to Hendon by Hurricane and from there drove by car to Whitehall.

It was very quiet, I remember, in the courtyard of Admiralty House which was watched by armed marines. In the entrance hall was a burly naval petty officer with a big revolver strapped to his waist. I went up in the lift and found the Prime Minister, as calm and cherubic as usual. Without undue waste of time he led the way into the dining room.

Lunch was very good. We ate lamb although this was at a time when food rationing was beginning to be severe. Churchill explained: 'This lamb was very unfortunate, Max,' he said solemnly. 'It broke its leg. Otherwise, of course, it would not be here.'

Naturally I knew that I had not been ordered to London in the middle of a battle simply in order to be given a good meal. Before I knew where I was I was being subjected to a severe cross-examination. The Prime Minister wanted to know how things were going with the fighter squadrons. He was not a man to be satisfied with official stories filtered through the various levels of the Air Ministry. I told him that things were going well. This was true although, as the Prime Minister knew even better than I did, we were in the early days of a long battle. The initial phase of the Battle of Britain was, in fact, no less than ten weeks ahead. In answering as confidently as I did, however, I was only expressing the absolute belief of the fighter squadrons. It simply did not occur to us that we could be defeated.

The Prime Minister's next question was, as he showed, backed by considerable technical knowledge. 'Are the German fighters better than ours?' I told him they were certainly not and he seemed to be satisfied with the evidence I gave him. At some opportune moment in the conversation I told him that I thought he should make a statement in the House about the RAF. In due course, as is well known, he did, although I do not know whether my suggestion had any connection with his speech. Then, quite suddenly, he got up from the table saying, 'Now I'm going to sleep.' The interview was at an end.

'Song for the Mess'
Norway, 1940

I've got Arctic Circles under my eyes,
And the Cold is making me blue
But the sun comes out and warms my heart
When I think of You.

The snow-clad mountains and the ice-bound fiords
Make summer at home seem untrue;
But I see the daffodils and smell the roses
When I think of You.

Buckingham Palace
Has Aurora Borealis
Beaten to a Frazzle.
And a crowd of gapers
At Norwegian papers
Would make me give a fiver for a peep at Razzle.

Yes, I've got Arctic Circles under my eyes,
And the Cold is making me blue,
But the sun comes out and warms my heart
When I think of You.

The NAAFI can carry
The Café de Paris
For only about two rounds
And tho' it sounds silly
The roar of Piccadilly
Would be the most welcome and the best of sounds.

Yes, I've got Arctic Circles under my eyes,
But I'm not yet by any means through,
For I'm inspired to grin and bear it,
When I think of You – my Darling,
When I think of You.

Thomas C. Harvey, Scots Guards, Norway, 1940

Death Strikes in the Arctic

If the news from France was bad, the reports from Scandinavia were certainly no better. The Allies' ill-starred Norwegian campaign was coming under gathering pressure. The six weeks' period from the last days of April 1940, to the disastrous sinking of the aircraft carrier, Glorious, *six weeks later on 8 June, with the loss of two Royal Air Force and two Fleet Air Arm squadrons on board, tested human endurance beyond the limit.*

263 Squadron, led by Squadron Leader J.W. Donaldson – 'Baldy' to the Service, the eldest of three exceptional Royal Air Force brothers, each of whom was to win the DSO and other decorations besides, was the first to land its Gladiators on the frozen waters of Lake Lesjaskog, near the port of Aandalsnes, in Central Norway. A makeshift runway cleared of snow, had already alerted the Luftwaffe to the squadron's imminent arrival. No sooner had the obsolete biplanes touched down on the ice than they became the target for repeated bombing attacks by more than a hundred enemy Heinkel 111s, Junkers 87s and 88s. The assault lasted for eight hours.

Nor was this the only problem. Group Captain Stuart Mills, then a flight commander in the squadron, and Donaldson's effective No. 2, has set down for the first time his story of events.

'The squadron had been warned in February 1940, when it was stationed at Filton, near Bristol, to ready itself for Finland, then fighting the Russians. A party of Finns had been to visit us during the hard winter when the country was covered with snow. They were after spares and equipment from the Bristol Aircraft Company.

Seeing our aircraft grounded by the thick snow, they told us that, in Finland, they simply rolled the snow down on the airfields. The packed surface made a good landing ground. Borrowing local machinery, Baldy had the squadron operating within a few hours – the only one in the south to be operational for nearly a fortnight.

Soon we were put on a fully mobile, expeditionary basis with additional personnel, vehicles, fuel tankers, insulated tents, ambulances, first aid – everything down to white cloaks and snow shoes and special intelligence data. We worked in shifts, twenty-four hours a day, to get the squadron ready. However, the poor Finns gave in and eventually all the assembled equipment had to be returned to maintenance units. We then reverted to our previous basis.

Within weeks, on 19 April, we were warned at short notice to move to Prestwick from where we should be embarking on an aircraft carrier. We guessed this meant Norway. Only minimum kit and equipment could be taken. Ground personnel, under Flight Lieutenant Tom Rowlands, the squadron's other flight commander, departed for Rosyth, while we flew up to Prestwick. There I was able to buy one *Daily Telegraph* war map which included part of Norway. We had no other maps of the country.

We boarded *Glorious* off Prestwick on 22 April. Much to our annoyance, Fleet Air Arm pilots, some of whom had had only limited experience of Gladiators, were detailed to fly our aircraft onto the carrier. One landed in the sea and was lost.

The crew of *Glorious*, which had been hurriedly recalled from a long spell in the Mediterranean, was in a poor, even mutinous mood. They had had no shore leave with their families after a protracted absence abroad.

D'Oyly-Hughes, the ship's captain, wanted the squadron to take off on 24 April at a point about 300–350 miles off the Norwegian coast and fly in from there to Lake Lesjaskog. He knew we had no maps of the landing area.

We felt this to be quite unreasonable. Baldy therefore asked the captain if we could be put off much closer in – 150 miles off Norway. Because we had no maps, he also asked that two navy Skua aircraft be put up to lead us to the frozen lake.

After 1½ hours flying, the Gladiators were landing on the ice. I flew on for another 60 miles or so to Dombås rail junction to reconnoitre the area. The main road and rail track between Aandalsnes and Dombås had recently been heavily bombed. On landing I reported to Baldy that there was intense air activity in the area and we must realize what we were in for.

Having got down on Lake Lesjaskog, the squadron found there were no refuelling tankers, only 4-gallon fuel cans and these were full of 100 instead of 87 octane spirit. This meant the engines would overheat and in due course seize up.

Moreover, there were no serviceable starter batteries so the aircraft had to be started by hand – difficult with a Gladiator. Then we found that the oil was the wrong (thin) grade. Instead of the thick, winter grade, the summer mixture would freeze in the Arctic temperatures. And a Gladiator needed a pint to a quart of oil after each flight.

On 25 April, our first morning, when the enemy attacked the lake, Baldy and I were stranded in our billets, 3 miles away from our aircraft, in deep snow. When eventually I got airborne, the enemy mounted another attack and two sticks of bombs fell close to Baldy. He was badly shaken and concussed.

After managing to refuel two Gladiators with the aid of two milk jugs acquired from a nearby farm house, I persuaded Baldy to join me on a sortie in the course of which we were able to engage two formations of six Heinkels which were approaching the lake. They were driven off by our attacks and did not bomb. We then engaged a lone Heinkel and, with a beam attack forced the German pilot to crash land the aircraft.

We returned to the lake after being airborne for over two hours. As I was touching I saw five Ju 88s beginning a dive-bombing attack on the lake. I at once climbed and engaged four of the 88s. The fifth attacked Baldy who had been covering me as I had started to land. He evaded by flying at nought feet down a narrow valley towards Aandalsnes, hotly pursued by the 88. He then performed a loop from ground level coming round onto the tail of the German aircraft as he pulled out. It was a remarkable manoeuvre. The 88 had to climb flat out to draw away and get out of range.

Meanwhile, the four Ju 88s I had been grappling with were now joined by the fifth. I was very much at a disadvantage because I had only one gun working. However, I managed to manoeuvre into a position from which I could make head-on attacks on the bombers with one gun firing. The German pilots were shaken by the tactics, re-formed and withdrew, leaving me in command of the lake.

After landing, I quickly went to our only Bofors gun and operated it against four attacking Heinkels; thereafter, I gave the order for the squadron to withdraw and prepare to move to Stetnesmeon, some 60 miles away, where Baldy had already landed safely. This was a large, serviceable field where we should have landed in the first place. The snow had melted and, with our normal camouflage, our aircraft would not have been so conspicuous to the enemy as they had been on the lake.

On 26 April, after flying a few more sorties over Dombås, Lille Hamaa and Aandalsnes harbour, the engine of one of the five remaining, serviceable aircraft seized up because the thin-grade oil had frozen; the pilot had been obliged to force land in the mountains.

The next day, after setting our four remaining aircraft alight, we were evacuated from Norway in the motor vessel, *Delius*, which was attacked for five hours by Heinkels.'

On the squadron's return, first to Scotland and then to Northolt, near London, Donaldson and Mills were subjected to detailed debriefing at the Air Ministry. Mills has recorded his impressions of these high-level meetings.

'When we pointed out the difficulties we had been confronted with – wrong aviation spirit, wrong oil, no serviceable starter batteries, unsuitable equipment, no maps, lack of ammunition – senior officers were unable to give any answers. We were simply told: "You appreciate the squadron was sent to Norway as a token sacrifice."'

The two were summoned to report to the Secretary of State for Air, Sir Samuel Hoare, at his room in the House of Commons. He was preparing to make a statement to MPs only hours before the fall of the Chamberlain administration and the formation, on 10 May 1940, of the great wartime Coalition under the leadership of Winston Churchill. Mills recounts the circumstances.

'We had been told by senior officers at the Air Ministry that on no account were we to relate to the Secretary of State any of the mistakes which had been made in Norway, but, rather, to paint as favourable a picture as we could. This Baldy had to do. I felt this whitewashing was all wrong and it was the reason why the reports were padded. The Norwegians have copies of Hansard and they know now that the statements did not tell the real story.'

The reception which these two officers received from Hugh Dowding, with whom they dined at Fighter Command, was in sharp contrast. The C-in-C, who was now the most senior serving officer in the Royal Air Force, was always ready to listen carefully to his junior officers. He prefaced the discussion by reading to Donaldson and Mills the citations which supported the decorations which each had just won – a DSO for Donaldson and a DFC for Mills. Thereafter, he absorbed the two officers' views and comments on their extraordinary experience. He expressed himself to be much in sympathy with their objective conclusions.

Back with the Air Staff at the Air Ministry, Donaldson and Mills were faced with the 64,000 dollar question. It is still etched deeply in Stuart Mills' mind.

'We were asked directly if we were prepared to go out to Norway again with the squadron in support of the forthcoming Narvik operations and, if so, what we considered our requirements to be.

We had no hesitation in expressing our willingness to go back, but we made the squadron's needs quite clear. We asked, basically, that our ground personnel and all necessary equipment be in position before the squadron arrived. There must be no repetition of our earlier experience.

I then submitted my forthright opinions regarding the use of the carrier, HMS *Glorious*. For these I was severely reprimanded although the sinking of the ship, with two battle-trained Royal Air Force fighter squadrons on board, on 8 June, during the evacuation, substantially vindicated my statements.'

The second supporting operation now went ahead. 263 Squadron, sailing this time with its Gladiators in the carrier, Furious, *was joined by 46 Squadron, under the command of Squadron Leader K.B.B. Cross (Air Chief Marshal Sir Kenneth Cross), 'Bing' Cross to one and all, whose Hurricanes were embarked in* Glorious. *Bardufoss, in the Narvik area, a Royal Norwegian Air Force station, was selected as the mainland operating base.*

Mills has described the take off from Furious.

'The weather was very bad, visibility was poor and it was snowing hard with intermittent blizzards. Baldy asked me to take Pilot Officer Richards and Pilot Officer Francis with me and try to get through. A Swordfish from *Furious* would lead us in.

'Baldy then took off with fifteen Gladiators and two Skuas to navigate, but after a short while decided it was too bad to continue and landed back on the carrier without incident.

'The carrier, which had not obtained a sighting of land all the days it had been at sea, was, in fact, 60/70 miles out of position. Instead of being opposite Bardufoss it was nearer Torsking Island and we were confronted by mountains rising to some 4000 feet.

'We pressed on, climbed to 1500 feet and, then, in the appalling conditions, there were three crashes almost simultaneously. The Swordfish hit a mountainside, but the crew escaped. Richards flew into the face of a mountain and was killed. My Gladiator hit 2000 feet up a mountain and at once burst into flames; I was able to get out. Francis, alone, got away with it and, with a remarkable piece of navigation, eventually made Bardufoss.

'I was taken, with Richards, whose body was a terrible sight, in a motor boat to Harstaad where he was buried. I was then moved, under medical escort, to Tromso, and was later evacuated in the cruiser, *Devonshire*, with the King of Norway, members of the Norwegian Royal Family and Parliament and a consignment of gold.'

(Group Captain R.S. Mills, Winchester, 1982.)

For a fortnight or so 263 Squadron and 46 performed heroics, destroying during the period some 50 enemy aircraft. On 7 June the order was given to evacuate. There would be no exceptional difficulty in flying 263's Gladiators onto the flight deck of HMS Glorious. *The Hurricanes, however, had never landed on a carrier before. The Air Ministry had concluded, after trials, that it was impracticable. Cross was, therefore, ordered to destroy his precious aircraft and deny them to the enemy.*

However, he had other ideas. He knew what the ten serviceable Hurricanes which remained would mean to Fighter Command if they could be flown onto the carrier and taken safely back to Scotland. In one of the most courageous flying decisions of the war, Cross elected to take off in the middle of a bright, Arctic night and have a go. He recorded the story three weeks after the event in a letter, dated 26 June 1940, written from Gleneagles where he was then in hospital, recovering from his horrendous experience. It was addressed to a friend in a law firm in the City of London, which had relations with one side of the Donaldson family. The original is retained in the Donaldson Archives.

'... I knew the chaps on the *Glorious* would be full out [to make the landing possible] and this was eventually arranged .... Of course, it never really got dark up there and it was as light at midnight as it was at midday....

'At midnight I called for volunteers for a shot at landing ... and of course the boys stepped forward to a man. There was nothing for it but to pick the senior ten. We had only ten serviceable aircraft by this time. We left at 0045 hours ... and were navigated out to sea by a Swordfish at 100 knots and the old Hurricanes had to do some fairly hearty zigzagging to keep behind.

'It wasn't a nice feeling knowing that if we couldn't get on the deck there was no way out.... However, we had taken the precaution of having a sandbag weighing 14 lbs put in the tail so we could use our brakes pretty coarsely.

'The navy were full out and there was a fresh breeze blowing [so] we had 33 knots over the deck and [the aircraft] all came on like birds. The last one landed at 0300 hours.... Most of the boys were pretty tired and after some ... welcome eggs, bacon and cocoa we all turned in ....'

The next day, 8 June 1940, death struck in the Arctic. *Glorious* was attacked by the two German battle cruisers, *Scharnhorst* and *Gneisenau*. Cross, one of the 39 survivors out of the ship's complement of 1400, had the dreadful experience starkly in mind as he concluded his letter.

'... All the boys went to their "Abandon Ship" stations and when the order came we went over the side. The whole thing was over in 45 minutes. I swam to a raft and a few minutes later young Jameson [Air Commodore P.G. Jameson, who was later to lead the Norwegian wing in Fighter Command] came swimming along. We eventually had 29 aboard, but after three nights and two days, when we were picked up, we had but 7 left of whom 2 died later....

'The boats that got away [after the sinking] were sunk by heavy seas, but in most cases they had been so badly holed by gunfire that they sank as they launched. When we were in the raft the Germans came up, had a look and then went straight away. I have a real hatred for Germans now....'

(Air Chief Marshal Sir Kenneth Cross. Letter dated 26 June 1940, The Donaldson Archives.)

Stuart Mills, by an unlikely quirk of fortune, was on the bridge of the cruiser, Devonshire, *acting as an air raid identification officer, when* Glorious *made her last two W/T messages. The staff officer to the Admiral, a naval commander, was standing beside him. When the plight of the carrier was known, Mills asked the commander: 'Why aren't we going to her rescue?' He recalled the reaction.*

'There was no reply. The effective answer was given when *Devonshire* was put on full emergency speed and headed out into the Arctic. Apart from the presence of the Norwegian Royal Family on board, the reasoning may well have been that the carrier, *Ark Royal*, and the rest of the Fleet covering the evacuation of British and Norwegian forces, were closer. But the Royal Navy have been silent on the issue ever since.

'So poor Baldy and my young and gallant friends, who had done so wonderfully well at Bardufoss with their slow aeroplanes against Heinkels and Junkers 87s and 88s, had a dreadful end.

'Bing Cross told me later, when he was in hospital recovering, that the last he saw of Baldy was lying on a bunk in his cabin. 'Come on,' said Bing, 'we're getting off.' But Baldy made no reply and didn't move. He was never seen again.'

(Group Captain R.S. Mills, Winchester, 1982.)

Prophecy

Air Commodore E.M. Donaldson, The Donaldson Archives, Gosport, 1982

There are three reflections to add to complete the poignant Donaldson saga.

When Teddy Donaldson, then on the threshold of his own success at the head of 151 Squadron in the Battle of France – and the winning of the DSO – said goodbye to his brother as he was about to leave for Norway, there was a prophetic parting.

'. . . Jack ['Baldy' was always 'Jack' in the family] said, and it really upset me, that he would never see me again and I must protect and look after Mummy. I said: 'What nonsense. The Donaldsons are indestructible.' But I was never able to cheer him up or get him to believe he would see the disastrous Norwegian campaign through.

'I was miserable until, some weeks later, 'Mac' [Air Chief Marshal Sir Theodore McEvoy] came down to see me at Selsey to give me personally the awful news. . . .'

Dowding's Despatch

The Air Officer Commanding-in-Chief, Fighter Command, in his Battle of Britain despatch, submitted to the Secretary of State on 20 August 1941, went out of his way to pay a tribute to Baldy Donaldson and his squadron.

'The fighting in Norway has only an indirect bearing on this paper. Certain useful tactical lessons were (however) gained, particularly with regard to deflection shooting, and I trust that the epic fight of No. 263 Squadron, under Squadron Leader J.W. Donaldson DSO, near Aandalsnes may not be lost to history....'

CAS Reflection

Marshal of the Royal Air Force Sir John Grandy, November 1982

To the wartime generation of Royal Air Force flyers, John William Donaldson, unlike his two brothers, Teddy and Arthur, was an unknown figure. What, then, were the characteristics of this legendary officer whose achievements, Dowding contended, should be retained for history.

None knew him better from the early days than Marshal of the Royal Air Force Sir John Grandy, who led 249 Squadron in the Battle of Britain and, later, became Chief of the Air Staff.

'I first met J.W. Donaldson on 11 September 1931 when we reported, as pilot officers on probation, for training at the Royal Air Force Depot, Uxbridge. There followed two weeks of some blood, much sweat and, as far as I remember, very few tears. This was known as the Knife, Fork and Square-bashing course and its object was to knock us all into some semblance of shape before we reported to No. 5 Flying Training School, Sealand, in Cheshire, to learn how to fly and ultimately how to become Royal Air Force officers.

We were there for the best part of a year until, proudly wearing our hard-earned wings, the coveted Royal Air Force pilot's flying badge, and dressed in stiff white collar, breeches, puttees and boots, with brown leather gloves and regulation cane, we reported in the late summer of 1932 to our squadrons, having dispensed with the "on probation".

We had made it: or so we thought. The great disillusion was yet to come.

As we all very soon discovered, pilot officers, despite being no longer "on probation" fell very much into the "to be seen and not heard" category in those early months of squadron life. Our training had, in truth, barely begun; that ultimate aim was yet to be reached.

And what of J.W.D. in all this?

Well, the first point to be made is that he was older than any of us. He looked older, too, having even then very little hair on his head. He had been a wireless operator in the Merchant Navy; this, of course, gave him an enormous cachet. He had been around, seen the world.

We were soon to discover the man. The overriding quality was, I think, an irrepressible self-confidence. I would imagine his flying instructors must have been hard put to cut him down to size because, and I have no doubt of this, he was a born aviator; it just came to him, and he had dash.

His aerobatics, beautifully flown, were usually just bordering on the height limits; his low-flying practice in the low-flying area over the upper reaches of the Dee estuary was just a bit nearer the mud than anyone else's. As the aircraft was somewhat underpowered, even our instructors could not always guarantee a truly vertical upward roll in the Siskin, the fighter on to which we converted from the Avro 504N. Not so Baldy: he got that old Siskin going in a dive and then eased it up into a perfect manoeuvre every time, which was delightful to watch. He would frequently land just that bit nearer the tarmac than authorized and in those clandestine, forbidden formation efforts we all had a go at, Baldy would get closer to you than any other ever dared and stick to you like a leech – disturbing. Even in those early days he just had the edge; he was a natural.

The reader will be thinking how insufferable this man must have been. But the extraordinary thing is that this was not the case at all. He was great company, a true and loyal friend, ever ready to help others, most generous, great fun to be with and highly popular.

I was very fond of him and despite going our different ways on leaving Sealand, he to No. 1 Squadron at Tangmere in Sussex to fly the Hawker Fury, that superb masterpiece of the designer Sydney Camm, and I to No. 54 Squadron at Hornchurch in Essex to fly Bristol Bulldogs, we continued to keep in touch. I have warm memories of him taking me in his very smart MG to meet his remarkable mother. He and his brothers were as justly proud of her as she was of them. We served together in 1936/37 as flying instructors in the London University Air Squadron, flying from Northolt.

J.W. Donaldson, never slapdash or foolhardy, was an outstanding fighter pilot. He was thoughtful, highly professional and an exceptionally good air-to-air shot. A determined character, his loss so early on with the sinking of the *Glorious* was a tragedy – perhaps the most profound and far-reaching result being the consequent denial to the Royal Air Force of the inspired leadership, already apparent, which would undoubtedly have developed.'

Escape

Lieutenant-Colonel Peter M. Klepsvik, Sola, Norway, 1982

The Nazi occupation of Norway was the precursor of many brave escapes. The tough Norges, like their counterparts among the Poles and the Czechs, the Dutch, the Danes, the Belgians and the French, were resolute in their refusal to accept subjugation. A chance to fight back, to take the contest to the enemy, was worth the acceptance of ultimate risks. Peter Klepsvik's get-away ranks high in the catalogue of human endeavour.

'Why a small rowing boat? Why not a larger vessel, preferably a motor or steamboat?

The answer was simple: there was no larger boat in our possession on the island.

The island – Lindøy in Finnås-Bømlo, approximately 3 miles long and one mile wide, housed my parents, five sisters, four cows, one horse, one pig, a couple of dozen sheep, some hens – all tied into a happy combination of a small farmer-fisherman's way of family life. Not much to live on, you might say. Perhaps so. But, thinking back, we were content and reasonably happy. We all helped to make ends meet.

I had already had experience of war, having been held back by the Russians in Arcangelsk in November/December 1939, when bound for the UK with timber. However, we were eventually released and made our way to Hull early in 1940. On my return to Norway, I signed off, after being called up to serve in the navy in May.

At twenty, I was, I considered, an experienced and grown-up sailor.

Came 9 April 1940. What a day! Our defences, having been totally neglected, had no chance whatever against so powerful an enemy. We had been spared wars for almost 130 years. So I hurried to report to our local county sheriff, whose duty it was to direct the local mobilization. I had no previous military training, so I was ordered to return home to await the call.

By 16 April, the western part of Norway had been overrun and had surrendered to the enemy. Fighting was still going on in the north, but a surrender there was inevitable also. One evening Johannes Baldersheim, a friend of the family, came to our island. Johannes had already been to the local sheriff protesting at the surrender. This was, he claimed, in contravention of the Norwegian Constitution. He suggested that we should escape to the UK. As we possessed only small rowing boats (the type is called Oselver in Norwegian), we soon agreed to use one of these rather than try to steal a larger vessel. Our country was at war. We knew what kind of enemy we were up against. There was no hesitation. Our duty was clear. My brother, Olav, a steward on a Norwegian ship, was already in the service of the Allies. We began to prepare the following morning.

We needed a sail. This, we bought from a neighbour for Kr 10 (about £1). We found another neighbour who had an old lifeboat compass. We bought it for Kr 30 (£3). We found some cork and nailed it underneath the seats of our rowing boat. This would keep us afloat if we were holed by machine gunning. We thought about the most important life-preserving item – water! My father had a couple of small old wooden kegs. They were dry after twenty-five to thirty years' storage. So we sunk them in a well to swell the wood and make them watertight – but they simply fell apart. The only alternative was to use regular zinc buckets. We filled two of these with the most wonderful well-water of the island.

We had time to visit Sheriff Henrik Robberstad. He was well aware of our intentions, and was later to become one of the key men of the local resistance movement. He issued a proforma, handwritten passport to Johannes and put his official stamp on mine which was already valid. It had recently been stamped in the UK.

The next item was food. Mr Skimmeland, who ran the local general shop, offered tinned fishballs, meatballs and various other consumables, insisting on post-war payment!

On 16 May 1940, we were ready to leave. My parents, sisters and the pig saw us off. My mother cried, blessed us and then prayed for a safe journey and a welcome back to a liberated Norway.

We had previously planned to set course west by south-three-quarter south, heading for Kinnairds Head, the nearest point in Scotland, some 300 nautical miles away.

17 May is the Norwegian day of liberty. It found us outside Norwegian territorial waters. Symbolically, we enjoyed the same kind of perfect freedom that our forefathers had experienced on 17 May 1814, after 400 years of Danish domination.

The North Sea was calm, just like a mirror. We were both rowing together all the time. Heinkels, Junkers and Dorniers passed overhead, flying north. They had nasty-looking swastikas on their tails. They seemed to ignore the little nutshell underneath.

We continued to row our four-oared little boat, keeping a steady course for 'Blighty'. By the following day the coastline of Norway was no longer visible, but the mountain, Sigjo, could be seen for well over 48 hours.

On the third night out we saw lights, but not knowing whether they were friendly or hostile we did not dare to try to attract attention. That night we passed a lot of horn mines and a floating life raft.

On the third morning the weather began to change and the wind from the northwest became increasingly uncomfortable. Rest or sleep was out of the question. During the day the northwesterly wind increased to a gale and we had to reduce sail to a little triangle. The rest of the journey was a battle for life with the sea and the wind. It was exhausting. Noticing that the current was running against the wind, we decided to change course to a more southerly direction. We did this three times, amounting to a variation of almost 30 degrees.

The waves breaking over the boat destroyed our water and bread. We had no time to open tins or get anything to eat; and there was nothing to drink. This situation lasted for about thirty hours altogether. At the end, we were so tired that the mind seemed to stop working. We continued sailing and baled automatically.

On the fifth morning the wind died and misty rain set in, but there was still a very, very heavy sea running. Suddenly, we heard a fog signal. It later turned out to be from exactly the spot for which we had been heading – Kinnairds Head, only 1 or 2 nautical miles away.

Then out of the mist loomed a grey shape. Johannes saw the ship's flag. He said at first he thought it was a swastika. But, as we drew nearer we realized we were being approached by a British ship, flying the White Ensign. We had a small Norwegian flag which we promptly hoisted on our own mast. As we passed one another, a hard Scottish voice shouted down: 'Where do you come from?' I shouted back 'We are coming from Norway.' Nothing more was said until the ship had turned round and come back to us. Then the same voice shouted: 'Where the hell do you say you've come from?'

I gave the same answer.

Things then started happening very quickly. The ship, an armed trawler, commanded by Skipper Lt. John Brebner, from Aberdeen, came alongside and, in a moment, half a dozen hands were dragging us over the rail and onto the deck of the trawler. They took our boat in tow.

The skipper offered us a drink. Baldersheim, being a teetotaller, refused. I took a drink of whisky. It just about knocked me out. We were treated like admirals, everybody trying to be of service. We were given food and put into bunks to sleep.

Then came the real trouble – CRAMP. My limbs would not function normally. Because of this, and despite our exhaustion, it took more than three hours to get to sleep.

The trawler which was escorting a convoy, took us to Kirkwall where we were accommodated in an old people's home. After staying there for two days, we joined a Norwegian merchant vessel. I sailed the Atlantic for a year, eventually signing off in London at the end of May 1941. I then joined the air force and was sent to Canada to train as a wireless operator mechanic.

Baldersheim spent the whole war in the merchant navy, having made a personal pact not to take a day off as long as the war lasted. Johannes kept his promise and I have remained in the Royal Norwegian Air Force ever since.'

'My Duty'

Capitaine François de Labouchère, 249 and 615 Squadrons, killed in action, 5 September 1942

Why did I go to England? Because I felt it was my duty. It was seeing where one's duty lay, not doing it, that had become so difficult.

DEFENCE OF AN ISLAND

'Their Finest Hour'

Winston S. Churchill, *World War Two, Volume II: Their Finest Hour* (Cassell, London; Houghton Mifflin, New York)

'... What General Weygand called the Battle of France is over. I expect that the Battle of Britain is about to begin. Upon this battle depends the survival of Christian civilization. Upon it depends our own British life, and the long continuity of our institutions and our Empire. The whole fury and might of the enemy must very soon be turned on us. Hitler knows that he will have to break us in this island or lose the war. If we can stand up to him, all Europe may be free and the life of the world may move forward into broad, sunlit uplands. But if we fail, then the whole world, including the United States, including all that we have known and cared for, will sink into the abyss of a new Dark Age made more sinister, and perhaps more protracted, by the lights of perverted science. Let us therefore brace ourselves to our duties, and so bear ourselves that, if the British Empire and its Commonwealth last for a thousand years, men will still say, "This was their finest hour."'

Comparison

Air Commodore E.M. Donaldson, The Donaldson Archives, Gosport, 1982

France was far worse than the Battle of Britain.

Engaging Baptism

Group Captain Edward Preston Wells, Marbella, Spain, 1982

Like others among that remarkable breed of New Zealand contemporaries, Edward Preston Wells, from Cambridge (they called him 'Hawkeye' because of his eyesight), earned the right to a place alongside the exceptional characters – and leaders – of the British and Commonwealth air forces. The Luftwaffe were beginning to turn the heat on Britain as Wells finished his operational training.

'Looking back, I see that it is almost exactly forty-two years ago that, at the end of June 1940, I, in company with about ten others, arrived in the UK as a pilot officer in the Royal New Zealand Air Force. We had already qualified in New Zealand as pilots on the, even then, obsolete Vickers Vincent biplane. This gentle giant with all its bracing struts slightly resembled a birdcage on stalks, but it could actually fly; take off was at 75 knots, cruising speed was 85 knots and top speed about 95 knots per hour. It hardly seemed a very suitable training machine for those of us who within a few weeks would be flying what was, at that time, the most modern high-speed monoplane with retractable undercarriage, flaps, variable pitch propeller, oxygen and many other sophisticated pieces of equipment, which together made up that marvellous flying machine, the Spitfire.

At the time of our arrival in the UK, France had fallen and there was a palpable feeling of tenseness and excitement in the air and I was, of course, mad keen to start flying operationally. However, even in such a crisis period, certain administrative requirements had to prevail and we were sent to RAF Station, Uxbridge, to get kitted out with all the proper flying equipment and to await a vacancy at one of the Operational Training Units. Here, in a few days, we would be shown how to fly a Spitfire. There was no time, as I later discovered to my cost, to learn how to operate these aircraft properly, which is a very much more exacting requirement than just to be able to fly them.

During our short stay at Uxbridge, we used most of our limited free time in the evenings and at weekends to go roaring up to London, in order to 'beat the place up a bit' as we used to say. These expeditions were carried out in decrepit old cars known as 'bangers', which were apparently owned by nobody but borrowed by everybody. Apart from shortage of free time, the other great limitation on these sorties was finance, or rather the lack of it, as in those days pilot officers received the princely sum of 10/6 per day (52½ new pence). However, in spite of these difficulties, we managed to have a really marvellous time and I had the additional good fortune to meet a charming Dutch girl, who some years later became my wife, which she still is.

After a couple of weeks some of us were posted to No. 7 OTU at RAF Station, Hawarden, where I gazed in awe and excitement at the rather battered collection of Mark I Spitfires which we were to fly. Various lectures on the technical features of these aircraft, plus some rather airy-fairy talks on what were alleged to be the enemy's tactics in the air and what was supposed to be the best way of dealing with them, filled the next few days. To us at that time, these talks were pure gospel and we drank in every word, as our lecturers were young pilots with operational experience. Some had just returned from France, where most had been flying Hurricanes and had seen quite a bit of action as the occasional DFC testified.

The next step was a few hours' dual instruction on a Miles Master aircraft to familiarize us with such things as retractable undercarriages and other sophisticated equipment, and also to give us the feel of a comparatively high-speed monoplane. I had 3½ hours of this dual before the great day came and I was at last sent off solo in a Spitfire. I will always remember, not only how slim and slight it seemed in the air, but also how it surged with a power such as I had never felt. When I actually landed this beautiful, but fragile, creature without damage, I was prouder than any two-tailed dog!

That evening proof came that this beautiful machine was also effective and deadly. It was the custom, after flying for the day had finished, to assemble at the tent which was our bar, for a glass of beer and a chat with the instructors. It was ordered that two Spitfires should be kept nearby fully armed and at readiness during all daylight hours. It was not yet dark and there was a low cloud base of some 2000 feet. Suddenly, while we were looking out of the tent, there was a loud burst of heavy machine-gun fire and we saw that a Dornier 17 had slipped out of the cloud, apparently without noticing us, and was machine-gunning certain factories, which stood at the far side of our airfield.

In a flash, the Duty Instructor, Squadron Leader J.R. Hallings-Pott (Group Captain J.R. Hallings-Pott) jumped into the nearer Spitfire and, within seconds, was roaring across our grass airfield, without regard for wind direction, straight towards the Dornier, which was just starting a second attack. The latter obviously never knew what hit him, because the Spitfire was still climbing towards him, firing as it came. We could all see the flashing of the de Wilde incendiary ammunition striking the Dornier and its nearer engine, which almost at once burst into flames. Within two or three minutes the enemy aircraft had crashed just off the airfield and the Spitfire was taxiing back to its readiness position.

I need hardly say that there were very hearty cheers for both the pilot and the Spitfire, and that night we drank rather more beer than we usually did. It seemed a very good end to a splendid day and one which I have never forgotten. I felt I had indeed come a long way in a short time from the peaceful green hills of New Zealand.'

Lord Beaverbrook
Minister of Aircraft Production, 14 May 1940
Member of the War Cabinet, 2 August 1940

Harold Balfour, *Wings Over Westminster* (Hutchinson, 1973)

On becoming Prime Minister, Churchill had set up the Ministry of Aircraft Production under Beaverbrook. Within hours of his appointment

Beaverbrook asked me to meet him at Stornoway House. First persons I met there were Sam Hoare, the newly-sacked Secretary of State for Air, 'Jay' Llewellin, the new parliamentary secretary, and a host of Beaverbrook's newspaper staff. Beaverbrook was carrying on three conversations at the same time, telephones were ringing and in spare moments he talked into a dictaphone. He asked for my help in what he described as 'a hell of a task Churchill had given him'. From that hour on Beaverbrook threw overboard every Whitehall custom and practice. At once conflicts started between Air Ministry and MAP.

In the Garden of Eden a rib was taken from Adam in a deep sleep to create woman. Beaverbrook took not only one rib from the Air Ministry to create his department but a whole flank. The first big battle was over whether the Air Staff or Ministry of Aircraft Production had the right to say what aircraft should be ordered. MAP claimed that while Air Staff needs would be met as far as was industrially possible, they could and would alter production schedules as they chose. Next, Beaverbrook claimed the right to deal with the Admiralty direct and supply their needs direct and not, as for years past, through the Air Ministry. Worse shocks were to come. Beaverbrook claimed the right to choose the air force officers to be attached to his department and, most shocking of all, to promote them out of term as he wished. This really put the cat among the personnel pigeons at Adastral House. Beaverbrook had Churchill's backing. He got his way in all these disputes.

A new spirit flowed through the world of aircraft manufacture and repair. Hurry, hurry, hurry, for we have so little time. Then the Battle of Britain was on us. The pirate of Whitehall really hoisted the Jolly Roger so far as service practice and methods. The old Air Ministry supply system said that if one Hurricane was at Cramlington, Northumberland, unserviceable with a broken airscrew, the damaged component had to be replaced by indent through its own group. If another Hurricane was unserviceable at Northolt with a broken undercarriage, the same local indenting procedure for replacement parts had to be gone through. Beaverbrook would have none of this. To the confusion of accounting and issuing departments now the two unserviceable Hurricanes would be thrown together to make one airworthy fighter. Technical experts laid down that each set of piston rings had to be fitted carefully to individual Rolls-Royce Merlin engines. Now any set of rings was to go into any Merlin and they all worked just as well as before.

Any hen roost was robbed to produce just one more fighter. Risks were taken. Experimental units were robbed of test aircraft. Training stations had to give up Service types. Factory management was bullied, praised, cajoled. Each week's effort was taken as the starting point of exhortation for a bigger and better next week. Bosses of great aircraft concerns learnt not to play golf on Sunday mornings but stand by the telephone for this was Beaverbrook's favourite hour for telephone talk with these men. His minions went round the country with delegated powers to requisition any and every building wanted for dispersed component manufacture or for storage. The Archbishop of Canterbury complained to Churchill. One of Beaverbrook's travelling staff served a requisition on a garage in Salisbury where the Bishop and other clerics of the Cathedral housed their cars. The Bishop protested whereupon the MAP official took out a blank requisition form and threatened the Bishop that any more resistance and he would requisition the Cathedral.

'The House That Max Built'

David Farrer, *G– For God Almighty: A Personal Memoir of Lord Beaverbrook* (Weidenfeld and Nicolson, 1969)

Beaverbrook's critics have always maintained that he was little more than a play-actor and that his main object was self-glorification. The same criticism might also be levelled at the magnificent and flamboyant rhetoric to which at the same time Churchill was treating the House of Commons. Certainly the personal touch which the Minister of Aircraft Production used with high and low made him for a time the second most popular man in the country. It was rough magic worked with a nasal twang. It was effective as never quite again. In any case his critics were wrong. In all the six and a half years I worked for him I never knew him so selfless as in the months of his adopted country's greatest danger.

Nor could play-acting have got the results which were achieved. They came in the last resort from a rapid mastery, by the quickest and sharpest brain I have ever known, of all the problems and intricacies of the aircraft industry, allied to an incomparable drive. The final ingredient was a ruthlessness bordering sometimes on sharp practice which was perhaps natural in a man who had made a million before he was thirty, but which now was applied in the national interest. The 'personal touch' was the ingredient which fused these qualities into success.

Lord Salter (then Sir Arthur Salter) tells a story of Beaverbrook at his most ruthless and agile in those days. Salter was then Second Minister at the Ministry of Shipping, and they sat together in the ante-room at 10 Downing Street awaiting appointments with the Prime Minister. It was just after the fall of France, and from scraps of conversation from within Salter could tell that the argument was over what to do with a consignment of arms previously destined for the French. At length Churchill's visitor was ushered out – and Salter discovered that Beaverbrook had vanished. The Minister of Aircraft Production, who was not then even a member of the War Cabinet, was already on the telephone to the port where the consignment was held up. Before anyone quite knew what was happening it had become the property of MAP. How he got away with it Salter does not know. He suspects that Beaverbrook claimed the authority of the Prime Minister. It may not have been necessary. By this time, with MAP barely six weeks old, it was widely felt not only that it held the chief clue to success or failure in the battles that lay ahead, but that its minister was Churchill's 'favourite son'.

For me, who in this doom-laden time of late spring and early summer saw him day in day out (including Sundays), and sometimes part of the night as well, the days had often the quality of a nightmare. The hasty scurryings between Stornoway House and Millbank, the increasing number of letters and documents for which I had to provide first drafts, the often brusque and always telegraphese instructions which had to be interpreted for onward transmission in the shortest form possible to the relevant individual or department, the absolute uncertainty as to what I was going to do next – all this put a strain on Beaverbrook's junior personal secretary that left its mark. Whenever I was not with him and my telephone rang I would jump in near panic out of my chair before I could bring myself to answer it. I had, after all, been catapulted into a strange alarming world for which nothing in my previous career had prepared me.

An even greater strain was placed on the senior personal secretary, George Malcolm Thomson. At that time and thereafter, he played a far more important role than I. I grew, I think, increasingly skilful at being 'His Master's Voice'. George was 'His Master's Ear'. He was his listening post, and where I observed, he gave advice which was listened to, and which on many occasions – though the Master never openly admitted it – caused a reversal of what Beaverbrook had originally intended.

It was a nightmare, but one in which, like other junior members of 'Operation Beaverbrook', I grew increasingly proud to share. For Beaverbrook it was a nightmare of a different kind, for if his nightmare turned into truth there would be a German occupation of Britain. This nightmare can be simply described. On the day he took office there were more trained pilots than there were aircraft in the front line of the RAF, and in Fighter Command there were only five fighter aircraft immediately available in reserve. The tensest, most nervous moment of the week was Saturday afternoon when Beaverbrook received the weekly production charts. From the start they rose steeply, and before the Battle of Britain was finally joined they showed a total reserve of 65 per cent of the Fighter Command's operational strength, which at the same time exceeded the number of trained pilots available to man them. Such was the achievement which by every manner of means, adding up to an inspired unorthodoxy, Beaverbrook had made possible. It came just and only just in time. In my view the Minister of Aircraft Production had his 'finest hour' even before Churchill had called upon the people of Britain to rise to theirs.

So, as far as MAP was concerned, by early August the stage had been set, the chief actors assembled, for the high drama of the Battle of Britain and the Blitz. There had been minor dramas already, with happy endings and Beaverbrook taking all the bows while distributing bouquets to all around him. And there had been comedies too, in one of which *in absentia* I took part. Beaverbrook had instituted a rule whereby letters written for his signature by Thomson had to have a small 't' in the top right hand corner, those written by me a small 'f'. One evening about half past eight, with his full council surrounding him, Beaverbrook pressed the bell which summoned me to his presence. I had gone home, so Thomson, with an air of conscious virtue answered the summons. Beaverbrook looked up as he entered the room. 'Where's little f?' 'Little f,' replied Thomson, 'has effed off.'

In the ensuing gale of laughter at this unexpected sally, my truancy was forgotten. It was a very human ministry to work in, led by a very human man.

Declaration

Harold Balfour, *Wings Over Westminster* (Hutchinson, 1973)

I do declare that from my first-hand knowledge, even though on the Air Ministry side of the fence, the margin for victory in terms of aircraft and pilots was knife-edged.

I do declare that if there had been no Beaverbrook and no MAP I do not believe the balance would have been tipped in our favour.

The Battle of Britain, a Polish View

The Polish Air Force Association, *Destiny Can Wait* (William Heinemann, 1949)

Prosaically, the Battle of Britain was the air engagement between the Royal Air Force and the Luftwaffe above the English Channel and southeast England in 1940 after the fall of France. It was the first air battle on a major scale ever to be fought, and upon its issue depended nothing less than the invasion of England. As it was Poland which had first fought the German menace, it was only right that Poles should play an important part in a battle which was one of the crucial battles of the world.

The essential condition for a successful Nazi invasion was mastery of the air over south and southeast England, and for this condition to be realized it was necessary to destroy the fighter strength of the RAF. Only then would the Germans have been able to launch their bombers against the ships of the Royal and Allied navies and the coastal defences of Britain in sufficient strength to prepare the way for actual landings, send aircraft at will to paralyse the transport system of southeast England and London, and so disorganize any relief forces the British sent to the invasion coast. The Battle of Britain was conceived by the Germans as the first stage – the air stage – of an offensive which would ultimately involve all arms of the German forces and end with the occupation of the British Isles.

The Germans had available for their plan about 3500 first-line aircraft, backed by reserves of about 60 per cent. This force included about 500 Junkers 87 dive-bombers, some 1500 bombers – Dornier 17, Heinkel 111 and Junkers 88 – as well as about 300 medium aircraft of the same types for reconnaissance. The fighter force contained about 850 single-engined Messerschmitt 109s and 300 Messerschmitt 110s.

The number of aircraft actually engaged in the battle by the Germans never exceeded 1750, of which not more than about 1500 could be engaged on any one day. Most of these were deployed in Luftflotte 2 and Luftflotte 3, between Amsterdam and Brest, under Field-Marshal Kesselring and Field-Marshal Sperrle respectively....

By July 1940 the strength of Fighter Command had risen to 52 fully operational squadrons – 20 Spitfire squadrons, 22 Hurricane squadrons, 2 Defiant squadrons and 8 Blenheim squadrons; a further 8 squadrons were forming or reorganizing after the French campaign. The effective daily fighting strength amounted to about 500 Hurricanes and Spitfires, as the

Blenheims, and later the Defiants, could not fight effectively in daytime.

This was the force that the Polish pilots joined. They proved a welcome and timely reinforcement, for Fighter Command was seriously short of trained pilots. There had been heavy casualties in France and Belgium and the training of fresh pilots takes time – normally about eighteen months, and never less than a year. Moreover, new pilots are sometimes more of a liability than an asset to a fighting force; they are more likely to damage equipment and they are an anxiety to more experienced pilots. All the Polish fighter pilots, however, had had at least two years of thorough training and an average of 500 flying hours. Many of them had already met the enemy in the air and in spite of flying inferior aircraft they had survived, and that in itself was a form of natural selection of the most skilful and resourceful pilots.

All these facts weighed in favour of giving the Poles a chance of meeting the Luftwaffe in good aircraft, backed by a good organization. They were to have for the first time an opportunity of showing what they could do, fighting the enemy on almost equal terms – although still outnumbered.

Into the Line

New squadrons were hurriedly being formed to meet the challenge from the Luftwaffe. No sooner were they trained and considered operational than they were pitchforked into the line to face the 'fury and might' of the enemy.

Typical of this process was 249 Squadron, re-formed, commanded and made operational within six weeks by Squadron Leader John Grandy (Marshal of the Royal Air Force Sir John Grandy). Its experience is specially significant because, from this momentarily-created base, it was to develop into one of the Royal Air Force's most successful wartime fighter squadrons. Indeed, Christopher Shores, the distinguished air historian, was to single it out, a quarter of a century later, for mention:

> 249 was to become *without doubt* [editor's italics] the highest scoring squadron of the British Commonwealth Air Forces....*

Also unique among its distinctions was the award of the Victoria Cross, the supreme wartime honour, to one of its members, Flight Lieutenant J.B. Nicolson. On 16 August 1940, Nicolson, with his Hurricane on fire and cockpit flames already torturing his body, drove home an attack on a German Messerschmitt 110 and shot it down before baling out. He was the only Royal Air Force fighter pilot to be awarded the Victoria Cross in World War II. Sadly, he did not survive the conflict, being killed in the Far East just before the end of the war.

Grandy has recorded his recollections of the squadron in those tumultuous summer and autumn days of 1940 and its involvement in the battle.

Marshal of the Royal Air Force Sir John Grandy, November, 1982

'On 16 May 1940, I received orders to re-form † No. 249 Day Fighter Squadron at Church Fenton, Yorkshire. Armed with Spitfires, we rapidly got

*Christopher Shores and Clive Williams, *Aces High*, (Neville Spearman, 1966).
†The squadron had previously existed in 1918/19.

going on these brand-new, exciting aircraft. We were not to enjoy them for long. The memory dims; Spitfire production may have lagged for some reason or, more likely, our aircraft were needed to strengthen already operational squadrons in the south. Whatever the cause, away they all went and we were sad – but not for long. There was no time for that. We were re-equipped immediately with Hurricanes, also brand-new and fitted with the Rotol three-bladed variable pitch propeller, and there soon developed a great enthusiasm for these excellent, sturdy fighters.

My orders from AOC 13 Group were that, despite the change of aircraft, I had to get 249 operational *within one month*.

We flew hard and we flew thoroughly. I was ably supported in our training task by my two flight commanders, Flight Lieutenants Kellett and Barton and a superb ground staff to keep us in the air; Flight Sergeant Bennett was a tower of strength in this work.

At the end of June, Air Commodore Nicholas, SASO, 13 Group, inspected the squadron and pronounced us operational. There was an important ingredient in this success. I knew that our squadron training task would be made infinitely the easier if I had good pilots. I wanted the best I could find. Fortunately, I had been Adjutant and Flying Instructor of the University of London Air Squadron in 1936/37. I knew all the University air squadrons and had a high regard for their members' abilities. Many of the undergraduates whom we had been training to *ab initio* standards on the Avro Tutor in 1936/37 had by 1940 joined up and were doing their advanced training on Hawker Harts (and I believe Furies) at Cranwell and other training schools. So, I managed to "fix it" that I got a very good share of the best that was going in terms of flying ability, guts and skill.

This was to stand us in very good stead. The reputation which the squadron built during the Battle of Britain was to remain with it throughout the war. New pilots, as they arrived both in the UK and, later on, when 249 Squadron had moved to the Mediterranean, very soon became aware that they had to fly and operate to a pretty high standard if they were to make the grade.

249 stayed at Leconfield, with little operational activity, until ordered south to Boscombe Down on Salisbury Plain, a Royal Air Force Experimental Station, also being used as a satellite in the Middle Wallop Sector of 10 Group. We left Yorkshire with a formation of twenty Hurricanes, our ground crews following in requisitioned Handley Page Hannibals. We flew to our new home on 14 August and the next day had our first major combat, B Flight destroying three Me 110s in operations over the south coast.

The game was on. We stayed, pretty actively engaged, for only just over a fortnight at Boscombe, before moving to North Weald, Essex, in 11 Group, to relieve 56 Squadron on 1 September. The squadron then became involved daily and continuously in heavy air fighting.

To my great chagrin and personal annoyance I stupidly allowed myself to be shot down (fire and bale out) during the morning of 6 September when attempting to re-form the squadron over Maidstone after an attack on a large formation of Ju 88s escorted by Me 110s and 109s ... Hospital.... But then, when able to hobble about a bit (damaged leg) and return to North

Weald, I could not even clamber up into a cockpit, let alone fly. I flew again in November; but it was too late. This was no good for 249, so command went to Flight Lieutenant Barton, my senior flight commander, who led the squadron from 7 September onwards with dash, conviction and flair until the end of the battle.

249 Squadron enjoyed with 303 (Polish) Squadron the distinction of being among the most successful of all the squadrons in the battle. It is of interest that Squadron Leader Kellett went straight from 249 on 19 July to command 303 at Northolt.

It is not easy to record accurate impressions of events, however momentous, which took place forty-two years ago. Of one thing one can be sure: all the emotions were there, excitement, boredom, eagerness, fear. We knew that an invasion was on the cards. We were, I believe, spurred on by this, and by anger at the Luftwaffe's attempt to fly in our sky, to bomb our country, our people. Victory was not in sight, yet the thought of defeat never entered our young minds.

On such foundations was the wartime history of 249 Squadron to be built.'

Ground Crews

Air Commodore A.C. Deere (New Zealand), *Nine Lives* (Hodder and Stoughton, 1959)

The cry for more pilots, give us more pilots, was being echoed by all squadron commanders at about this time. There was a desperate shortage throughout the Command. Not so fighter aircraft. The miracles worked by Lord Beaverbrook at the Ministry of Aircraft Production were being felt at squadron level where replacement aircraft arrived virtually the same day as the demand was sent. This achievement was due in no small measure to the sterling work of the maintenance unit which serviced the Spitfires allotted to operational units. On-the-spot repairs of damaged aircraft were carried out by our own ground crews, who were magnificent. All night long, lights burned in the shuttered hangars as the fitters, electricians, armourers and riggers worked unceasingly to put the maximum number on the line for the next day's operations. All day too they worked, not even ceasing when the airfield was threatened with attack. A grand body of men about whom too little has been written but without whose efforts victory would not have been possible.

The pilot picture was not so rosy. Not only was the replacement problem serious, but this growing strain on those who had been in action continuously, with only brief rests, was also beginning to tell. Small things which earlier would have been laughed off as irrelevant, now became points of bitter contention. At this stage of the battle pilot losses far outstripped the replacements and it was only a question of time before the serious position became a grave one.

Gaps at the Table

Harold Balfour, *Wings Over Westminster* (Hutchinson, 1973)

One lovely summer day I went on to the Air Ministry roof. I watched a formation high in the blue sky coming up the Thames. For twenty-five years I had watched formation flying but now I was looking up at a formation of German bombers invading our air, over our own countryside, on their way to drop bombs on a target right in the heart of Britain. I was overwhelmed by a sense of complete and detached unreality.

I flew to Hornchurch, landing just as a formation of Hurricanes were returning from a patrol over the Channel. I joined the boys at tea in the mess. There were gaps at the table. Suddenly all heads turned to the door. Into the room swayed a young man, very drunk, wearing long sea boots, a filthy sailor jersey and perched on top of his head a far too small naval petty officer's cap. His appearance drew a roar of welcome. His story was that he had been shot down over the sea about 7 p.m. the previous night but not before he had accounted for two certain and a possible third German aircraft. He had baled out and luckily hit the sea fairly near one of our destroyers which picked him up, wet, cold and shaken. Naval rum in large and frequent doses was prescribed by the ship's doctor. As he thawed out between blankets the patient happily and steadily absorbed his medicine. He had been landed that morning, lorry-hopped back to Hornchurch but still clutching a bottle of his medicine which he had not neglected to take. By the time he reached the mess the bottle was just about empty and he was just about full. In a bunch of casualty telegrams I looked through a week later this boy's name was marked 'Missing, believed killed.'

'For Johnny'

Reprinted by permission of David Higham Ltd from *Dispersal Point*, John Pudney (Bodley Head)

Do not despair
For Johnny-head-in-air;
He sleeps as sound
As Johnny under ground.

Fetch out no shroud
For Johnny-in-the-cloud;
And keep your tears
For him in after years.

Better by far
For Johnny-the-bright-star,
To keep your head,
And see his children fed.

John Pudney

'Proem'

Richard Hillary, *The Last Enemy* (Macmillan, 1942)

... Then, just below me and to my left, I saw what I had been praying for – a Messerschmitt climbing and away from the sun. I closed in to 200 yards, and from slightly to one side gave him a 2-second burst: fabric ripped off the wing and black smoke poured from the engine, but he did not go down. Like a fool, I did not break away, but put in another 3-second burst. Red flames shot upwards and he spiralled out of sight. At that moment, I felt a terrific explosion which knocked the control stick from my hand, and the whole machine quivered like a stricken animal. In a second, the cockpit was a mass of flames; instinctively I reached to open the hood. It would not move. I tore off my straps and managed to force it back; but this took time, and when I dropped back into the seat and reached for the stick in an effort to turn the plane on its back, the heat was so intense that I could feel myself going. I remember a second of sharp agony, remember thinking 'So this is it!' and putting both hands to my eyes. Then I passed out.

When I regained consciousness I was free of the machine and falling rapidly. I pulled the rip-cord of my parachute and checked my descent with a jerk. Looking down, I saw that my left trouser leg was burnt off, that I was going to fall into the sea, and that the English coast was deplorably far away. About 20 feet above the water, I attempted to undo my parachute, failed, and flopped into the sea with it billowing round me. I was told later that the machine went into a spin at about 25,000 feet and that at 10,000 feet I fell out – unconscious. This may well have been so, for I discovered later a large cut on the top of my head, presumably collected while bumping round inside.

The water was not unwarm and I was pleasantly surprised to find that my life-jacket kept me afloat. I looked at my watch: it was not there. Then, for the first time, I noticed how burnt my hands were: down to the wrist, the skin was dead white and hung in shreds: I felt faintly sick from the smell of burnt flesh. By closing one eye I could see my lips, jutting out like motor tyres. The side of my parachute harness was cutting into me particularly painfully, so that I guessed my right hip was burnt. I made a further attempt to undo the harness, but owing to the pain of my hands, soon desisted. Instead, I lay back and reviewed my position: I was a long way from land; it was unlikely that anyone on shore had seen me come down and even more unlikely that a ship would come by; I could float for possibly four hours in my mae west. I began to feel that I had perhaps been premature in considering myself lucky to have escaped from the machine....

The water now seemed much colder and I noticed with surprise that the sun had gone in though my face was still burning. I looked down at my hands, and not seeing them, realized that I had gone blind. So I was going to die. It came to me like that – I was going to die, and I was not afraid. This realization came as a surprise. The manner of my approaching death appalled and horrified me, but the actual vision of death left me unafraid; I felt only a profound curiosity and a sense of satisfaction that within a few minutes or a few hours I was to learn the great answer. I decided that it should be in a few minutes. I had no qualms about hastening my end and, reaching up, I managed to unscrew the valve of my mae west. The air escaped in a rush and my head went under water. It is said by people who have all but died in the sea that drowning is a pleasant death. I did not find it so. I swallowed a large quantity of water before my head came up again, but derived little satisfaction from it. I tried again, to find that I could not get my face under. I was so enmeshed in my parachute that I could not move. For the next ten minutes, I tore my hands to ribbons on the spring-release catch. It was stuck fast. I lay back exhausted, and then I started to laugh. By this time I was probably not entirely normal and I doubt if my laughter was wholly sane, but there was something irresistibly comical in my grand gesture of suicide being so simply thwarted....

It is often said that a dying man relives his whole life in one rapid kaleidoscope. I merely thought gloomily of the squadron returning, of my mother at home, and of the few people who would miss me. Outside my family, I could count them on the fingers of one hand....

I began to feel a terrible loneliness and sought for some means to take my mind off my plight. I took it for granted that I must soon become delirious, and I attempted to hasten the process: I encouraged my mind to wander vaguely and aimlessly, with the result that I did experience a certain peace. But when I forced myself to think of something concrete, I found that I was still only too lucid. I went on shuttling between the two with varying success until I was picked up. I remember as in a dream hearing somebody about: it seemed so far away and quite unconnected with me....

Then willing arms were dragging me over the side; my parachute was taken off (and with such ease!); a brandy flask was pushed between my swollen lips; a voice said 'OK, Joe, it's one of ours and still kicking'; and I was safe. I was neither relieved nor angry: I was past caring.

It was to the Margate lifeboat that I owed my rescue.... Owing to wrong directions, they were just giving up and turning back for land when ironically enough one of them saw my parachute. They were then 15 miles east of Margate....

At the hospital they cut off my uniform, I gave the requisite information to a nurse about my next of kin, and then, to my infinite relief, felt a hypodermic syringe pushed into my arm.

I can't help feeling that a good epitaph for me at that moment would have been four lines of Verlaine:

> Quoique sans patrie et sans roi,
> Et très brave ne l'étant guère,
> J'ai voulu mourir à la guerre.
> La mort n'a pas voulu de moi.

Spitfire versus Messerschmitt 109
1940

Air Commodore A.C. Deere, *Nine Lives* (Hodder and Stoughton, 1959)

There was no more experienced or persistent exponent of air fighting in World War II than Al Deere, the New Zealander. He was in the thick of the contest in the Battle of Britain and he was still at it five years later. None is therefore better placed to assess the merits of the Spitfire against the Messerschmitt 109 in the great battle of 1940.

Here, in a piece he wrote for his old friend, Douglas Bader, in Fight for the Sky *(Sidgwick & Jackson, 1973) he describes at first hand one of the original, prolonged, hand-to-hand duels between these two fine aeroplanes and then draws his own authoritative conclusions from it.*

... Thus the stage was set for the first sustained dogfight between a Spitfire and a Messerschmitt....

At first the 109 pilot continued in his turn but he soon realized that not only was I closing the range but I was able to hold the vital inside position. Quickly he rolled into a reverse turn and at the same time he pulled up into a steep climb. Momentarily, I lost distance, but as the climb continued I closed steadily causing him to adopt a new tactic, and one which really caught me by surprise. One moment I was watching his tail and the next I was presented with a full view of his pale-blue underbelly, as pushing his stick abruptly forward he went from a steep climb into a steep dive. Involuntarily, I pushed hard forward on the control column at which the engine cut dead as the carburettor, momentarily starved of petrol under the force of negative 'g', failed to function. Quickly, I rolled onto my back, and pulling hard back gained speed and engine power before again rolling out into the chase. In a matter of seconds the 109 had gained valuable distance, putting it well out of firing range. Slowly, but obviously noticeably to the German pilot, I again closed the range but before I could bring my sight to bear, the German wheeled upwards into a steep climbing turn, puffs of black smoke pouring from his exhausts as he did so. Throttle fully open I followed, and, nursing my Spitfire around the turn I was able not only to hold him in the climb, but also keep inside his turn and yet slowly but surely close the range. Again he went into a dive, this time by turning over on the inside of his turn, and just at that moment I gave him a quick burst, in desperation really because I couldn't bring my sights to bear, but I hoped it would warn him that I was determined to fight it out. And I was. By now, charged with the adrenalin of battle, spurred on by increasing confidence in the superiority of my aircraft, I knew that the end result would be dependent on my ability as a pilot not only to stay with the German but to get into a decisive position for the kill. No doubt the German was thinking likewise; he certainly reacted as if he was. From a steep dive into a steep turn and almost immediately into a steep climbing turn in the opposite direction, each manoeuvre following in quick-changing succession. Grimly, determinedly, I hung onto his tail dragging, pulling, bucketing my Spitfire from one impossible firing position to another, rarely being able to bring my guns to bear even for a split second. Sweating from exertion and greying from excessive 'g' I nevertheless managed to keep the 109's heaving belly in sight. A quick, hopeful burst now and again I did manage but with no apparent effect. Whenever I thought I'd got into the ideal position to make a kill, he reacted violently causing me to fire another wasted burst, until eventually there was no answering response from my eight Brownings. I was out of ammunition. At this juncture I was really close to the 109, so close that I could clearly see the pilot's head darting to the left and to the right as skidding, turning and weaving he tried to keep me in view. He was not to know that I had run out of ammunition, but I had the feeling that he would soon do so, and after savouring the enjoyment of being the hunter, I decided to break off the engagement on my terms before I became the hunted....

As a result of my prolonged fight with the Me 109, it was possible to assess the relative performance of the two aircraft. In the early engagements between the British Hurricanes in France and the 109s, the speed and climb of the latter had become legendary and were claimed by many to be far superior to those of the Spitfire. I was able to refute this contention and indeed was confident that, except in a dive, the Spitfire was superior in most other fields and, like the Hurricane, vastly more manoeuvrable. The superior rate of climb was, however, due mostly to the type of Spitfire with which my squadron was equipped. Aircraft of 54 Squadron were fitted with the Rotol constant speed airscrew on which we had been doing trials when the fighting started. Other Spitfires were, at that stage, using a two-speed airscrew (i.e., either fully fine pitch or fully coarse) which meant that they lost performance in a sustained climb. There was a great deal of scepticism about my claim that the Spitfire was superior to the Messerschmitt 109; there were those who frankly disbelieved me, saying that it was contrary to published performance figures. Later events, however, proved me to be right.

Messerschmitt versus Spitfire 1940

Generalleutnant Adolf Galland, *The First and the Last* (Methuen, London, 1955; Bantam Books, New York, 1978)

Finally, (Göring) asked [us] what were the requirements for our squadrons. Mölders asked for ... more powerful engines. The request was granted.

'And you?' Göring turned to me.

I did not hesitate long. 'I should like an outfit of Spitfires for my group, Herr Reischmarschall.'

After blurting this out, I had rather a shock for it was not really meant that way. Of course, fundamentally, I preferred our Me 109 to the Spitfire, but I was unbelievably vexed by the lack of understanding and the stubbornness with which the (High) Command gave us orders we could not execute – or only incompletely....

Such brazen-faced impudence made even Göring speechless. He stamped off, growling as he went.

Fatal misjudgement 7 September 1940

Laddie Lucas, *Flying Colours: The Epic Story of Douglas Bader* (Hutchinson, 1981; Granada, 1983)

I put the question to Adolf Galland.... It was 7 September 1980....

Forty years ago on this very day, the German High Command, with Hitler and Göring pressing down upon it, had, in a fatal blunder, switched the direction of its attack from Fighter Command and the Essex and Kentish airfields to London and other major cities. The decision, politically quite as much as militarily motivated, was an undisguised consequence of Bomber Command's attacks on Berlin and the heart of the Third Reich. To appease their own population, the Nazi leaders had been stung into retaliation. The precipitate deviation of aim was to cost Germany the Battle of Britain and, with the invasion of Russia which followed, eventually the whole war.

'Tell me, Dolfo,' I asked, 'why did your High Command allow itself to be diverted from the attacks upon our airfields in September 1940, when it just about had the battle "made"? Looking back, now, it seems an incredible misjudgment.'

Galland, with his suave courtesy and impressively good manners, threw up his hands and lifted his big round eyes to the heavens. A shrug of his broad shoulders and a slow shake of the head made an answer superfluous.

No. 303 (Polish) Squadron's Day 7 September 1940

Jean Zumbach, *On Wings of War* (Opera Mundi, Paris, 1973; André Deutsch, London, 1975)

7 September was a hot, clear day and I sat baking in the cockpit with the rest of the stand-by flight. Suddenly the earphones vibrated with the familiar call – '303 scramble! 303 scramble!' The red flares soared. In thirty seconds I was airborne from Northolt in west London.

The controller sent us north at first, then southeast. Height 25,000 feet. Maximum boost. Soon I could see the cotton-wool shell bursts of the British ack-ack over the Port of London. Beneath us, to the right, a flock of German bombers escorted by swarms of Me 109s was bearing down on the docks. I expected us to fall on the enemy before they could drop their bomb-load, but our British commander didn't seem to realize where they were headed. I was writhing with frustration when a Polish voice on the RT broke the suspense:

'Attack! Follow me!'

It was Paszkiewicz, rocking his wings to rally the others as he broke formation. He swooped to the attack with the other sections streaming after him together with the commander, now alive to the situation. At the same time I saw two flights of Spitfires pounce on the Messerschmitts.

Something told me it was now or never. Two Dorniers were already ablaze in front of me, parachutes drifting downwards. The black mass of bombers flashed towards us as we dived to get in a glancing frontal attack before turning to take them from the rear. My leader opened fire to the left. My turn. I pressed the button. Nothing happened. Shit! Shit!

It was too late to break off, and the tracer was flying in all directions. Suddenly I noticed the safety-catch – it was still on! I wrenched the Hurricane into a tight turn, doubled up over the stick by the crushing centrifugal force. Now I was behind my group, and a Dornier 215 sprang up ahead, growing until it filled my sights. I could see the tail gunner firing at me. At last I pressed the button and the Hurricane shuddered with the eight-gun recoil.

Smoke streamed out of the Dornier's port engine, I let him have another squirt and saw a banner of flame. Just like that, it all seemed easy.

A babel of English and Polish voices came through the earphones. Another Dornier 215 to my right had dropped slightly out of formation. Again I dived and watched him fill my sights. This one was a sitting duck, his tail gunner already out of action. It blew up with my first burst, and I was lucky to come through the flying debris unhurt, though I felt something hit the Hurricane.

A voice on the RT shouted 'Watch out, Messerschmitt!' and in the same instant I spotted a Hurricane with a Messerschmitt on its tail. I was too late to help the other pilot, and was left to admire his agility in baling out of the burning plane.

The dogfight went on, and I found myself in the middle of half a dozen Messerschmitts drawn by my attempt to intervene. With bullets coming at me from all sides I steep-turned violently to the left, and blacked out.

Luckily I had the plane at full throttle and nose up. It only lost 12,000 feet of altitude before I came round, quite alone in the sky, and with a few holes in my port wing. Only when I checked my equipment did I notice that my oxygen had come adrift. Now I knew why I had passed out. A few deep breaths cleared my head . . . then I was touching down at Northolt.

Everybody was dancing with excitement, and with good reason. The squadron had the phenomenal score of 11 Dorniers downed and 3 damaged, plus 2 Me 109s downed and 3 damaged.

We had our own bill for the victory. Forbes (Flight Lieutenant Athol Forbes, soon to command 66 Squadron) added 2 Dorniers to his score, before being shot down. In spite of being wounded in both legs, he managed to bale out. Lieutenant Daszewski lost half his left buttock to a Messerschmitt bullet. He was the man I'd seen baling out. Lieutenant Pisareck got a Messerschmitt, but another got him. His boot jammed in the cockpit as he started to jump, and the burning plane took him down with it for a few seconds, but he managed to wrench himself loose and parachuted into a suburban rose garden. The owner proceeded to give him a formal reminder that he was standing on private property! Then he brewed a pot of tea. What made Pisareck feel small was the discovery of a large hole in his exposed left sock.

Total debits: 3 planes and a boot.

With those two kills, 7 September was a milestone in my flying career.

Wing Commander Jean-Eugene Ludwig Zumbach had an exceptional air force career. A professional officer in the Polish Air Force before the Nazi invasion, he moved to France in 1939, fought with the French Air Force until the collapse and then found his way to England. He rose to become the commanding officer of 303 Squadron before being appointed to command

133 Wing in 1944 with its one British and three Polish squadrons equipped with long-range P-51 Mustangs.

University Contribution

Air Vice-Marshal F.D.S. Scott-Malden, Icare, Paris, 1965

David Scott-Malden, scholar of Winchester and King's, Cambridge, with a first in Classics, was twenty, and a product of the University Air Squadron, when he played his reservist's part in the battle. In this short narrative he paints again the ethos of those days, the spirit of the Auxiliaries, and the selflessness that animated the people.

'I saw the Battle of Britain through the eyes not of a fighter "ace" but of the newest of new recruits to a Spitfire squadron – one of the few hundred from university squadrons and other reserves who completed their training after the outbreak of war. In June 1940, at an Operational Training Unit hastily formed with instructors from the Hurricane squadrons in France, I first flew the Spitfire. To boys, this was a glimpse of a man's world. In the evenings, we plied our instructors with beer and made them talk of the fighting retreat against overwhelming odds, the constant moves to makeshift airfields; the way to turn inside the Me 109, to watch the sun, to stay alive. An unfinished diary records: "17 June: The French cease hostilities. Cannot yet conceive the enormity of it. I suppose it will not be long before we start defending England in earnest."

I opted to join an Auxiliary squadron, of so-called "weekend fliers". It seemed natural to seek a home with this improbable collection of business men, lawyers and farmers, rather than in a regular squadron. I never regretted it. For the non-professionals, there was solace in the variety of interests, the interplay of personality, the after-dinner talk, as the menace across the Channel grew larger day by day. "30 June. Looks as though the blitzkrieg on England is coming soon. Whole squadron at readiness most of the time."

In retrospect, it was always fine, that summer and autumn. Sound, smell and touch were vivid, as well as sight. They linger still: the freshness of a September morning outside the dispersal hut, with mist clinging to the grass undulations of the airfield; the crossing white trails in the blue sky, showing the first squadrons already in action, the puffs of blue-black smoke as the Merlin engine started; the exhilaration of climbing in tight formation. London, spread out like a map, hazy against the sun and partly obscured by the burning oil tanks of Thameshaven; the feel of the bevelled ring on the stick, carefully turned from "safe" to "fire". The enemy is a thing, not a man. First a string of tiny silhouettes against a distant cloud. Then suddenly all round you, so close that you note with a sense of shock the rough metal and grey-green paint, with sometimes a vivid yellow nose. You are turning savagely, trying to stay with your leader, trying to keep track of the crossing enemy, pressing the gun-button with a fierce exultation as the sight comes to bear. Then – just as suddenly – the utter solitude of the sky. How can you be so alone, where only seconds before was the heart of a battle? As that autumn wore on, fatigue dulled the impressions on the senses. Sleep was hastily snatched between sorties, in a chair or on the grass at dispersal, on a sofa in the Mess. At last, as darkness fell, there was relaxation. Ancient cars were packed with pilots bound for their favourite pub. Often we were not allowed to pay for our drinks; though meat was scarce, steaks would appear in a back room. Sturdy, cloth-capped figures in the bar would come up, shake hands, and return in silence to their seats.

How does the Londoner express his feelings at such a time? In the language of another age, we were their champions, sallying forth from the walls of the beleaguered city. They would not have recognized such language; but they showed, by a proffered drink or a silent handshake, that they understood and wished us Godspeed.

We had the glory. They went back to work in office or factory, railway yard or dock, and gathered round their radios for the news. Strength came from the man who offered them nothing but blood, toil, tears and sweat. I saw something which I have never forgotten. If the heart of a nation is sound, danger brings out unexpected kindliness. As on the battlefield, men facing the unknown become considerate to each other. Friendly greetings were exchanged between strangers, as each morning brought the threat of new attack. Small services were done for neighbours, without thought of thanks or reward. For one moment in history, the nation felt like a family. Un-British, often comic in its embarrassed emotion, this was the true meaning of "Their finest hour". It is only in the aftermath of victory or defeat that greed, meanness and self-interest flourish, like weeds in a forgotten garden.'

Victory – and Loss
The Closeness of a Family

John, and Hugh Dundas were brothers in two Yorkshire Auxiliary squadrons. Each fought in the battle.

John, an Exhibitioner of Christ Church, Oxford, with a first in History, was a promising young journalist on the *Yorkshire Post*. Hugh, known to the Service as 'Cocky', was to become in 1944, the youngest group captain in the Royal Air Force, commanding 244 Wing, with its five squadrons, in the drive northwards through Sicily and Italy. Between them they won two DSOs and three DFCs.

Each could write, and selected passages from private letters which passed between them and their mother, tell, as well as anything can do, the thoughts of a family facing mortal danger and the closeness and concern of brothers, one for the other. They show, too, the affection and regard that pilots had for each other, and the levity of spirit which animated squadron life.

Royal Air Force Station, Drem, near North Berwick 25 November 1939
John Dundas, the journalist, to his mother:

I say, aren't the papers dull these days? Writing for the papers must be a wretched job when there's nothing to say except how well the boys are doing, even when they're not. I shouldn't care to be a war correspondent.

Royal Air Force Station, Northolt, West London 2 June 1940
John to his mother:

Just a line to say that for the time being things have quietened down and that I'm still alive and kicking and in one piece. The last three days have been very hectic, patrolling Dunkirk most of the time. After one patrol I force-landed with zero gallons of petrol on a cricket pitch in the middle of Frinton-on-Sea. The poor aeroplane came to rest on its wheels and its prop in a kneeling attitude in some iron railings. No one was hurt....

I'm afraid we've had some losses, and may go up north for a bit to recuperate and make up our numbers.

You never enclosed Hughie's letter, but I knew all along where he was and hear that he scored over a Heinkel some days ago. Nice work.

Royal Air Force Station, Kenley, Surrey 20 August 1940
Hugh Dundas to his mother:

Nothing much has been doing today, which was quite welcome as it has given us the chance to settle down and find our feet. Our flying quarters are a bit odd at the moment, as, the day before we arrived, a Boche, together with bombs, effected an unwilling landing which rather devastated everything, including our hut. However, one Hun is considered to be worth quite a lot of huts and discomfort, and things are getting straight again pretty quickly.

With luck and good management I should be able to make contact with John now that we are both in the same part of the world – that is if he hasn't moved: I haven't heard from him for some time, and I don't know exactly where he is. I have written to him today, suggesting a well-coordinated movement to the same place.

Kent and Canterbury Hospital 23 August 1940
Hugh to his mother:

This is a quick letter to let you know that all is well with me, despite the ominous change of address. I had to jump out yesterday evening when two explosive shells at 12,000 feet put my machine out of action and set fire to it. Please don't worry as I really had quite an enjoyable parachute descent! Moreover, my injuries are very slight; the machine was blown about quite a bit, and the jar dislocated my left shoulder, and a lot of little splinters got into my leg; but there is only one puncture worth speaking of, the rest being mere scratches.

I saw John yesterday when he flew over for luncheon: it was a surprise, and a very pleasing one. He looked surprisingly well, and was full of life.

Kent and Canterbury Hospital 24 August 1940
Hugh to his mother:

George (Moberley, close friend in the squadron) came over yesterday evening and spent the night at Canterbury. He tried to ring you up, but he couldn't get through. He tells me I was shot down by Me 109s, which came out of the sun and attacked our rear section. I was the only one they got out of action before George saw them, and swung the section round. The other two in the section routed the 109s, while George flew down to protect my tail as they continued to attack as I fell; he followed for about 7000 feet and as I still hadn't baled out, and was burning merrily, he gave me up and rejoined the squadron. Actually I was having some trouble getting clear, and I didn't get out till about 800 or 1000 feet; when at last I managed it I really quite enjoyed the jump, which showed up the world at an entirely new angle!

Royal Air Force Station, Middle Wallop, near Stockbridge 25 August 1940
John to Hugh, his brother:

Very sorry to hear that a 109 – or rather twelve of them – inflicted grievous bodily harm on you over Dover two days ago. Mummy sent me a wire yesterday, and you were mentioned as wounded in an 11 Group Intelligence Summary this morning. I haven't heard any details, but I do hope the damage isn't too bad. Write and tell me about it as soon as you're well enough to do so. Anyhow, you'll now get a nice spell of sick leave which I rather envy you.

The 109s nearly made hay with us over the Isle of Wight yesterday. They sent us off too late to do anything about the bombing of Portsmouth and too low to do anything about the myriads of 109s who were hovering around the scene and who, when they saw poor old 609 painfully clambering into the sun, came down on us. The result was that one of our machines was shot to hell, two more damaged and not one of us succeeded in firing a round. I was reduced to the last resort of a harassed pilot – spinning. It was most humiliating. But fortunately we didn't lose any pilots.

Cawthorne, Barnsley, Yorkshire (the Dundas home) 29 August 1940
Hugh to John:

You will have heard, I suppose, that George Moberley was shot down and killed last Monday – an overwhelmingly powerful force of 109s. It is no good trying to express my feelings about it; the bottom has just fallen out of things for the moment. He was the best friend, and the best person, I have ever known, or, I feel, I am ever likely to know.

I came home today. Life is likely to be very peaceful. I only wish that you could be here to enjoy it with me. I walked up to Cannon Hall with Mummy and Daddy tonight, and any village people who saw me more or less fell on my neck; everyone seems to have heard of my contretemps, and they seem to think it was rather a good show, instead of just damned clumsiness. Life is quite embarrassing! Some of these people take the *Daily Mirror* too literally.

Best of luck and watch your tail.

Royal Air Force Station, Middle Wallop 29 August 1940
John to his mother:

It almost looks as though, touch wood, August may peter out fairly quietly. Only two and a half more days to go now. Our last engagement was on Sunday when we got in among a gaggle of Me 110 bombers and shot them down in considerable numbers. 609 shot down thirteen of them and altogether fifteen crashed on land near Warmwell, the aerodrome in Dorset which they were trying to bomb, and where 609 has spent quite a lot of time on and off.

I felt very pleased about it because (a) we lost no one and (b) I was leading that time and managed to bring the chaps in against the leaders of the Jerry formations which broke up the raid completely. It was a most gratifying sight!

As from Royal Air Force Station, Kirton-in-Lindsey, Lincolnshire
20 September 1940
Hugh to his mother:

I am sitting in a hut on the edge of a field in Cambridgeshire [it was Fowlmere, a satellite of Duxford], where we come at crack of dawn every day. We patrol Piccadilly as a wing of five squadrons, superbly led by Squadron Leader Bader of wooden legs fame. It's the best thing I've seen since this war started; instead of being almost invariably practically alone among many Huns, we are a large, concentrated and formidable-looking formation. We have done several patrols in the last six days, since we joined the wing and rarely get near the Hun as he turns and belts off home as soon as he smells us. As I say, it's the most heartening thing I've seen since this war began; and at last someone has got things organized.

John told me about his leave. It won't be easy to get time off as we are short of experienced pilots and, so far, I've been on every time. Things look up every day for the squadron. The more I see of the CO and Macfie, my flight commander, the more I think of them.

John wrote urging me to get posted to 609; much as I would like to be with him – and I know you would like us to be together – there is no question at all of my pushing off just now. Were John in my position, he would feel the same.

For heaven's sake don't worry about me now. We're on a grand job, and very lucky and honoured to be included in this wing, which I think is keeping Jerry off more than anything else: we are in company with some pretty crack squadrons.

As from Royal Air Force Station, Kirton-in-Lindsey 25 September 1940
Hugh to John:

I will be on forty-eight hours leave from 30 September – 2 October so, if you will let me know where you will be then between those times, I'll try to organize myself to be there, too.

There isn't much time for recreation just now. It gets quite tiring – flying down here (Duxford sector) every day before breakfast, doing perhaps two or three patrols at 25,000 feet or more, then flying back at dusk; by the time you've washed and changed you don't feel like doing more than drinking a spot of dinner and going to bed.

Royal Air Force Station, Middle Wallop 6 October 1940
John to his mother:

Good news! When I got back here I found I'd got B Flight and have been given the DFC. The Station Commander had the ribbon tucked away in a drawer in his desk and produced it immediately, so I am now wearing it and feeling rather a fool. Not that it's worth much these days.

Royal Air Force Station, Kirton-in-Lindsey 17 October 1940
Hugh to John:

I wonder whether you would sound your CO about his views on the subject of having two Dundases in his squadron. I can't stand this business much longer, and while I have a great loyalty towards 616 as a squadron, I have no particular loyalty towards it as a number and a dead-beat OTU. I have never felt in such good form or so keen to get into the air, and fail to see why I should stooge around up here wasting my time when there are jobs of work to be done.

Shortly after receiving his brother's letter, John visited Hugh at Kirton. It was the occasion for some revelry by night. Hugh made a reference to the goings-on when he wrote to his mother on 7 November:

John is a great success in the Kirton mess, and is much inquired after – 'Where is that other mad Dundas?' and 'When is your crazy brother coming again, we want to lock up the fire extinguishers next time?!'

Then came the terrible tidings. On 19 November a telegram arrived at Cawthorne from the Air Ministry:

Regret to inform you that your son, Acting Flight Lieutenant John Charles Dundas,

DFC, is reported missing as a result of air operations on 18 November 1940. Letter follows.

Two days later, Michael Robinson, Commanding Officer of 609, wrote to John's father:

It is with very deep regret that I have to inform you that John has been posted as missing since the afternoon of 28 November....

John was, from every point of view, the ideal and perfect fighter pilot, and although we had only known each other a short time, we had become very great friends and I relied on him more than anyone else to keep up the prestige of 609 Squadron.

He was by far our most popular and courageous pilot, and had only the day before chased a Ju 88 from Winchester to Cherbourg, finally shooting it down over French soil. His loss is as keenly felt by the NCOs and men of the squadron as it is by the pilots and myself.

More than three months elapsed before the Air Ministry, on 17 March 1941, removed all further hope:

... I am directed to inform you with deep regret that all efforts to trace your son, Acting Flight Lieutenant J.C. Dundas, DFC, have proved unavailing.... It is feared that all hope of finding him alive must be abandoned.

I am, Sir,
Your obedient Servant.

Hugh, now a Flight Commander in 616, was flying with the squadron in Douglas Bader's Tangmere wing when his mother sent him a copy of the Air Ministry's letter. He replied on 1 April:

... I hope that you have not been too upset by the finality of the ... statement. I'm afraid that I personally never really expected anything else. You have had the most terrible winter of suspense, but the way you have stood up to it has been a greater inspiration than I can possibly tell you....

Under a new, and very categorical, order which has recently been circulated re: operational hours without a rest, I may not be allowed to go on with this job very much longer before being pushed into a staff or training job: that wouldn't suit me at all, and I want to make my presence felt a bit more before I get kicked out as a dead-beat. (Dead-beat, aged twenty!)

There was a sequel to John Dundas's loss. The daily monitoring service of German broadcasts to the Americas recorded a relevant statement in its digest of 8 February 1941:

Marshal Göring is still lauding the bravery of Major Helmut Wieck, Commander of the Richthofen Squadron, who was shot down over the Isle of Wight on 28 November. Wieck was regarded as one of Germany's greatest fighter aces. He had fought with the Luftwaffe in Spain and Poland and was credited with 56 victories in air combat. Hitler had decorated him with the Knight's Cross with Oak Leaves.

Pilots flying with him on the day he was killed have been interrogated again and again by the Chiefs of the Luftwaffe, broadcast accounts of his last flight have been made from many German stations, but they have never been able fully to substantiate the true manner of his death.

Shortly afterwards, a British broadcast gave the Germans the answer they had been seeking. It added that, when Wieck was shot down, his victor, John Dundas, then had twelve German aircraft to his credit.

New Zealander Before the Monarch

Air Commodore A.C. Deere, *Nine Lives* (Hodder and Stoughton, 1959)

On arrival at Hornchurch we found the station buzzing with excitement. The distinguished visitor was none other than His Majesty The King. A small parade was to be held at which 'Sailor' Malan, Bob Tuck and I were to be presented with our decorations. The ceremony was carried out with a minimum of fuss.... The parade was held on a small square of tarmac between two hangars and, appropriately, adjacent to the parked Spitfires of 65 and 74 Squadrons. For me it was a memorable occasion. As a New Zealander brought up to admire the Mother country and respect the King as her head, it was the honour of a lifetime, an ultimate milestone of my flying ambitions – the Distinguished Flying Cross presented by the King, in the field of action.

Deere on Malan

'Sailor' Malan ... was, to my way of thinking, the best fighter tactician and leader produced by the RAF in World War II....

Allen on Malan
Wing Commander H.R. 'Dizzy' Allen, *Battle for Britain* (Arthur Barker, 1973; Corgi Books, 1975)

...The other squadron (No. 74) was completely differently organized and was under the command of the redoubtable South African ace, 'Sailor' Malan. He made his pilots live, by comparison [with 66 Squadron] like Boer farmers in their wagons. He even made sure they went to bed at 10 p.m. each night although, presumably, he did not actually close their eyelids for them.

He was tough, strict, a martinet and, operationally, it paid off. I doubt if there was a more successful squadron commander in 1940 than Malan. He had been a merchant seaman – hence his nickname 'Sailor'. He was so long-sighted he could have seen a fly on the Great Wall of China at 5 miles.

He was a crack shot, a brilliant aerobatic pilot, but, above all, he was utterly determined. He even won a dogfight against the German ace, Mölders.... If Malan had had decent ... armament instead of the puny battery of eight ·303-inch Brownings we were equipped with, Mölders would never have got back to his base at Wissant....

Free French ...

Lieutenant-Colonel Henri Lafont, 247 and 615 Squadrons, Icare, Paris, 1965

'...I don't really know what happened. I was "weaver" when I heard a loud noise, at the same time as my cockpit was filled with glycol vapour. There was a big hole above the engine from which a not-too-thick stream of white smoke was coming. Detling airfield was just below. Life was fine. But I stopped appreciating the joke when I saw flames instead of smoke. I was, however, low enough to land, undercarriage up....

My beautiful brand-new uniform, covered with oil, was irrecoverable. I immediately put in ... for another.... The Quartermaster General didn't agree. The captain to whom I told my misfortune said he was an old soldier and that it was useless for me trying to pull a fast one. If you had a fire (he said) you must certainly be to blame in some way...!

That was "my" battle of Britain. A succession of never-ending alerts, of take offs in 3 minutes to try to intercept an elusive enemy; and the superior performances of the adversary who always had the advantage ... (of height and) sun....

When high up, at several tens of degrees below zero, we suffered from the cold. I remember one day forgetting my gloves during a scramble. My fingers stuck to the throttle and to all metal parts I touched.

We were, it is true, a little jealous of the Spitfire squadrons which were better armed than we against the Messerschmitts, and therefore their missions were more likely to result in victories.

Despite all our prayers to the Gods of Battle, our score at the end of 1940 was a blank. Some of us were to catch up later on, but I remember that period as one of exultation mixed with a little bitterness and regret that, despite my efforts, I arrived a little late in this Battle of Britain....'

Regia Aeronautica Over Britain, November 1940
'My Personal War With Mussolini'

Group Captain Karel Mrazek, Jablonec, Czechoslovakia, 1982

Karel Mrazek, one of the outstanding Czech officers in the wartime Royal Air Force (he was to become a highly-decorated Group Captain), followed the classical train of his determined countrymen after the rape of Czechoslovakia in 1938 – Poland, France (the Foreign Legion), North Africa and thence, in 1940, to Britain.

Living now in Czechoslovakia, Charles Mrazek has recorded his recollection of the field day his Squadron, No. 46, enjoyed, in company with 249 and 257, with the Regia Aeronautica during Mussolini's token (and abortive) intervention at the tail-end of the battle for the daylight air over Britain.

After arriving in England and doing a conversion course onto Hurricanes at No. 6 OTU at Sutton Bridge, beside the Wash, I was the only Czech pilot to be posted to No. 46 Squadron, then based at Stapleford Tawney, a satellite airfield of North Weald, just east of London.

At about 1300 hours on 11 November 1940, the squadron, led that day by Flight Lieutenant A.C. Rabagliati, a South African, to whom I was flying No. 2, was ordered by 11 Group to patrol a convoy of ships off the coast at Foulness, height 1000 feet. We had been on patrol for no more than fifteen minutes, when the operations controller gave our leader fresh instructions:

'Climb to angels 20 (20,000 feet) over the Thames Estuary. 30 plus bandits over the Channel approaching Brighton and heading for Kent and London.'

It was just my luck that, on this day, the engine of my Hurricane wasn't developing its usual power. I pushed all the levers forward, but there was not the proper response. Try as I might, I couldn't keep up with the formation. Dropping 100, 200, 500 yards behind, I became highly vulnerable to attack.

At 18,000 feet, and still heading south towards the incoming raid, the squadron was above me. I then sighted a formation of twin-engined bombers flying towards us of a type I hadn't seen over England before. As I flew under them I could see the Italian insignia on their wings and fuselage. I identified them as Fiat BR 20s. They were flying in tight formation in five sections of three aircraft. Apparently, they had no fighter escort.

As Rabagliati manoeuvred the squadron up-sun for the attack, two other squadrons from the North Weald sector, 249 and 257, were also hurrying to get their chance at this rare quarry. Engagement took place over Maidstone. All I could do with my lazy engine was to stay where I was hoping to pick off any stragglers.

As the attacks developed, the bombers soon began to veer eastwards towards Southend and Harwich and then making off in a slanting dive for Margate, the Straits and Calais, on the French coast. As they turned away, I saw three BR 20s going down in flames followed by the crews in their parachutes, floating earthwards like white mushrooms.

At that moment I saw, just above, thirty or forty unknown biplanes. They looked like Avia B 534s, the standard fighter aircraft in my country. Then I realized that here was a gaggle of Italian CR 42s, supposedly protecting their bombers. A fat lot of good they were doing!

I struggled to get my lazy beast of a Hurricane up to their level. As I did so two of the fighters were falling in flames and then two more broke away from the main formation, looking as if they might have had enough of the fighting. As they crossed my path without seeing me I gave the second aircraft a short burst with my machine guns, allowing full deflection. It exploded and went down like a fireball.

To my surprise, the other CR 42 turned to fight using all the aeroplane's manoeuvrability. The pilot could get on my tail in a single turn, so tightly was he able to pull round. I did my best to keep up my attacks, getting in a short burst whenever I could until I saw smoke and flames coming from the fighter.

By now I was over the sea and then, holy mackerel, six CR 42s started diving at me out of the sun to come to the aid of their comrade. I was now the prey. It was time for me to quit so I turned away putting my nose down at about 6000 feet. I quickly left the Italians behind, recrossed the coast and eventually landed at Rochester, in Kent, at 1305 with my fuel tanks almost empty. I had ten bullet holes in the wings and fuselage of my aircraft.

After refuelling I returned to Stapleford, claiming one CR 42 destroyed and another probably destroyed from my personal war with Mussolini.

My claims were never credited. In the excitement, I had forgotten to switch on my ciné camera and, because I had been on my own, separated from the squadron, there was no other confirmation.

All the same, it makes one of my most cherished memories of my days with the Royal Air Force in England where, despite my bad English, I made so many friends. I look back on them now with a lot of affection.

AFTER THE BATTLE

Post Mortem on the Battle
The German Verdict
Reasons for Defeat

Those who questioned Adolf Galland after the war were in no doubt about the views of the German operational commanders.

PLAN NOT FEASIBLE

... [When] it was decided to destroy the RAF by bombing its air bases, the plan was not regarded by Galland, Mölders and other fighter [leaders] as feasible; nevertheless it was carried through at the insistence of Göring, who was from time to time at Kesselring's headquarters.

For these attacks the size of bomber formations was increased from sixty to eighty aircraft with roughly three times as many escorting fighters.

Losses of both fighters and bombers increased, but remained bearable. Escort at this time was still confined to close escort and top cover and fighters were not permitted to leave the bombers to engage RAF fighters positioning themselves for attack. Real damage was wreaked on the Luftwaffe as a result of this ruling.

The Luftwaffe could see that they were pounding the RAF fighter fields with great effect on the physical installations, but the actual flyers also knew that, despite the damage, the RAF was still coming up to engage. It was nevertheless the practice at Kesselring's headquarters to cross off the Order of Battle each British fighter unit whose airfield was bombed; Göring refused to listen to his fighter commanders' protests that such claims were not realistic.

It was soon decided that the Luftwaffe was ready to begin the great bombing assault on London and other strategic points. Göring and others actually believed that Fighter Command was stricken and that ordinary battle attrition would soon destroy the few remnants of RAF fighter units which remained.

ATTACKS ON LONDON

The heavy bombing attacks on London and southern England were carried on by two main forces of the Luftwaffe, one based in the St Omer area and the other between the Seine and Cherbourg. The entire planning and conduct of the offensive was dominated by bomber experts supported by Göring and Kesselring, who had no experience in such matters (no one had such experience, for that matter), and who consistently refused to listen to the objections and observations of fighter men like Mölders and Galland.

Galland's own views on why the daylight bombing offensive failed are given below in the form of a detailed analysis of the conduct of the missions from a tactical standpoint. It should be said that he still believes that the Luftwaffe could have won this phase of the Battle of Britain under different conditions, as stated below.

So many bombers were to take part in the mass attacks on London that escort strength barely equalled bomber strength, giving a lower margin of safety than during the three previous phases of the Battle. The routes flown by the bombers were limited by the endurance of the fighter escort.

Since the escort had to accompany the bombers from the French coast to the target and back at least as far as the Channel, and since the Me 109 could at that time barely fly from French bases to London and back in a straight line, leaving a fuel margin for combat en route, the whole formation of bombers and escort fighters had to fly in an almost straight line from the French coast to London and then straight back. This greatly reduced the possibility of varying the route of approach and made RAF interception work easier.

WRONG FORM OF ESCORT

After assembly, each fighter formation was supposed to pick up its bombers as they crossed the coast. This was very difficult because radio contact was uncertain and because the missions were so planned that several bomber and fighter formations were supposed to rendezvous in a space of minutes over a stretch of coast a few miles in breadth.

The result was that a fighter formation often arrived at the right time and picked up the wrong bomber formation, or that two fighter formations attached themselves to one bomber formation while another bomber formation came a few minutes late and had to fly over England unescorted. Unescorted flights usually ended in calamity, thanks to the RAF fighters. Often a fighter formation was sent on ahead to sweep the route the bombers would take, but it was limited in its freedom and could not detour to attack enemy fighters. Moreover, Fighter Command did not usually send up much of its strength against this sweeping force, but instead saved its bullets for the real battle.

Based on experience gained in the Spanish War, the German fighter units recommended an escort system with a close escort, a top cover, and freelance patrol to fly high above the bomber formation and engage any and all RAF fighters which they sighted. The advice of the more timorous bomber men prevailed, however, and fighter escorts were limited to close escort in twos and fours and a form of wider escort and top cover by Staffeln which was far too close to the bombers and which was strictly forbidden to move away in anticipation of attack to engage enemy fighters....

In addition to the disadvantages of the straight routes along which the bombers approached, and the false philosophy of the fighter escort, the altitude chosen by the bombers was very bad for the fighters. To minimize the danger from British anti-aircraft fire, the bombers flew at an altitude of from 21,000 to 23,000 feet, a height too great for their power and loading. As a result they were very slow, and the fighters had to weave continually and closely to stay with them.

Each weave carried the fighters away from the immediate vicinity of the bombers and made the bomber crews more nervous and more insistent in their demands that the fighter escort stay close to them. When the fighters objected and pointed out that the extremely poor formation flying of the bombers resulted in bomber formations 10 km long and impossible to protect, they were met with more rebuffs and Göring actually ordered them to cease weaving and to fly straight, level and close to the bombers.

Göring's order caused Galland to explode and declare that it was impossible for Me 109s to fly straight at the speed and altitude of the slow bombers....

OPPONENTS' ADVANTAGE

The worst disadvantage of this type of escort was not aerodynamic but lay in its deep contradiction of the basic function of fighter aircraft – to use speed and manoeuvrability to seek, find and destroy enemy aircraft. The Me 109s and Me 110s were bound to the bombers and could not leave until attacked, thus giving to their opponent the advantage of surprise, initiative, superior altitude, greater speed, and, above all, fighting spirit, the aggressive attitude which marks all successful fighter pilots and units.

To Galland and the other German fighters the tragedy was all the more marked because in Spain, Poland and France they had all learned these lessons by bitter experience. Now they were blocked by Göring and the bomber men.

Under these conditions the carrying out of effective escort duties was almost impossible, but Galland claims that his unit, JG 26, as well as JG 54, were able to carry out their function with fairly low losses. He claims that usually the bombers he escorted had lower losses than did JG 26 itself.

GÖRING PINS THE BLAME

Throughout the Battle of Britain the formation flying and planning of bomber missions became worse. As losses rose, the bombers began to fly in to the target and out again along crooked routes, frequently overstaying the limited flying time of the Me 109 escort. On one mission seven aircraft of JG 26 landed in the Channel, and Galland himself, bringing up the rear, decided he would have to land at Manston (on the Kent coast) and surrender for lack of fuel, but instead he flew on and landed on the beach at Cap Gris Nez.

The decisive factor in causing the breaking off of daylight bomber operations over England was, to Galland's mind, the weather. Towards the end of September, the weather over England and northern France started to close in. Fighter assemblies and the meetings of fighters and bombers became even more difficult to effect and when unescorted missions were flown by accident, losses rose alarmingly. The insistence that bombers should fly despite bad weather made escort conditions so bad and losses so high that day missions were discontinued and night missions substituted....

The German fighter force was exhausted mentally and physically by this time and was glad of the temporary halt that was called. For over two months they had flown several missions a day. Replacements of men and aircraft had kept pace with losses but the stamina of the superbly trained and experienced original establishment was down to a point where operational efficiency was being impaired.

The stopping of the daylight offensive also followed from the realization that Fighter Command had not been, and could not be, eradicated with the means at the disposal of the Luftwaffe. British fighters, at the end of the daylight offensive, were more effective than they were at the beginning, and the range of German fighters did not permit a deep enough penetration of bombers to wipe out British fighter production. The inferior armament of German bombers made them especially vulnerable.

Despite all the above reasons adduced ... for the failure of the daylight bomber offensive, Göring believed that the failure was due solely to the cowardice and poor flying on the part of the German fighters. His attitude was that the fighters had failed in their clear duty to protect the bombers....

Post Mortem on the Battle
The Bader View
Mistakes in Victory

Group Captain Sir Douglas Bader, 'The Bader Papers', 30 November 1969

*'Who cares, anyway? We won, didn't we?'**

It was Douglas Bader's telling retort when old controversies about the conduct of the 1940 battles were raised.... Yet the arguments aroused his feelings intensely. This was manifest in a private paper he circulated among a few trusted friends, among them Harold Balfour, the wartime Under-Secretary of State for Air, when, at the end of the 1960s, the publication of Robert Wright's book, Dowding and the Battle of Britain, *and the showing of the film,* The Battle of Britain, *had stimulated controversy afresh.*

Selected extracts bear witness to the strength of Bader's conviction.

Much has been written on the so-called controversy between Air Vice-Marshals Park (of 11 Group) and Leigh-Mallory (of 12 Group) on the subject of the Duxford Wing during the Battle of Britain.... It is reasonable to mention that none of the post-war writers, with the exception of Air Vice-Marshal J.E. 'Johnnie' Johnson, has taken the trouble to discuss the matter with me. This is relevant in the context of the Duxford Wing because I led it and my Group Commander at the time, Leigh-Mallory, died in a crash at Grenoble in 1944. The authors of *The Narrow Margin*, Derek Wood and Derek Dempster (Hutchinson, 1961), which was widely acclaimed as an accurate documentary, failed completely in their views about the 12 Group Wing because they did not seek the facts. They merely identified themselves with 11 Group strategy....

*Laddie Lucas, *Flying Colours: The Epic Story of Douglas Bader* (Hutchinson, 1981; Granada, 1983).

INTRIGUE SUGGESTED

It has been suggested that the whole period of the Battle of Britain represented a continued and sustained intrigue by Leigh-Mallory to undermine the Commander-in-Chief, Dowding, using the big Wing formation, led by Squadron Leader Bader as his spearhead. It followed [from this] that Leigh-Mallory had won when Stuffy Dowding was replaced at the end of 1940 by Sholto Douglas as C-in-C and Keith Park handed over to Leigh-Mallory. Nothing could be further from the truth....

A further suggestion that has been accepted over the years that Keith Park should have operated wings like the 12 Group one is also nonsense. You cannot operate large formations from close to an attacking enemy. At no time – and I say this with certainty – was it in Leigh-Mallory's mind....

As regards Keith Park, it must be remembered that he was a tired man since he alone conducted the Battle of Britain. Neither Dowding nor Leigh-Mallory fought the battle; the sole weight lay squarely on Park's shoulders. To suggest that he was removed from 11 Group because he failed is not borne out by the facts. He won the battle. He was rested by going to Training Command in the same way that tired pilots were rested. He (later) continued a distinguished career by commanding Malta and, finally, as an Air Chief Marshal, Southeast Asia.

DOWDING'S CONTRIBUTION

At the end of 1940, Dowding had been C-in-C, Fighter Command, for four years. He was due to retire in 1938, but was kept on. He was finally retired after the first, major air battle (indeed, the vital one) of World War II and Sholto Douglas, a man thirteen years younger, succeeded him.

Dowding made two vital contributions to the defeat of the Luftwaffe in the summer of 1940. Firstly, during his pre-war command, he had laid down radar coverage of the south of England so that we had early warning of the Germans' intentions. Secondly, he had persuaded the War Cabinet against sending more RAF fighters to a defeated France in May 1940.

Park fought under the disadvantage of being too near to the enemy effectively to deploy the strength that Dowding had given him. His problem was compounded by his operations room displaying a map of 11 Group territory only (plus a tip of the Pas de Calais). In other words, Park was fighting an 11 Group battle that should have been a Fighter Command battle.

A map of the whole of England lay on the plotters' table at Fighter Command. It showed every fighter airfield with the location and state of readiness of every squadron on the board above it. The difference is paramount.... A controller at Fighter Command operations room table, with the whole picture spread out in front of him, would instantly have realized the need to scramble squadrons from the further away airfields first against the enemy.

Admirer . . . and admired. Winston Spencer Churchill, East Anglia, 1943

Facing page:
Blitzkrieg! Poland overrun, September 1939. German troop-carrying JU 52s on the airfield at Zelechov as Polish prisoners were interrogated (p. 40 *et seq.*)

Right and below: The unquenchable spirit of Poland. Polish aircrews, making their hazardous escapes to Britain, were soon training and operating with the Royal Air Force. 303 (Polish) Squadron became a stand-out in Fighter Command (p. 81 *et seq.*)

Facing page: Dunkirk, June 1940. Where, they asked, was the Royal Air Force? Churchill gave the answer in the House of Commons (pp. 49 and 50)

Left and below: The biter bit. German bombardment of Namsos, Norwegian campaign, May 1940 and battleship *Tirpitz,* in Alten Fjord, brilliantly photographed by Frank Dodd and Eric Hill on 12 July 1944 in an Mk 16 Mosquito of 544 Squadron, before sinking by Bomber Command.

Leaders all. As the air war over Britain and France developed, personalities, old and new, caught the public's gaze. For the defence, 'Boom' Trenchard, *(top left)* 'father of the Royal Air Force' (left with Dickie Barwell and Michael Robinson), and Douglas Bader, Duxford wing leader *(top right),* were arresting figures. For the Luftwaffe, while the Kampfgeschwader crews *(bottom left)* were planning their bombing attacks, Adolf Galland *(bottom right)* was quickly scaling the Jagdgeschwader ladder. Here, in a captured British flying jacket, he is with 'Schweinebauch', a favourite Laverak Setter which he bought in Spain in 1938 while serving with the Condor Legion. The picture is taken at Audembert, near Wissant, northern France, the HQ of JG 26, which, as Kommodore, 'Dolfo' was commanding

Walking at Göring's side is Werner Mölders (right) the Luftwaffe's outstanding tactician, killed in a flying accident in 1941 (p. 42)

Left: Exceptional brothers. John (left) and Hugh Dundas at Cawthorne, autumn 1940. Their letters (p. 84 *et seq.*) tell the story

Below: On 17 May 1940, during the battle for France, Douglas Douglas-Hamilton (Duke of Hamilton) flew to France, at Dowding's request, to make an on-the-spot assessment. He picked an 80 m.p.h. training aircraft for the job. Here are his log-book entries

Year 1940 Month	Date	Aircraft Type	Aircraft No.	Pilot, or 1st Pilot	2nd Pilot, Pupil or Passenger	Duty (Including Results and Remarks)	Single-Engine Aircraft Day Dual (1)	Single-Engine Aircraft Day Pilot (2)	Single-Engine Aircraft Night Dual (3)	Single-Engine Aircraft Night Pilot (4)
—	—	—	—	—	—	— Totals Brought Forward				
May	16	Master		Self	—	Grantham - Northolt		40		
"	17	Magister	L5925	"		Northolt - Hawkinge		50		
"	17	"	L5925	"		Hawkinge - Arras		50		
"	17	"	L5925	"		Arras - Abbeville		30		
"	18	"	L5925	"		Abbeville - Arras		30		
"	18	"	L5925	"		Arras - ~~Bethune~~ Bethune		15		
"	18	"	L5925	"		Bethune - Merville		15		
"	19	"	L5925	"		Merville - Northolt		1·20		
"	19	"	R1822	"		Northolt - St Inglevert		1·10		
"	20	"	R1822	"		St Inglevert - Merville		30		
"	20	"	R1822	"		Merville - Croydon		1·00		
"	20	"	R1822	"		Croydon - Northolt		10	15·25	
"	21	"	R1822	"		Northolt local		10	15 35	
June	1	Proctor ~~Magister~~	P6130	Self	AC G.F. Glynn, AC Drury	Northolt . Hornchurch		35		
"	1	"	P6130	"		Hornchurch . Rochford		10		
"	1	"	P6130	"		Rochford . Northolt		35		
"	15	Magister	P6405	"		Northolt . Middle Wallop		35		
"	15	"	"	"		M Wallop . Warmwell		35		
"	16	"	"	"		Warmwell . M Wallop		30		
"	16	"	"	"		M Wallop . Northolt		30		
"	18	"	N3881	"				10		
"	18	"	P1822	"		Northolt . Warmwell		40		

GRAND TOTAL [Cols. (1) to (10)] 1993 Hrs. .... Mins.

Totals Carried Forward (1) (2) (3) (4)

Carnage by daylight. 2 Group's low-level Blenheim losses in 1940 and 1941 were horrific. Len Trent, VC, from New Zealand *(right)*, remarkably survived. The Rotterdam attack *(below)*, in July 1941, looked like this

Battle winners. Lord Beaverbrook *(facing page, top),* who fed the aircraft to the squadrons, (right in picture), with Max Aitken, his son, CO of 601. The ground crews *(below)* who, marvellously, kept them flying. The aircrews of 600 Squadron *(bottom)* who flew them by night and *(facing page, bottom)* six of Fighter Command's best – left to right: Tony Bartley, Desmond Sheen (Australia), Widge Gleed, Max Aitken, Sailor Malan (South Africa) and Al Deer (New Zealand)

Same game, different sides, 1941-2. For Spitfires and Hurricanes read Messerschmitt 109s, and a Heinkel 115c float plane. Luftwaffe fighter pilots at Wissant *(below)* and ground crews *(bottom)* servicing the aircraft. Armourers *(facing page, top)* feeding the 'ammo' belts for the 109s and loading a torpedo for a strike against British shipping in the North Sea *(facing page bottom)*

'When the odds were even, he towered over the Desert. . . . We had no general with his flair or ability . . .'
Peter Atkins leading SAAF navigator, on Rommel (p. 268), here flanked by Italian top brass *(top)*
The General took his own picture of three smashed British tanks *(bottom)*

Top of the Desert pops. Clive Caldwell, RAAF, outstanding CO of 112 Squadron, with his ground crews at Antelat *(bottom)*, January 1942, and high life for the Luftwaffe's Nos 1 and 2 *(top)*, the exceptional Hans-Joachim Marseille (at the wheel) and squadron commander, Werner Schroer (sitting on Marseille's right)

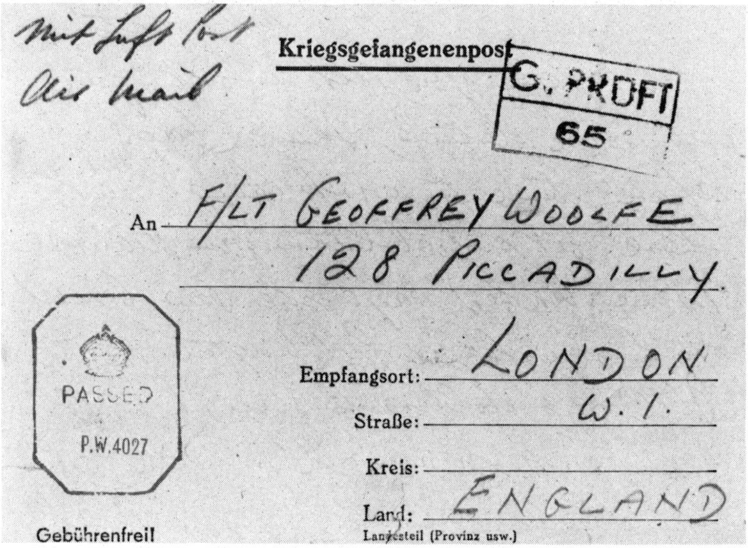

Mit Luft Post
Air Mail

Kriegsgefangenenpost

G. PRÜFT
65

An F/LT GEOFFREY WOOLFE
128 PICCADILLY

PASSED
P.W.4027

Empfangsort: LONDON
W. 1.
Straße:
Kreis:
Land: ENGLAND
Gebührenfrei!
Landesteil (Provinz usw.)

Left: In the 'bag' with 27 aircraft destroyed, the irrepressible Bob Stanford-Tuck (centre) and, left, George Harsh, journalist, from Atlanta, Georgia – Stalag Luft III, March 1943. Another inmate, Paddy Barthropp, floored by an FW 190 over St Omer a year earlier, wrote to Geoffrey Woolfe at the Royal Air Club, Piccadilly, on a POW letter-card *(below)* as follows:

Dear Geoff,
You will be surprised to get this, no doubt . . . . No drink and nonsense here, but I'm quite OK and not badly wounded. Remember me to the boys and get hold of Fisher and tell him the news. Any cigarettes would be more than welcome, and especially a letter. I haven't much else to look forward to . . . . We will have the hell of a party after it's over. I shall be pissed on about ½ pint. Sex life is bad here; saw a cat but couldn't catch it. Keep smiling and have one on me. Yours, Paddy.

This would have provided the classic air defence in depth so desperately needed to make life easier for controllers and less costly for pilots....

However, let there be no mistake, the battle was won (under Park) by the efforts of tired but resolute controllers and the immense courage of 11 Group's pilots.

Nevertheless, the 12 Group Wing, properly exploited, could have provided the spearhead against the enemy formations, creating havoc among them and giving 11 Group's pilots time to gain height and position to continue the destruction....

Over Twenty-six? Too Old!

Air Chief Marshal Sir Hugh Dowding, Air Officer Commanding-in-Chief, Fighter Command in his Battle of Britain despatch, 20 August 1941

'Only exceptionally should officers over twenty-six years of age be posted to command fighter squadrons.'

Dowding

Group Captain W.G.G. Duncan Smith, *Spitfire Into Battle* (John Murray, 1981; Hamlyn Paperbacks, 1983)

At the end of November 1940, Air Chief Marshal Sir Hugh Dowding ... handed over his command to Sholto Douglas. The news came as a shock to most of us because he had schemed the victory, held Fighter Command together in the Battle of Britain ... and proved himself an all-time great among air commanders.

Though he was relatively unknown personally to many pilots, a great bond of affection and respect flowed between them.... He was expert in making decisions of far-reaching effect and would go to any lengths to support his subordinate commanders and squadrons. It seemed, therefore, a strange turn of fate, particularly since he had defeated with resounding success the German Air Force in its attempt to destroy Fighter Command....

Park

Keith Park, an Essay, written by Dr Vincent Orange

Dr Vincent Orange, head of the Department of History at the University of Canterbury, Christchurch, New Zealand, has recently made a penetrating study of the life and commands of Air Chief Marshal Sir Keith Park. He has the advantage of having had access to Park's private papers.*

Here, in a specially written essay, Dr Orange quotes correspondence revealing some of the thoughts that troubled the old 11 Group commander's

*A biography of Sir Keith Park by Dr Orange is to be published by Eyre Methuen.

mind about his 'unjust dismissal' in the aftermath of victory in the most critical defensive air battle of all time.

Shortly before 25 November 1940 Keith Park learned that he was to be relieved as Commander of No. 11 Group, Fighter Command. Richard Saul (Commander No. 13 Group) wrote to say how sorry he was 'in view of the magnificent achievements of your group in the past six months; they have borne the brunt of the war and undoubtedly saved England. My sympathy is entirely with you.'[1] Park replied: 'It was a shock to be told that I was to be taken away from my Command after only seven months but it was a far greater shock to hear the name of my successor (Trafford Leigh-Mallory, Commander of No. 12 Group), in view of the little support that No. 11 Group has had from No. 12 Group ever since away back in May. Your Group and No. 10 Group (Sir Quintin Brand) have always sent properly trained squadrons to relieve war-weary squadrons from the front line. Moreover, your Group has always cooperated in helping out with junior leaders and providing properly trained pilots required to replace wastage. On the other hand, a number of pilots provided from No. 12 Group were rejected by squadrons because they had not been trained after leaving OTU.'

Park agreed that if a change had to be made, the job should have gone to Saul who knew the stations and the country. But for himself he flatly refused a position in the Air Ministry: 'As I was told that the only reason for my leaving No. 11 Group was because I had carried the baby long enough for one man and was due for a rest from the responsibility, I do not quite see why I should be stuffed into a very busy office job at the Air Ministry.'[2] He went instead to a Flying Training Group, where his precious experience was passed on to those most in need of it, but that had not been the Air Ministry's original intention.

Although dismayed at his removal from Uxbridge, Park did not repine. He spent his last days there energetically pursuing two campaigns against the Air Ministry: one to improve the standard of comfort at his stations, the other to win recognition for those who had not earned immediate gallantry awards. 'The senior members of my group staff deserve the lion's share of the credit for our successes but neither they nor many devoted officers and men around my twenty-six stations have had anything.'[3]

He himself was given a CB on 4 December and as soon as the news was announced, congratulations poured in. Sir Cyril Newall (Chief of the Air Staff during the Battle) wrote: 'It was good to see you singled out and you deserved it thoroughly,' to which Park replied: 'I sincerely hope that the powers that be will not leave me indefinitely doing a training job, after having demonstrated a certain flair for beating up the Boche under varying conditions. There is a saying that once in Training always in Training during war, and I sincerely hope this will not be my reward for my efforts in 1940.'[4]

Leslie Gossage (Park's predecessor at Uxbridge and later Air Member for Personnel) wrote: 'The consequences of what you have achieved may prove wellnigh incalculable in the history of the world and my admiration for the way in which you have met and overcome the serious and novel problems which have so frequently confronted you is unbounded. I know that you are concerned about awards for those who have served you so well and before leaving the Air Ministry I took special pains to impress upon the Secretary of State for Air the importance of their publication without delay. He was holding them up until you received yours.'[5]

Douglas Evill (Park's successor as Senior Air Staff Officer at Fighter Command Headquarters) was also quick to congratulate Park: 'You know, I think, that I have admired immensely the way in which you have commanded and led your Group through the battles of this summer and autumn – and regret your departure.'[6] Park's response to this letter shows his resilience. He swallowed his hurt at what he considered an unjust dismissal, knowing full well that to sulk in his tent would lead him swiftly to the retired list, and began to fight back. 'Maybe I shall be able to help Fighter Command,' he told Evill, 'by raising the standard of the FTS pupils we send to the OTUs. To that end I wish one of your staff could write to me direct at No. 23 Flying Training Group and let me know where our weakness lies. I cannot promise an immediate cure for every fault, but I should like to know in what direction I can assist Fighter Command to defeat the enemy in 1941.'[7]

Nearly six years later, on 30 September 1946, Viscount Stansgate (Secretary of State for Air) wrote to Park on his retirement to convey to him the thanks of His Majesty for his long and valuable services: 'he recalls with gratitude your distinguished services in the Middle and Far East.' Stansgate added his own tribute to Park's efforts: 'during the grim days of the Battle of Britain and later in command in Malta you played a great and active part in the organization of victory against odds which were at times overwhelming. No one did more to deserve to be raised to the position of C-in-C, as you eventually were and in the final stages of the war to be the only airman to hold a major command against an active enemy.'[8]

References:

1. Saul-Park, 25 November 1940: Park Papers (Lucerne Road), Item 386 (folder entitled 'Copies of correspondence re: the Battle of Britain').
2. Park-Saul, 27 November 40: Park Papers (NZI), Item 44.
3. Park-Gossage, 27 Nov 40: Park Papers (NZI), Item 45.
4. Newell-Park, 15 Dec 40: Park Papers (NZI), Item 56. Park-Newall, 17 Dec 40: Park Papers (NZI), Item 79.
5. Gossage-Park, 11 Dec 40: Park Papers (Lucerne Road), Item 386.
6. Evill-Park, 16 Dec 40: Park Papers (NZI), Item 57.
7. Park-Evill, 16 Dec 40: Park Papers (NZI), Item 68.
8. Air Ministry Historical Branch.

PART THREE

In the Teeth of Adversity

If the events of the summer and autumn of 1940 could be seen as some kind of divine deliverance from the oppressor, the next twelve months were to tax to the limit the resolve and resources of the Allies. Everywhere, it seemed, there were reverses: in the Balkans and Greece; in the Mediterranean; in the Western Desert; and then, as the year's end approached, in the Pacific and Southeast Asia, as the Japanese let loose their fury against the US Fleet in Pearl Harbor and, immediately after it, against the Royal Navy's capital ships, *Prince of Wales* and *Repulse* off Malaya.

It was a sombre period of the war, brightened only by the knowledge that, with the United States and Russia in the fray, the long-term outcome could hardly be in doubt. Extending well into 1942, it was a time when the Allies had to try to stand where they were and fight.

However, there were still a few postscripts to add to the year just passed.

Honoured Czech

Wing Commander H.R. 'Dizzy' Allen, *Fighter Squadron: A Memoir 1940–42* (William Kimber, 1979)

The outstanding fighter pilot on the British side in 1940 was a member of 303 (Polish) Squadron; he was a Czech, Sergeant Pilot Josef Frantisek.

Jean Zumbach, *On Wings of War* (Opera Mundi, Paris, 1973; Andre Deutsch, London, 1975)

One of our best pilots was Sergeant (Pilot) Frantisek, who had patented a special tactic of his own. After the first onslaught by the whole squadron, he would profit by the confusion to slip away and hang about over Dover, where he kept a patient vigil for homeward-bound raiders. There were always a few of these – planes which had run out of ammunition or (were) short of fuel. He placidly put paid to them. The CO kept telling him off for breaking formation, but Frantisek went on adding the black crosses to his scoreboard. . . .

Frantisek Fajtl, *Vzpomínky na Padlé Kamarády*, (Mlada fronta edice Třináct, Praha, 1980)

Josef Frantisek was the son of a carpenter and car-body repairer in Otaslavice, near Prostejov in Moravia. From childhood his interest was in motor cars. This led him to an engineering apprenticeship and he qualified as a mechanic. Immediately afterwards, he joined the Czechoslovak Air Force as a student pilot. After qualifying he was posted to No. 5 Squadron of the 2nd Air Regiment in Olomouc.

Lance Corporal Frantisek was a good looking, pleasant young man. He was particular about his dress and his personal appearance. He laughed heartily and was a friendly man. He did not like discipline, but he loved flying. He lived his life to the full and he was often absent from military barracks when he should have been in his billet. His many friends covered up for him, but this did not always stop him from being reported as a defaulter.

The real problem for Frantisek was his friendship with another pilot named Kokes. People tended to tease Kokes because he was so short of stature. They called him names like 'Small Beer'. This greatly annoyed Frantisek and it led to fights, often involving a crowd in a pub. It was all the same to him whether he faced one or twenty opponents – he would fight, disregarding danger or damage. His CO warned him that this might lose him his lowly rank or even cause grounding. Josef wasn't bothered about the prospects of demotion, but permanent grounding was something else.

When a new CO took over he called the eternal lance corporal in and said: 'Frantisek, I command you to stop being a lance corporal and become a corporal!' He added an aside which surprised the young pilot. 'For Christ's sake at least behave for six weeks so I can promote you.' Josef came sharply to attention. 'Understood, sir,' he said, saluting smartly.

When Frantisek came to England in 1940, he travelled mainly with Poles. The Czechoslovak Air Force didn't reclaim him so he flew with 303, the Polish fighter squadron, in the Battle of Britain. The environment of the brave and generous Poles was greatly stimulating. After warnings from the considerate CO about leaving the squadron to attack by himself, he eventually came to be regarded as a kind of guest performer with the right to hunt on his own.

He was a brilliant pilot and marksman. His ability to shoot from any angle and attitude was matched by his thirst for combat. His victories mounted. When he was killed, aged twenty-seven, on 8 October 1940, in a mysterious landing accident, he had shot down 17 enemy aircraft during twenty-seven days' operations with 303. Some Polish and British sources claim that to this total should be added the 10 or 11 aircraft he was believed to have shot down during the fighting in France.

Once Josef Frantisek decided to serve with the Poles, he did not wish to change, but he was never a renegade; he did not become alienated from his own country. The squadron respected him for it.

His name is at the top of the pilots who fought for the freedom of the Czechoslovak Republic. Granted, posthumously, the rank of flying officer, Frantisek will always be the pride of Czechoslovak military aviation.

Somebody's Sitting On My Parachute

Pat Ward-Thomas, *Not Only Golf: An Autobiography* (Hodder and Stoughton, 1981)

In common with many another in time of war I nursed the delusion that disaster was no part of my immediate future. Men might vanish, be killed or taken prisoner but I was immune. It cannot happen to me was the sustaining thought, but one moonlit night in November that year [1940] it certainly did so.... We were trundling back from Berlin, having deposited our little load of bombs which may have mildly inconvenienced a few citizens. All was serene until a German gunner, who had not been taught that it was unsporting to shoot sitting ducks, scored a hit on one engine. An aircraft doing 130 m.p.h. at 12,000 feet looks almost motionless from the ground.

The engine fluttered, as if slightly irritated, but settled down for another hour and more.Then, as we were about to cross the Dutch coast, both engines went completely out of control, alternately cutting and screaming at maximum power. We lurched about the sky like an elderly gull in a hurricane but nothing we tried would work and we lost height rapidly....

There was some confusion over the process of baling out. The flight commander, Pritchard, and I, the second pilot, had different types of parachutes and he was sitting on mine. When he suggested that I might like to depart first we realized that we would have to change. This was not easy with the aircraft reeling all over the place and Pritchard striving to keep it steady. It was well for us that there was no fire aboard.

I dropped through the hatch in the nose of the aircraft with a feeling of despair familiar to thousands. For perhaps a second I did not care if the parachute failed to open but this did not prevent me from instantly pulling the release with no thought of counting three before doing so, which I think, was the instruction in those days. Then came the tug on the harness and, of course, profound relief that the parachute had opened. Suddenly all was peace. The flat polder land north of Amsterdam lay silver under a brilliant bomber's moon but it seemed that I would never reach it. Hanging there, at about 5000 feet, I distinctly remember feeling foolish and hoping that no one was watching. Later I discovered that others had felt the same.

Judging the distance can be difficult enough on a golf course in daylight. It is much harder when making a vertical descent by moonlight. Aware that I was falling towards one of the canals that make a checkerboard of that part of Holland I tried to sideslip but was far too late. Without any warning I was immersed. Luckily the parachute did not fall on me but drifted laterally and helped my floundering scramble up the steep sides of the dyke.

I sat on the bank tearing up various documents that should not have been with me. All was silence, not a soul was abroad in a desolate scene, but a quarter of a mile away I saw a faint light in a small farmhouse....

My hosts, humble smallholders, spoke only Dutch of which I had not a single word, but I conveyed that I wanted to sleep. The morning could take care of itself. I was put in a vast bed, which several warm bodies had recently left, within a sort of cupboard, and had been asleep for an hour or so when awoken by a light in my face and a German saying in English that I must go with him....

The Germans in the police station were quite amiable, gave us beer and insisted that for us 'Der Krieg ist fertig', as if we needed reminding. You will be home by August they said, but they were almost five years adrift in their estimate. We were taken to Amsterdam, where I was hardly at my best clattering in clogs and pyjama trousers across the lobby of the old Carlton

Hotel, then the Luftwaffe headquarters. We were taken towards a room whence emerged a screeching little German major. Behind him we could see a party in progress, women, drinks and cigar smoke, but we were not invited to join. Eventually the major stopped showing off; we were taken into a room and left until a minibus took us across Holland to Dulag Luft, a reception centre for air force prisoners near Frankfurt.

Awakening the next morning, refreshed after a long sleep, and realizing what had happened remains the worst single moment of my life. I looked round the plainly furnished single room and faced the full impact of my plight, thinking how, in God's name, can I stand living in a room like this for years to come, little dreaming that it was the equivalent of a first-class hotel compared with what was to follow.

Fleeing to Fight

Jan Cĕrmàk, Brno, Czechoslovakia, 1982

By 1941, most of the brave fugitives from Nazi-occupied Europe and Scandinavia had made their devious ways to Britain to be blended by the Royal Air Force into a wonderfully spirited and aggressive cosmopolitan machine. Jan Cĕrmàk's story was typical.

'On the airfield at Toulouse, in the Unoccupied Zone, there were dozens of new Devoitines. The Station Commander, a French colonel, was very good to me. "Take what you like," he said, "if you can fly one you can have it."

Taking all sorts of risks I flew from Perpignan, across the Mediterranean to Maison Blanche, the airfield near Algiers. I had no maps, no ammunition. Near the island of Mallorca, three Me 109s appeared. They didn't see me. I was worried about my navigation and also about having enough fuel. But I was lucky.

After landing at Maison Blanche and rejoining the French air command with French pilots, we were briefed for the first sorties against the English ships in Oran. I was so astonished I refused to fly. I was then put in jail. The Zouave, French colonial troops, smuggled me out at night after three days and I escaped, spending a month wandering about by foot, by train and by camel caravan with the Arabs. Eventually, I reached Casablanca. To my astonishment the Gestapo were there, too! I was in the depths of despair.

I was interrogated repeatedly by the French Secret Service who told me I would perish in England. With great difficulty I was finally able to board a mysterious Portuguese ship, bound for Lisbon.

With the help of the British Embassy there, I was flown to England in a DC3 Dakota. Years later I wrote to the Embassy to thank for the help which a little letter gave me at a moment when I was very downhearted. The reference was AA 22/2 and it was dated 16 August 1940:

To Whom it may Concern,
The bearer of this letter, Lieutenant Jan Cĕrmàk, has been sent by me to England to join the Royal Air Force. Any British authorities to whom he addresses himself should facilitate his onward journey to London and inform him how he can put himself in contact with the Air Ministry.

It was signed "Group Captain Chamberlayne, HM Air Attaché".

I met my comrades again on the friendly, welcoming soil of Great Britain, and soon joined the Royal Air Force.

The Battle of Britain was virtually over by the time I was ready. However, I later got the chance to take part in the Battle of the Atlantic and then, later still, in the Invasion of France. But what a relief it was to feel that the Gestapo had been removed from my life. I realized then that I was in the right place.'

Reinforcing the Middle East

As eyes now turned to the Middle East, the Western Desert, the Mediterranean and Malta, the cry was for aircraft, more aircraft – and quickly.

Of the earliest seaborne, reinforcing operations to Malta the first, on 2 August 1940, when a dozen Hurricanes were successfully flown off the carrier *Argus*, raised hopes about what could be achieved with a combination of Royal Navy and Royal Air Force enterprise. The second, however, on 17 November, was a monumental mess-up. A further twelve Hurricanes, navigated into the island by a pair of Fleet Air Arm Skua aircraft, were put off *Argus* at the extremity of the fighters' range – around 450 miles west of Valletta.

The Royal Navy worked to nautical miles, the Royal Air Force to statute miles, and on other counts the planning was inexpert and imprecise. Out of the formation of fourteen aircraft, five – one Skua and four Hurricanes – reached Malta. The rest were lost, having run out of fuel before they reached the island. It was an uncharacteristic, combined-Service reverse.

However, for ineptitude, maladministration and culpable negligence, an earlier land-based ferry of twelve Blenheims and twelve Hurricanes from Aston Down and Tangmere to the Middle East took the prize. Fifty per cent of this precious force was lost in circumstances which should, justifiably, have landed the planners at home in court-martial proceedings. Had it not been for the ability, persistence and courage of the survivors, the operation might well have become a total write-off.

Two of the Hurricane pilots on that ferry have, from their homes in Zimbabwe, provided details of the disaster. W.R. 'Dick' Sugden, who, more than three years later, was to take part in Pickard's famous Amiens prison raid, and Roger 'Jock' Hilton-Barber, who was subsequently to make his mark on photo reconnaissance and other varied operations, have records to substantiate their recollection of this ill-starred venture. Hilton-Barber is clear about the background to the operation which, altogether, covered six days from 16–21 June 1940.

Roger 'Jock' Hilton-Barber, Harare, Zimbabwe, 1982

'The whole ferry was organized in quick time because of the urgent need for reinforcements in the Middle East. The aircraft were in the hands of No. 4 (Continental) Ferry Pool which was responsible for ferry operations. It had been supplying aircraft to the BEF in France before Dunkirk.

The long-range tank system for the Hurricanes, being used for the first time, had never been properly tested. In the briefing, such as it was, before we left Tangmere, I do not recall that any mention was ever made of what we were supposed to do in bad weather. The Blenheims were to navigate and the Hurricanes were to formate on them, but what if they got separated?

The flight plan was practically direct – Tangmere, Marseilles, Tunis, Malta and Mersa Matruh. The leader of the Blenheims was to be Squadron Leader "Scotty" Pryde. It was impressed upon the Hurricane pilots that, to conserve fuel, they were to cruise at around 160 m.p.h. with weak mixture, the engine revs kept right down at 1200–1300 per minute, and the airscrew set in coarse pitch.

We didn't realize – and certainly we weren't told – that, at this level of cruising, the engine revs were too low for the generators to cut in. All pilots, during the ferry, experienced flat batteries – and, with them, the inability to switch over from the long range to the wing tanks. All that was needed was for the pilots to be told to select fine pitch and raise the engine revs while transferring fuel.

The weather was the other factor. Small attention had been paid to the met. forecast and, in any case, it was brushed aside. When we set off from Tangmere there was beautiful sunshine, but shortly after crossing the French coast we flew straight into solid, black cumulus cloud. I tucked in tight with my Blenheim leader who gently let down. This was most frightening because of the near misses we had. We broke cloud at a few hundred feet and carried on independently of the main formation. We never saw another aircraft. Eventually, the weather deteriorated still further and, with the high ground ahead, it would have been crass stupidity to have gone on – although another Hurricane, without oxygen, climbed up through the overcast to something close to 20,000 feet and made it.

My leader turned back at Tours and we landed at Tangmere again in time for lunch. We were given a very cool reception by the briefing officer who, for his own sake, shall still remain nameless. Having accused us of Lack of Moral Fibre, he then told us to get cracking. In the meantime, the weather had cleared somewhat although we were dodging heavy rain squalls most of the time.'

If Hilton-Barber's troubles were only just beginning, Dick Sugden, on ahead, had already had a taste of what was to come. His diary of the ferry leaves little to the imagination.

W.R. 'Dick' Sugden, The Sugden Diaries, Ruwa, Zimbabwe, 1982

'18 June 1940. Day of departure from Tangmere. Before some of us had even started refuelling, and without having a final briefing, a "gaggle" of Blenheims and Hurricanes took off and set course. Panic. There were three or four Hurricanes and two Blenheims also in the same boat – not yet refuelled and not properly set to go.

Took off and formed up in an untidy crowd. Set course over Poole, crossed Channel Islands, hopefully aiming for Marseilles. Weather fine, but gradually deteriorating until, after about three hours, cloud had thickened almost down to the deck. The leading Blenheim turned and those of us who still had him in sight followed. Suddenly a small airfield appeared, and we thankfully followed the Blenheim in to a very tight landing. Now pouring with rain. F/O Smythe's aircraft tipped over on its nose. We later righted it, ran it up and found it was OK; but, unluckily, the next day the same thing happened; this time the prop had really had it.

There was a large hangar with a petrol dump, some French airmen and ancient aircraft on the field. Our contingent now consisted of six Hurricanes – with pilots – F/O Haddon-Hill, F/O Smythe, P/Os Beardon, Collins, McAdam and myself, plus one Blenheim IV with F/O Parker and crew. Haddon-Hill managed to get some ham and cheese from a nearby village. We then set about refuelling by hand. Still raining heavily with the aircraft sinking deeper and deeper into the mud. General depression. This place was called Ussel, some 50 miles SE of Limoges, and 200 miles NE of Marseilles, on high ground. Nobody had a clue about the war situation. We half expected to see German tanks appear any moment. Eventually we got on a bus, and, just as it was getting dark, we were deposited at a village hall, full of French officers who gave us a great welcome and a huge meal with plenty of vin. Feeling much more cheerful we dossed down on the floor and got a little sleep.

19 June. Fine and hot. We found our way back to the airfield where the Blenheim was now well and truly bogged. After several fruitless attempts to get it out, Parker decided that he and his crew would try and thumb a lift south to the coast. Smythe's Hurricane also had to be abandoned after nosing over again; he, too, took to the road. Collins and MacAdam took off successfully downhill and downwind. I followed but lost sight of them. I circled several times waiting for the other two; but as I had no maps and they were having taxiing troubles, I (apprehensively) landed again.

Eventually we all got off and set course for Marseilles/Marignane. The weather then closed in, and, over the mountains, Beardon got separated. Wretched visibility, but we spotted the airfield at Toulon. Landed after a 2 hr 30 flight. Here, there was a French Air Force mess. The Italians had raided the place a few days before.

Refreshed, we then set off for Marseilles in lovely weather. With the airfield in sight, we were amazed to see a large formation of Blenheims and Hurricanes setting course southwards. If we had had more fuel we might have joined them. What a blessing we didn't! We were later told that only one man out of that bunch reached North Africa alive – P/O Mansell-Lewis – and he had to swim the last couple of miles. Apparently they were overtaken by darkness, ran short of fuel, and were shot at by "friendly" anti-aircraft fire. Poor Beardon had reached Marseilles on his own, refuelled, and joined the formation. (Most of my kit and logbook were in his Hurricane.)

On landing we found that Collins and MacAdam had arrived safely. In one of the hangars "Jock" Hilton-Barber was putting the finishing touches to a home-made tail skid on his Hurricane. His was a hair-raising story.

With fuel transfer system not working, and tanks reading zero, he had put his aircraft down in a tiny field in mountainous country near Monde. With extraordinary help from the local population, two 12-volt batteries, borrowed from the local garage, were lugged up the mountain. Connecting them to the electrical system, Jock pumped fuel into the wing tanks. Eventually, he got off, squeezing between a hedge and telegraph wires. With the aid of a local blacksmith in Istres and the leaves of an old motor car spring, he had fashioned a tail skid to replace a sheared tail wheel.

There was also a Blenheim pilot at Marseilles with his crew. Apparently Scotty Pryde had crashed on take off, and had promptly requisitioned this chap's aeroplane. His name was Miller, a New Zealander, and he was hoping to find a serviceable French aircraft in which to get over to Tunis. Subsequently we heard that Scotty, who was commanding the whole operation, had crashed in high ground in Sardinia.

20 June. A scorching day. The French had charged all our batteries, but we were worried about our long-range tanks, and whether the pumps were working properly. We took all the fuel we could squeeze in, I still hold a crude, pencilled invoice for "960 litres Essence B, 12 kgs thick Intava 120 pour Hurricanes P 2623, P 2629, P 2614, P 2625, reçu de SPR Marignane". And so, at about noon, we set off for the long Mediterranean crossing with Haddon-Hill leading, followed by Hilton-Barber, Collins, MacAdam and myself. Apart from one heavy rainstorm the weather remained fine, the only land we saw was Sardinia. "Jock" Hilton-Barber took over the lead when we crossed the African coast, as Haddon-Hill appeared to be lost. After a weary 4-hour flight, we all landed at Tunis. A few Frenchmen, standing around, told us that the French Air Force had left, and they invited us to help them finish up the bar stocks at the Flying Club – mostly delicious champagne off the ice. Never did a drink taste better. Sadly, we then learned that none of the previous day's flight had made it, intact, to North Africa.

21 June. Up early, and back to the airfield. No fuel there. We would have to fly west to Medjez-el-Bab where there was plenty. On take off, Haddon-Hill's engine cut and he finished up in the fence, a write-off. Fortunately he only had a few cuts. Nothing to do but leave him behind. At Medjez we were surprised to find two Blenheims, flown by two very experienced warrant officers, and two more Hurricanes, with P/Os Carter and Glenn, none of whom we had seen since Tangmere.

The Blenheims soon took off, heading for Mersa Matruh, and then the two Hurricanes got airborne for Malta. At last we were topped up with fuel, and Collins and Glenn took off. My battery was flat again, and Jock nobly stayed with me to hand-crank my engine. For an hour and a half we cranked and cranked, in a temperature of well over 100 degrees. The engines were boiling hot. At last mine fired and we got away. Jock had our only map – an Admiralty chart of the Med. with the depth of the sea measured in fathoms. Very helpful. But he steered a wonderful course, and after passing Linosa and Lampedusa, we hit Malta bang on the nose, in one hour and forty-five minutes. I literally hit Malta as on my landing run my port wing struck the side of an old bus; Luqa was covered with obstructions. Total flying time since leaving Tangmere was 14 hours. It seemed more like fourteen days!

(Instead of being back in England within a week, I was to remain in Malta for another sixteen months. It was certainly a balls-up from beginning to end!)'

Postscript by Hilton-Barber:

'My Hurricane, No. P 2653, survived many sorties over Malta which made me feel the blacksmith and tail-skid episode had been worthwhile. After eight months of intense activity over the Island, I saw it shot down on 23 February 1941 when being flown by Flying Officer J. "Terry" Foxton.

After the war, the Air Ministry wrote and said that the French Maquis had returned my passport which had been found in a Blenheim which had crashed in the Alps. It had been carrying all my kit. The Maquis wanted details; they were mystified, to say the least! The accident had happened on 18 June 1940.'

'Faith, Hope and Charity'

Lord James Douglas-Hamilton, *The Air Battle for Malta* (Mainstream Publishing, 1981)

At the beginning of the Battle of Malta, the Gladiator pilots, flying 'Faith, Hope and Charity', had led the resistance. One of these pilots, Wing Commander Burges, later said:

> My lasting impression of Malta at war is of an island whose strength was its unity and singleness of purpose. This produced an invincible defence and a formidable offence.

The Gremlins

George Burges, Cranwell-trained, and an exceptionally popular young officer, was a flight lieutenant and personal assistant to the Air Officer Commanding just before the Regia Aeronautica began to attack Malta. As the AOC's close aide he mixed easily with the other Services and, in particular, with the Royal Navy's Fleet Air Arm whose aircrews, when the fighting started, quickly came to respect his special talents as a pilot. From their mess at Halfar, Burges obtained these anonymous verses in 1938. He believes the author was a Fleet Air Arm pilot.

The Gremlins are waiting just over the wall,
They're waiting and watching to give you a fall –
When least you expect it they're out in a flash
To trap the unwary and laugh at his crash.

They've been in the island since flying began
And try to prevent it as much as they can –
Their method is always the same – more or less –
They make your arrival a hell of a mess.

The wall to the south is the favourite lair
Of these little Gremlins who jump in the air,
Without any warning they make a bee-line
For any young chappie who's cutting things fine.

When coming in slow with a few feet to spare
The Gremlins will spot you and creep from their lair
They'll shoot out their arms with lashings of mirth
Will seize your contraption and drag it to earth.

So if you approach with your height running short
Keep plenty of speed or you'll surely be caught;
The man who depends on his flaps and his slots
Is grabbed in a jiffy and tied into knots.

Beware of the Gremlins who live by the wall,
They're waiting for fellows who flirt with the stall –
Remember my warning and give her the gun,
Or else you'll arrive as some others have done.

Anonymous, Malta, 1938

'In a Monastery Garden'

W.R. 'Dick' Sugden, Ruwa, Zimbabwe, 1982

Having ferried two aircraft to the island, 'Jock' Hilton-Barber, an 'old' Rhodesian, and I were now part of the air defence of Malta – all four Gladiators and four Hurricanes. The aircraft were seldom serviceable all at the same time; spares were a problem. This often meant a 4-hour stand-by, sitting in the cockpit in the hot sun, waiting for the hooter from the control tower to scramble us. Sometimes it might be just one Gladiator and one Hurricane against all comers....

I had never had any fighter training, being destined for bombers. I knew nothing about aerobatics or front gunnery. I suppose I was lucky to last a month, being shot down only once by CR 42s. I found the Gladiator tricky to land straight ahead; after a few ground-loops, the AOC decided it would help the war effort if I was posted elsewhere.

On 25 July I reported to No. 3 Ack-Ack Cooperation Unit, commanded by Squadron Leader 'Rosie' Houlbrook at Hal Far. The unit was quartered in 'The Monastery'. This had been a genuine monastery, situated to the southeast of the airfield. It now accommodated eight pilots, the ground crews and gunners. Three Swordfish were parked nearby. Our duties were mainly to cooperate with the AA and Searchlight units, met flights etc. In the event of a seaborne invasion we were expected to sell our lives dearly while 'spotting' for the coastal guns. Every day we provided a stand-by aircraft and crew under the control of 830 (Fleet Air Arm) Squadron, who lived on the other side of the airfield. Our duty telephonist would toll the monastery bell when we were needed.

I was duly initiated into the mysteries of the Swordfish, and after a few circuits and bumps was considered 'fully operational'. Actually, I found the aircraft delightful to fly, a real old gentleman's aeroplane, capable of 100 knots flat out. It could land itself. The only hazard in landing at Hal Far was to avoid all the obstructions – old buses, bowsers and lorries etc. These were intended to deter the enemy from landing.

The unit had recently lost an aircraft near Sicily, the crew now being POWs. Since then, comparative calm had prevailed which not even three or four air raids a day seemed to disturb.

So it was that, Gerald Bellamy and I were taking our ease on the roof after lunch on 31 July, when the silence was broken by the tolling of the monastery bell. 'God, what does that mean?' I asked. Gerald, putting aside his saxophone, was objective, 'You'd better get moving,' he said, 'and find out. *You're* the duty tit.'

I tore down the stone stairs and found my gunner, LAC Parke, already strapping on his mae west. Beside him was an FAA observer who rapidly briefed me. An Italian cruiser had been reported about 100 miles to the northeast, steaming towards Italy. We were to shadow it and send back any news which might help 830 Squadron to make a night attack. I was stunned; it was almost as though I had been told I was to ride in the Derby that afternoon.

The observer also wore a rather stunned look. He was Petty Officer Pinkney, a small, regular RN type who, obviously, had reservations about his pilot-to-be. As we walked over to our aircraft which was already being run up, he asked me how much Swordfish flying I had done. 'About three hours,' I said. 'God Almighty!' His reaction was understandable. 'How many hours altogether then?' 'Oh,' I said, casually, 'just over two hundred.' He again invoked the name of his Maker, but more loudly.

The three of us got strapped in and staggered into the air. 'Steer 055 magnetic,' said Pinkney through the Gosport speaking tube. The pilot's compass had one of those curious mirror affairs on top of it, which I had never seen before. It took me a minute or two to set it. Actually I had caught a glimpse of Pinkney's face in the mirror. He was half-leaning over my shoulder to see what was going on. Neither of us was encouraged by what he saw.

Once under way, however, we climbed to 5000 feet. There wasn't a cloud in the sky nor a ripple on the sea. There was no small talk, except when we had to drop a smoke float to check our drift. After an hour and a half we found our cruiser! Very pretty it looked, too, in the evening sunlight. Pinkney came on the intercom. 'Now use your bloody sense,' he said, 'and don't get too close. Let's go down a bit and keep circling.'

He was busy coding his messages to base, although he couldn't quite identify the ship's class. For one awful moment I thought he might want us to dive and read its name. When we saw the flak coming up we decided to skip that.

We continued circling for half an hour. Then Pinkney gave me a course for home. 'And for God's sake,' he said 'steer a good one!' All the time we had been praying no fighters would appear.

It was dark by the time we reached Malta. Pinkney was anxious to know whether I'd ever landed a Swordfish at night. 'Once,' I replied. But I didn't tell him that on that occasion my wheels had taken a bit of the boundary wire with them. He'd suffered enough.

We made a beautiful landing on the primitive flarepath. After debriefing the three of us had a drink or two together. Pinkney chose rum. I made sure they were large ones, after which we were all loud in our praise of each other's performance. Even our gunner, Parke, said how much he had enjoyed the 3½-hour trip.

830 Squadron went out to make an interception; but no luck, our cruiser had got clean away.

I was relieved to see my monastic couch again, and listen to Gerald's appalling saxophone practice.

First Overseas Radar

Flight Lieutenant R.T. Townson, Reading, Berkshire, 1982

How did a handful of pilots in their biplanes manage to position themselves so successfully, and hold their own, against the early Italian air attacks on Malta?

Few know that when Italy came into the war in 1940, the island already possessed the first transportable radar (or RDF – Radio Direction Finding – as it was then called) station to be sent overseas by the Royal Air Force. It was this early warning system which helped the gallant pilots to meet daily the numerical odds which the Regia Aeronautica stacked against them.

The Air Ministry, with a lot of foresight, dispatched to the island in January 1939, a small party of Signals tradesmen with instructions to set up, within six months, No. 241 AMES (Air Ministry Experimental Station) and then come home. Hitler and Mussolini, however, had other ideas and this tiny group of technicians, in the charge of Flight Sergeant James, under whom I served as No. 2, stayed for more than a year and a half.

Setting the equipment up, and making it work, would not have been accomplished without the expertise of a one-time BBC engineer, H.T. Roberts. When he was required in another theatre, nine of us were left to keep the station at Dingli running, and man it under what became very active operational conditions.

Our transmitter (Type MB1), built by Metropolitan-Vickers, and the Cossor receiver (Type RM2: Serial No. 1) were driven by a Meadows-engine, 12 KVA generating set. Heat was a problem – until we rehoused the equipment in buildings of local limestone. Another difficulty was the high rate of consumption of transmitter output valves (NT 77). However, Alitalia kindly came to our aid by bringing a dozen valves on the last civilian service they flew into Malta before the outbreak of war – a consignment which, I suspect, would never have reached us had they realized its significance.

Our little unit received cooperation – and much personal kindness – from the Command Signals Officer, Squadron Leader A.D. Messenger (Air Commodore A.D. Messenger), and from our Signals colleagues throughout the island. One cannot disguise, however, the Air Staff's lack of understanding of what it was that we were trying to achieve. Thus, when we had amply demonstrated what RDF could do, and it was plain that some form of 'Fighter Control' would have to be established, all wing commanders and above were roped in as 'Controllers'. The idea of anyone below the rank of flight-sergeant being entrusted with the job of putting tiddley-winks on the plotting table showing aircraft tracks or plots supplied by our solitary RDF station, was unacceptable!

We weren't surprised at the reaction when we suggested that we should all be promoted to Air Rank to read the cathode-ray tube; but we certainly were amused when we found that the (untrained) controllers with the rank of wing commander and above would not themselves use the R/T to talk to the Gladiator pilots, but, instead, fetched the Duty Signals Officer from his office to do the job, thus wasting vital time.

This extraordinary attitude became much more serious when Malta's RDF network was expanded. The battle to persuade the War Room in St John Cavalier to adopt a simple grid system when four RDF stations were in use (instead of the original range rings and bearing lines from the first station at Dingli) made little headway. An early morning mix-up, however, when eight aircrew were unnecessarily killed at Luqa, and two Sunderland flying boats were destroyed on the water because a telltale 25-mile plot or track of Me 109s was woefully ignored, quickly changed the Air Staff's stance.

The reliance which came to be placed on this warning service by the island's defences was demonstrated when the tall Merryweather tower carrying our transmitter aerials was blown down in a gale. It was a 'naked' 48 hours that the Gladiators had to face until, with a combined operation embracing the Royal Navy and the Royal Air Force, the resources of the two Services had the station 'back on the air'.

When Flight Sergeant James was invalided back to the UK I was left in the driving seat and, with eight others, we carried the RDF banner until eventually relieved by No. 242 AMES and the three low-looking 'beam' stations sited at Maddelena, Ta Silch and Dingli, which provided a pretty effective RDF coverage for the whole island.

The transformation was completed when a new Operations complex was carved out of the Baracca Cliff in Valletta and, with a Filter Room added, a more experienced staff took control under the new AOC, Air Vice-Marshal H.P. Lloyd (Air Chief Marshal Sir Hugh Lloyd).

Thus was 'radar' brought to Malta and additional 'eyes' given to a bunch of very gallant and devoted pilots to help them strike at the invading bombers. This island experience was soon to bring benefits to other theatres of war. The pioneers of Dingli had lit the way.

The Fleet Air Arm Observer's 'If'
(with apologies, etc.)

If you can keep your track when all about you
Are losing theirs and setting 'mag' for 'true'.
If you can trust yourself when pilots doubt you
And get back to the ship out of the blue:
If you can keep control of your dividers
And Bigsworth board and Gosport tube and pad,
Or listen to the wireless and the pilot
Talking in unison – and not go mad....
If you can fill the unforgiving minute
With sixty seconds' worth of ground-speed run,
Yours is the Air – and everything that's in it,
And – what is more – you'll be an 'O', my son.

Taranto
Fleet Air Arm's Epic Attack
Night of 11–12 November 1940

Commander Charles Lamb, *War in a String Bag* (Cassell Collier Macmillan, 1977; Arrow Books, 1978)

The *Illustrious* was to be detached on the evening of the 11 [November 1940], with the cruisers *Gloucester, Berwick, Glasgow* and *York* and an escort of four destroyers, to make for the island of Cephalonia, opposite the Gulf of Corinth, and about 70 miles south of Corfu. We were to steam to the north, towards Corfu, and the two squadrons were to take off to attack the Italian fleet in their harbour, at intervals of an hour. Each squadron was reinforced by the aircraft and crews from 813 and 824 Squadrons, from *Eagle*. 815 was to lead the first attack, followed by 819. As the first strike, we were scheduled to take off at 8.30 in the evening and attack immediately on arrival, with 819 Squadron, the second strike, following an hour afterwards. We had 170 miles to fly, across the Ionian sea; and the second strike an hour later, would have 150 miles of sea to cross.

At the final briefing in the wardroom a large-scale map of Taranto and a magnificent collection of enlarged prints of the photographs I had brought out from Malta were pinned to cardboard backings and were on display. It was possible to study every aspect of the harbour and its defences, and the balloons; and, of course, all the ships in detail. In the outer harbour, called the Mar Grande, there were six battleships moored in a semi-circle: four of the *Cavour* class with ten 12·6-inch guns, and two *Littorio* class with 15-inch guns. All these ships were protected with weighted anti-torpoedo nets, suspended from booms, which reached down into the water as far as the ships' keels; but the Italians had a shock to come, because our aerial torpedoes were fitted with Duplex pistols, a magnetic device which exploded the torpedo's warhead when it passed underneath the ship. These attachments had been invented at HMS *Vernon* when Captain Denis Boyd had been in command. They were called Duplex because they performed a dual function: the 'fish' would explode either as it passed underneath, or on contact, if it struck the ship's hull.... The eleven torpedoes which were being used that night were set to pass under the hulls, to avoid the nets.

To seaward of the six battleships, and between them and the harbour entrance, were three 8-inch gun cruisers, the *Zara, Fiume* and *Goriza*; and stretching right across the harbour, from side to side, were eleven moored balloons. Another eleven encircled the harbour to the south and east.

In the inner harbour, called the Mar Piccolo, were two 8-inch cruisers moored in the centre, the *Trieste* and *Bolzano*; and alongside each other, stern-to, in true Mediterranean fashion, were four 6-inch cruisers and seventeen destroyers.

The harbour defences at Taranto were designed to protect one of the biggest fleets in existence, if not the biggest. The Italians possessed all the necessary skills to make it into the impregnable fortress that it should have been: the guns, placed at strategic points on all the breakwaters, and all over the harbour, were expected to safeguard all the ancillary installations ashore which combined to make this their most important port. It had to be impregnable for a huge fleet to be able to rest, and to carry out repairs in complete security.

Promptly at 8.30 in the evening Williamson, the CO of 815, and Scarlett our Senior Observer, took off, followed by the rest of the first strike of twelve Swordfish. The second strike, led by Lieutenant-Commander 'Ginger' Hale, the CO of 819 Squadron, and a Navy and England rugby player as unshakable as the Rock of Gibraltar, were due to take off an hour later. Owing to the ditchings and collisions, Hale's flight had been reduced to eight aircraft. Six in our flight and five in the second were armed with torpedoes, and the remainder with six 250-lb armour-piercing bombs. Kiggell and I, the flare-droppers, were armed with sixteen parachute flares apiece, and four bombs.

Cruising along quietly at about 5000 feet, waiting for Kiggell to begin flare-dropping, I realized that I was watching something which had never happened before in the history of mankind, and was unlikely to be repeated ever again. It was a 'one-off' job. 815 Squadron had been flying operationally for nearly twelve of the fifteen months of war, and for the last six months, almost without a break, we had attracted the enemy's fire for an average of at least an hour a week; but I had never imagined anything like this to be possible. Before the first Swordfish had dived to the attack, the full-throated roar from the guns of six battleships and the blast from the cruisers and destroyers made the harbour defences seem like a sideshow; they were the 'lunatic fringe', no more than the outer petals of the flower of flame which was hurled across the water in wave after wave by a hot-blooded race of defenders in an intense fury of agitation, raging at a target which they could only glimpse for fleeting seconds; and into that inferno, one hour apart, two waves, of six and then five Swordfish, painted a dull bluey-grey for camouflage, danced a weaving arabesque of death and destruction with their torpedoes, flying into the harbour only a few feet above sea level – so low that one or two of them actually touched the water with their wheels as they sped through the harbour entrance. Nine other spidery biplanes dropped out of the night sky, appearing in a crescendo of noise in vertical dives from the slow-moving glitter of the yellow parachute flares. So, the guns had three levels of attacking aircraft to fire at – the low-level torpedo planes, the dive-bombers, and the flare-droppers. The Swordfish left the Italian fleet a spent force, surrounded by floating oil which belched from the ships' interiors as their bottoms and sides and decks were torn apart.

The Italians were faced with a terrible dilemma: were they to go on firing at the elusive aircraft right down on the water, thereby hitting their own ships and their own guns, and their own harbour and town, or were they to lift their angle of fire still more? Eventually they did the latter, because all the other five attacking Swordfish managed to weave their way under that umbrella to find their targets. Had the arc of fire been maintained at water level, all six would have been shot to pieces within seconds, instead of two;

but the guns would have done even more extensive damage to the ships and the harbour itself.

All the way back to our rendezvous with the ship off Cephalonia, the moon was on my starboard bow, which helped me to relax. The clouds had all dispersed and the shimmering path of watery gold, lighting up the sea's surface from the horizon to the water below us, made night flying simple.

After the last thirty minutes of bloodcurdling flying, over a man-made volcano, we both needed a breathing space to regain our sense of proportion. About an hour had elapsed before I had to hail Grieve on the voice-pipe: I was uneasy.

'I'm a bit worried,' I said. 'We may be the only survivors, I shall be very surprised if we are not. I doubt whether any of the torpedo or bombing pilots got away with that, and I saw nothing of Kiggell's aircraft after he had dropped that last flare on the far side of the harbour.'

'I'm afraid you are right,' said Grieve, 'but we can't do anything about it now.'

'No, but we should be thinking about what we are going to say. All the top brass will want to know exactly what happened and whether the attack was a success, and now many hits were scored, and so on....'

When, back on the carrier, the aircraft was struck down to the hangar on the lift, I wondered what we were going to say that would explain why the hangar was empty of all Swordfish but ours. When the lift reached the hangar deck, Burns and Brown leaped on to the stubplane, as the aircraft was being pushed aft. Their faces were expressive with relief.

'When nearly all the others had got back without you, we began to think you had bought it!' Brown said, shaking his head reprovingly; then he jumped down to help push.

I craned my head round in astonishment to look aft. I saw that the hangar was stacked with aircraft, in neat rows, swarming with men, and I stood up on my parachute to stare in disbelief. Grieve and I exchanged mystified glances; it was nothing short of a miracle.*

At noon, when we had all slept, and bathed, and dressed, the news began to come through; the RAF Glenn Martins had done a good job at first light, and had taken some startling photographs from which they deduced that one *Cavour*-class battleship had been sunk, the huge *Littorio* had been hit by three torpedoes and would be out of action for many months to come, and a battleship of the *Duilio* class had been beached to prevent her sinking. Three cruisers and one destroyer had been badly damaged, and both harbours were a mass of black floating oil....

...The rest of the Italian fleet was moved to Naples, and Taranto ceased to be a military port. As the captain said in a talk to the ship's company, in one night the ship's aircraft had achieved a greater amount of damage to the enemy than Nelson achieved in the Battle of Trafalgar, and nearly twice the amount that the entire British fleet achieved at the Battle of Jutland in the First World War. But what was more important, it was the first good news to reach the bomb-weary British since the war began. 'It will cheer the entire free world,' he said.

*Only two aircraft were lost in the attack.

Mediterranean Power

Winston S. Churchill, *World War Two, Volume II: Their Finest Hour* (Cassell, London; Houghton Mifflin, New York)

By this single stroke [Taranto] the balance of naval power in the Mediterranean was decisively altered.

'The Illustrious Blitz'

Air Commodore R.C. Jonas, Fowey, Cornwall, 1982

Wing Commander R.C. Jonas was commanding Luqa, Malta's principal air base, when, in the opening days of January 1941, the damaged aircraft carrier, Illustrious, *limped into Grand Harbour and, within hours, came under aggressive and determined enemy attack. In the island, they still talk about 'the Illustrious Blitz'. It lasted for a fortnight.*

Jonas and his wife, Gina, were then living in a flat overlooking the harbour, a few hundred yards from where Illustrious *lay. None is better placed than Luqa's former Station Commander to describe and assess, in retrospect, the form and strength of the Axis onslaught.*

'10 January 1941 had been clear and sunny, typical of Malta at that time of year, when the sun was warm and the nights were still and cool.

Just before eight o'clock that evening my wife, Gina, and I were standing on one of the balconies which overlooked the harbour. Dusk had fallen and bats fluttered to and fro among the trees below and over the water. The sharp outline of houses and churches had changed into sombre, formless blocks. Above a few stars twinkled in the heavens. Behind us, beyond Luqa, a pale amber light still glowed low down in the fading western sky. A light, cool breeze was moving the little twigs and pine-needles in the gardens.

As we watched, we could see small ships were moving towards the harbour entrance, their lights looking like glow-worms, their wakes moving like white snakes across the water. These were the tugs on their way to meet *Illustrious* and guide her to her berth. The next moment, the dark, ungainly shape of the carrier appeared between the breakwaters, a shadowy, crippled giant, dwarfing the dark, Lilliputian shapes of the tiny ships below her.

Half an hour later, the ship had nosed her way into French Creek, one of the long, narrow creeks exactly opposite our flat, and was tied up alongside. Lingering on the balcony, long after night had fallen, I listened to the sounds of ceaseless activity on her decks and watched the shaded arc-lights as the dead and wounded were gently carried ashore down the sloping gangways. Seven hours of enemy bombing at sea had taken its toll. My thoughts must have been typical of others' in the island that evening. Tonight all was peaceful. Tomorrow, we knew, Hell would be let loose.

Early next morning, air-raid sirens heralded the arrival of a single, high-flying photographic reconnaissance mission over Grand Harbour and the dockyard. A brief period of quiet followed – the lull before the storm – and then *Illustrious* was viciously and recklessly attacked by the Luftwaffe. I watched the raids – and there were many in the next days – from the natural grandstand formed by the rising ground of Luqa camp.

The attacks were pressed home by about a hundred dive-bombers – practically all of them Junkers 87s – accompanied by small fighter escorts. As the raiders approached, above the broken cloud layer, they sounded like a distant, and immense, swarm of bees.

As the first dive-bomber plummeted earthwards through the clouds, the ground defence opened up in a continuous, percussive thudding, the shells bursting, in black puffs of smoke, over the large target area. Soon the sky over the harbour, at a height of about 5000 feet, was covered with a plate-like circle of bursting shells and drifting smoke. Through this hail of shrapnel, hopefully and fearlessly, dived the Junkers.

The succession of attacks on the aircraft carrier made, undeniably, a magnificent spectacle. Here, spread over nearly a fortnight was the embodiment of modern, twentieth-century technical skill, prostituted in the cause of death, havoc and destruction. Here was ability, aggression and reckless daring on the part of the dive-bomber pilots, spurred on to sacrifice by a cause in which they implicitly believed. Here, also, was the determination and will of the men manning the guns, both on the shore and on board the wounded carrier.

A number of the attackers received direct hits from shells as they drove on through the inferno to the target. Either they disintegrated in the air or, with the pilots dead or helplessly wounded, they dived vertically into the houses below. Others, disappearing momentarily into the smoke, swept through to continue their onslaught against the ship. Then, with bombs released, sometimes at mast-head height, the marauders snaked away, zigzagging below the level of the houses, over the breakwaters and out to sea. Disaster overtook one escaping Junkers after another. Low down over the water, and turning and twisting as they might, guns of every type and calibre opened up at them from either side of Grand Harbour. It was no longer necessary for the gunners to elevate their barrels; it was enough to fire horizontally or even downwards at the fleeing targets with a full deflection aim. As the force headed home for Sicily, we watched the Hurricane pilots coming down on the enemy from behind, and heard the muffled hammering of machine-gun fire as it ripped into the ill-protected bombers. Running for home, with their fangs drawn, and no longer able to maintain their diving speed, the 87s became sitting targets for the aggressive fighters.

As for *Illustrious*, obscured by black smoke and yellow dust, it seemed impossible, after each attack, that she had not been utterly destroyed.

Standing on the balcony of the flat in the evening, as dusk fell, the long, grey shape of the aircraft carrier still lay a few hundred yards across the water. Each evening, as Gina and I watched, the ship looked just the same; but each evening the skyline beyond had changed. Tonight there was an ugly white gash in the cliff side; last night, a church had crumbled and a familiar line of roofs and turrets had disappeared. By tomorrow, some other landmark would have gone.... Vittoriosa, Senglea and Conspicua – the three ancient and picturesque cities were now broken into rubble. They were cities of the dead, inhabited by ghosts and memories.

When, eventually, *Illustrious* sailed out of Grand Harbour, miraculously under her own steam, our flat was still standing. Not even a pane of glass had been broken. Only a handful of shrapnel and the nose cap of a German bomb were found on the roof to remind us that we had been residents in the target area.

By now, however, we had decided to move elsewhere. Providence might not be so benevolent next time.'

Spirit of Malta

Air Commodore R.C. Jonas, Fowey, Cornwall, 1982

'One of the toughest and most skilful of the island's Hurricane pilots during the early days of 1941 was a young flight lieutenant named J.A.F. Maclachlan. Maclachlan had, with other reinforcements, flown into Malta from an aircraft carrier lying some 450 miles to the west of the island. He had been in France flying Fairey Battles during some of the hardest operations of the war and had already been decorated with the Distinguished Flying Cross.

I see from my diary that it was 16 February when I stood watching Mac and the flight he was leading getting into a stupendous, milling dogfight high up in the sky midway between Hal Far and Valletta. The whine and howl of over-stressed aircraft and engines, and the intermittent chatter of machine guns, told us what was happening.

Then, from all that chaotic sound and movement, a Hurricane, pilotless and doomed, was seen to be spinning earthwards, while its pilot followed slowly, dangling from his white parachute. Maclachlan landed, faint and wounded, on the roof of a house close to Kalafrana. He was rescued and taken immediately to hospital.

It was the end of Mac as a fighter pilot in Malta. A cannon shell from a Messerschmitt 109 had penetrated the side of his cockpit and almost severed his left arm. He was weak and faint from the pain and loss of blood but, with remarkable endurance, he was still conscious and still had the strength to pull the rip-cord. A few days after his arrival in Imtarfa and, in spite of the efforts of the hospital doctors and nurses – and even Mac himself – his arm had to be amputated.

Despite this physical and mental reverse, the patient was still as keen as ever to get back into the air. Two weeks from the day of his operation, Mac flew a Magister, solo, from Ta Kali aerodrome.

From Malta, he went on to Egypt and South Africa, doing ferry work and flying anything on which he could lay his remaining hand.

The last time I saw Maclachlan was quite recently when he visited me in London. He gleefully showed me his latest acquisition, an artificial arm, which he promptly took off and placed on my desk, to demonstrate its simple efficiency.

The spirit of Malta still wasn't so far away.'

How Two Girls Sank the *Tarigo* Convoy

Group Captain E.A. Whiteley, 'The Whiteley Papers', Croydon, Australia, 1977

'Titch' Whiteley, an Australian, was one of the most resourceful squadron

commanders in the Middle East during the early fighting in the Mediterranean and the Western Desert. As a cadet, he had won the Sword of Honour at Point Cook in 1935. When, some five years later, he was promoted to command No. 69 Squadron, based in Malta, he quickly turned it into the hardest working and most versatile reconnaissance unit in this theatre. With Adrian Warburton, George Burges and others he brought aggression to this operational role.

Whiteley's sighting of the famous Tarigo *convoy, in the rare circumstances here described, led to a brilliant naval victory. After retiring from the Service in 1956 with the rank of group captain, he pursued a career in the aircraft industry, becoming a director and general manager of the Bristol Aeroplane Company in Sydney.*

This story, written towards the end of his life, was picked out by his widow, Yvonne Whiteley, from among his private papers.

I

What follows will apprise two United Kingdom girls – for the first time – that they are the first women since Cleopatra to play a major role in the destruction of enemy maritime forces in the Mediterranean.

The *Tarigo* convoy – four German merchantmen and one Italian – with an escort of three Italian destroyers (*Tarigo, Lampo* and *Baleno*) left Naples on the 13 April 1941, with a cargo of petrol, ammunition and vehicles for Tripoli in North Africa, where Hitler's forces, under Field Marshal Rommel, were trying to drive the British back into Egypt. The convoy proceeded via the western tip of Sicily, down the coast of Tunisia, keeping as far away from Malta as possible. In Malta, about 220 miles from the convoy route, were four British destroyers commanded by Captain P.J. Mack, RN in HMS *Jervis*, and a handful of RAF aeroplanes including a few reconnaissance aircraft of No. 69 Squadron (myself in charge). Also in Malta were two English girls who, quite innocently, were to lead me and that convoy into trouble.

I was the first to run into trouble. Shortly after 14.00 hours on 15 April, flying a Glen Martin reconnaissance aircraft, I almost hit the mast of one of the *Tarigo* ships. Because of the low visibility I had not expected to find an enemy convoy – or even sight an enemy ship. All the way out I had flown in mist and rain at reduced speed – barely able to keep below the clouds and avoid flying into the sea. The signal I sent to Malta about 14.30, reporting the position of the enemy convoy, would have one certain consequence. Next morning, I would be 'on the mat' at War Headquarters to explain to the Royal Navy how I came to be off the Tunisian coast three hours ahead of the time laid down in the joint navy/air orders. When a diligent [*sic*] air force officer lands himself in such a situation there is one likely explanation – feminine distraction.

Meanwhile, as I and my crew shadowed the convoy sending further reports every half hour or so, my navigator, Flight Lieutenant Arnold Potter, relieved his boredom by tuning the radio compass into some dance music broadcast from Bizerta, while I contemplated the social engagement I had planned for that evening. Let us now describe what was happening in Malta and introduce my ingenuous and true blue Cleopatras.

II

By April 1941, I had been a pilot in the besieged island for some seven months. A few single English girls had been caught there when the siege started, but I had not met them. Then, one day, in Valletta another air force officer, Squadron Leader Hipwood, introduced me to his wife and daughter. Miss Hipwood certainly was attractive, but I was en route to a meeting at War Headquarters and the contact was all too brief.

As time passed Miss Hipwood kept returning to my thoughts. Increasingly I felt that it would be good to talk with her, and escape for a while from our depressing environment. But she lived some miles away at Kalafrana and there were no private cars, taxis – or any social occasions that I knew of when we might expect to meet.

Then, in April, another officer's family provided the opportunity. Melanie Scobell and her mother invited me to a cocktail party at their home. My friend and deputy, Squadron Leader George Burges, pressed me to take a night off duty and go to Melanie's party. Miss Hipwood might be there. I accepted – for the evening of 15 April 1941. Note that date.

To have the evening off duty, I gradually adjusted 69 Squadron's flying roster so that I and my crew would be rostered for the first operational sortie on the morning of the 15th. We would be briefed on the night of the 14th, take off early the next day and be back in Malta by about lunch. There were, however, those four destroyers in Grand Harbour, and I had instructions to send one aircraft on a search patrol late each afternoon to find an enemy convoy – for the destroyers to attack under cover of darkness. I decided that the second or third duty crew on the roster for the 15th could look after that routine task.

My first setback came on the evening of the 14th at the briefing conference at War HQ. The Senior Air Staff Officer announced that only one sortie was required on the 15th – the afternoon search for a convoy. As first crew on the list I was stuck with the search – and would be back too late for the party.

The only way out was to carry out the search earlier in the day. While I was trying to think up some excuse for doing so, an enemy air raid appeared on the War-Room plotting table, and the Senior Air Staff Officer lost all interest in what my squadron should do on the 15th. I was dismissed with 'No, Whiteley, just the search tomorrow. Nothing else. Write it up on the board, will you?' I slowly chalked on the forward operations blackboard for 15 April – '69 Squadron, Glenn Martin AR 714, Naval Search No. .... Take-off time *11.30* hours'. I should have written 14.30 hours, but no one took his eyes off the radar plots of enemy aircraft on the War Room table.

Unless my fiddle was detected, I would take off at 11.30 hours and be in the search area off the Tunisian coast three hours earlier than ordered. Unknown to me, so would the *Tarigo* convoy. The navy's estimate of the enemy convoy's timetable was just that far out. Miss Hipwood and Melanie Scobell had struck their first blow.

III

In London that same evening (14 April), Winston Churchill may have noted that four of his front-line destroyers had been lying inactive in Malta since the 10th. Whatever the reason, he decided to do a little prodding. A directive went that day to C-in-C Mediterranean Fleet. 'Every convoy which gets through (from Italy to Africa) must be considered a serious naval failure.'

On 14 April, things were also happening at Supermarino, Italian Naval HQ. On the previous day, German reconnaissance aircraft had detected the four British destroyers lying to in Grand Harbour. Admiral of the Fleet, Romeo Bernotti, reveals in his book, *Storia della Guerra nel Mediterraneo, 1940–43* that arrangements were therefore at once made for the *Tarigo* convoy to be provided with a strong air escort on 15 April as it moved southwards from Sicily towards Tunisia and Tripoli.

On the morning of the 15th I was standing by with my two crew at Luqa half expecting a telephone query about the 11.30 take off – but no call came. That was some comfort, but I had another problem – weather. Malta was covered by low cloud and drizzle. If I took off it might be impossible to land back in the island. We had no blind approach aids and every alternate airfield within 1000 miles was hostile. But if I did not complete that sortie, I and my crew would be 'first for duty' the following morning – and that would mean another briefing at HQ, instead of cocktails.

In Sicily, the German and/or Italian squadrons standing by to provide the air escort for the convoy were faced with the same weather problems. Admiral Bernotti relates what happened. 'Our aircraft failed to provide the escort on the 15th *owing to weather conditions*.' He then adds a little tersely 'but a British reconnaissance aircraft took off from Malta'. If the Admiral had seen my navigator's surprised face when I announced about 11.30 that I would take off, his doubts about the judgement of the Luftwaffe would have been dispelled. Mark up a second point to the girls.

Within seconds of becoming airborne, I realized that I had been impulsive. Apart from uncertainty about landing on return, there was another worry. Flying under the cloud base and just above the sea, Arnold Potter, in the nose, could see nothing on either side of the aircraft even when we were 100 miles out; and the search had been planned on the assumption that we would sight any ships within 15–20 miles (either side) of our route. At the very least, I should have postponed take off until after lunch – the proper time, anyway. Potter, unaware of my social commitment, let out a polite hint about giving up. But, with no forward visibility, I doubt whether I could return to Malta without serious risk of flying into the cliffs. The weather, if it changed, could only improve. So on I went towards Tunisia. I will not enlarge on the two hours' mental misery which followed.

It was at this moment, at the extreme limit of the search, that I almost flew into a ship's mast – or a wreck – or a buoy. I was not sure what it was. Neither Potter, in the nose, nor the wireless operator behind me, noticed anything in the rain. I snapped on a stop-watch and flew a timed run to the familiar, low sandy coast of Tunisia. From there we did a run back to the same spot and turned straight into a search. In no time we were reporting eight enemy ships.

As Admiral Bernotti tells us, our sighting report which had been intercepted, was quickly decoded. It caused quite a panic at Italian Naval Headquarters.

'The enemy aircraft,' he writes, 'had accurately estimated the convoy's course and speed of *8* knots.... In consequence our convoy found itself heading towards a disastrous fate.'

Unfortunately for my ego, everyone in 69 Squadron knows that Potter and I always estimated the speed of *every* enemy convoy as 8 knots – a safe middle figure.

Supermarino in Rome promptly asked Superaereo, the Air Command, to despatch fighters to dispose of the shadowing aircraft. But, as we know, because of the weather, the enemy fighters 'were unable to take off'. Belatedly, I commend those enemy pilots on their decision. Sound airmanship. Better than mine.

So I and the crew of the Glenn Martin AR 714 spent a relaxed afternoon in company with the convoy – sending regular radio reports back to Captain Mack in the British destroyers. Later, we landed back in Malta with less difficulty than I had expected and in good time for the party.

As I was dressing, Captain Mack in HMS *Jervis* led his destroyers out of Grand Harbour. The weather was still miserable. I had a nasty feeling that his fuel, as well as mine, would be wasted. However, it was a perfect night for a cocktail party – plenty of low cloud with, consequently, little risk of air-raid interruption.

IV

In January, on returning from a short visit to the Air Ministry in London, I had smuggled into Malta some of the latest gramophone records and a few lipsticks – precious commodities in a siege. An army officer had also solved my transport problem – a pony and trap. So off to the Scobells' I went with a well-groomed horse, polished harness and, for Miss Hipwood, the last three of the smuggled lipsticks.

I suffered two further setbacks. The intended recipient of the lipsticks, Miss Hipwood, was not there. After waiting hopefully for latecomers, I decided to make the best of a very bright function. About that time a senior RAF officer let fly a piercing arrow. Having observed my arrival by pony and trap, he complimented me on setting an example in saving petrol. He probably wondered why I blushed.

But you, Miss Hipwood, will want to know who ultimately received those three lipsticks. I gave them to my second love. My Glenn Martin aircraft would not fly without Lockheed hydraulic fluid – and, of course, the supply in Malta had given out. Medicinal castor oil, properly dispensed, was the only substitute in the siege but the army doctors refused to surrender what they had. So George Burges visited some nursing sisters with the last of my lipsticks – and came back with the island's entire supply of castor oil. My second love in Malta – my squadron's aircraft – were satisfied.

As the evening progressed the weather improved. About midnight, Captain Mack and his destroyers made a perfect interception at Kerkenah Bank, about 70 miles from where I had left the convoy. In a confused night battle, the entire enemy force – five merchant ships and three destroyers – was sunk or driven ashore, within fifty minutes. When the first blew up, pieces of ammunition landed on HMS *Jervis*, a mile away. HMS *Mohawk* was sunk in a spirited torpedo attack by *Tarigo* before the Italian destroyer, in turn, went down.* The remaining British destroyers rescued most of the *Mohawk*'s complement of about 167. Next day the Italian hospital ship *Orlando* picked up over 1200 enemy survivors.

On 16 April, in this period of almost endless reverses, when Britain was 'going it alone', Mr Churchill addressed a message to President Roosevelt. In it he included an account of this 'noteworthy success'.

The Royal Navy was so excited by its success that no one ever asked the expected question about the timing of my air search....

I never did catch up with Miss Hipwood. She was evacuated from Malta (probably before 15 April) to join the Women's Auxiliary Air Force in England. She was determined to do some useful war work. But Miss Hipwood and Melanie Scobell already had the following ships on their scoreboard – the destroyers *Tarigo, Lampo* and *Baleno*; and the merchant vessels *Adama, Arta, Aegina, Iverhold* and *Sabaudea*.

Desert Air Force

Group Captain W.G.G. Duncan Smith, *Spitfire Into Battle* (John Murray, 1981; Hamlyn Paperbacks, 1983)

After the Germans invaded the Soviet Union and thrust into the Balkans there never were the numbers of Luftwaffe fighter squadrons in France and the Low Countries as [there had been] previously. Many of these squadrons were sent to the Eastern front and, more particularly, the Germans reinforced Luftwaffe units in the Mediterranean with about 25–30 per cent of the total strength available. The fighting over Malta and the Western Desert influenced the conduct of operations far more than those in France and the Low Countries. But little account was taken of this in the UK presumably because an attack against England of 1940 proportions could not be ruled out.

No less than seventy-five fighter squadrons with the best aircraft were kept in the UK ... while in Malta and the Western Desert the Luftwaffe were playing ducks and drakes with the outmoded Hurricanes and Tomahawks.... Not a single Spitfire squadron was sent to the Mediterranean area until the spring of 1942 – a strange state of affairs....

... Forged into a unique brotherhood with the 8th Army, Desert Air Force operated beyond the reach of traditional and accepted Service regulations when it came to dress and a way of life, but this in no way detracted from the efficiency and operational capability of its members. It was a happy family which grew in stature with every success until, by the end, it had become a legend in its own time....

Desert Songs

The songs the aircrews sang in the Desert (and elsewhere) often exposed the

*The Commanding Officer of *Tarigo*, Captain Pietro de Christofero, of Naples, was posthumously awarded the Italian Gold Medal for Valour.

roles upon which their squadrons were engaged and the environment in which they lived.

Typical of the mundane, yet essential, character of the work 70 Squadron was doing with its Wellingtons, towards the end of the first great British advance across Cyrenaica to Benghazi, Agedabia and El Agheila in the opening months of 1941, before the demands of the Grecian campaign turned everything sour, was the song its members sang to the tune of 'Clementine'. Based in the Canal Zone, the squadron was ranging widely across the Mediterranean to the Italian ports and then, westwards, down the Desert to Benghazi and Tripoli beyond. The Benghazi detail, for its hazards and monotony, soon became known as 'The Mail Run'.

Down the Flights each ruddy morning,
Sitting waiting for a clue,
Same old notice on the Flight board,
Maximum effort – guess where to.

Seventy Squadron, Seventy Squadron,
Though we say it with a sigh,
We must do the ruddy mail run
Every night until we die.

'Have you lost us, navigator?
'Come up here and have a look;
'Someone's shot our starboard wing off!
'We're all right, then, that's Tobruk.'

Seventy Squadron, Seventy Squadron, etc.

Oh, to be in Piccadilly,
Selling matches by the score,
Then we should not have to do the
Blessed mail run any more.

Seventy Squadron, Seventy Squadron, etc.

In another vein and a different operational role, there was the song 229 Squadron used to sing. Group Captain G.B. 'Johnny' Johns recalls it was an 'adaptation of the old Indian ditty which started "Down a road so dark and dusty...." with a chorus of head-shaking wails....'

In the Desert after loot
Came the TwoTwo Nine pursuit
Wails
By its side, who should I see?
Two Three Eight and Two Fifty
Wails
Sixteen months we had these Is*
Now the IIs* and their twelve guns
Wails
Six machine guns in each wing
Still can't hit a f—g thing
Wails

*Hurricane Is and IIs.

Alexandria? Very fine place
Wish we had it for our base
Wails
To the Pongos we will shout
'Pull your f—g fingers out.'
Wails etc., etc.

Was South Africa's 'Pat' Pattle the Greatest?

Comparing the great bomber and fighter pilots of World War II, and trying to establish, forty years and more on, who was 'the greatest' is not usually a rewarding process. It's much like comparing games players of different generations. There were too many discrepancies in operating conditions, too many differences in war theatres, in opportunities, in opposition, in aircraft flown, to allow authoritative comparisons to be made. But none of that has stopped opinions from being freely aired.

There is now, for those who have a mind for it, certainly a case to argue for the great South African, Squadron Leader Marmaduke Thomas St John Pattle – Tom to his family and 'Pat' to the Royal Air Force – being the outstanding Allied fighter pilot of the war. One thing is irrefutable; of all the 'greats', 'Pat' Pattle was the least well known. He had, in effect, rather less than a year of active operations in the Desert and Greece before, sick of a fever and with a high temperature, he was shot down and killed over Eleusis Bay, near Athens, on 20 April 1941, while leading No. 33 Squadron in battle. Yet what he accomplished in the time he was allotted (he was finally credited with 41 aircraft destroyed, the highest Allied scorer; and the total could well have been significantly higher had his squadron's records not been lost) was nothing short of astonishing.

Born at Butterworth, in Cape Province, on 23 July 1914, he came from a close, well-regarded family with a strong military background. On 22 March 1936, while still a cadet with the Special Service Battalion at Roberts Heights, he wrote to his mother ('mom dearest') telling of his plans to go to England and apply for a commission in the Royal Air Force, where, he had been assured, he would get 'the finest training in the world'.

Gee!! Mom, I simply must go over. I am writing to Uncle Sam to see whether he can manage to fix me up with a job on a boat. If I could get a job I reckon about £30 ought ... to see me through ... Anything I have got over I could return. Mom, couldn't you possible manage it? I know how hard pressed we are and if it can't be managed, I'll give up the idea. But it wouldn't take me long to get on my feet ... and I could return it all.

Well, cheerio, Mom and God bless you
Love
Tom

PS. I'm going to the RAF even if I haven't a penny in my pocket. I'm going to put the Pattles in the limelight again.

His mother responded, Pattle went to Britain and, five months later, after doing his initial flying with Scottish Aviation Ltd at Prestwick, in Ayrshire, he wrote to tell his mother that he had been presented with his 'A' Licence.

The letter is dated 2 August 1936.

Mom, darling, you said you haven't given me a birthday present this year. Well, you're wrong, entirely wrong, for [the 'A' Licence] was my birthday present from you. Without you, I would not have come here and I treasure it above everything I possess. It's the fulfilment of a long, long dream, and I owe it to you....

Pattle's progress in the Service was uninterrupted. He joined 80 Squadron, to his intense pride and delight, less than a year before the outbreak of war. Then came the Desert and, finally, Greece. Success seemed inevitable. On 16 January 1941, he wrote to his family, but the letter wasn't received until 18 March, just a month before he died. The desperate fighting and, with it 'Pat's' golden days, were yet to come.

... I have been in quite a number of engagements and, at the moment, hold the Middle East record for the number of aircraft brought down. I have fourteen brought down without aid, and have participated in several scraps – together with other pilots – which have resulted in the enemy being destroyed.

As one's experience grows, it is amazing how one's reactions during a fight change. At first, I was plainly nervous about my own safety. I got over that and entered the reckless stage, when I did lots of silly things.... Then I became nervous about other pilots who were less experienced than I, and I lost a lot of opportunities through playing the protective role to too large an extent.

Now I regard it as a science.... I make full use of clouds and sun to make an unseen approach, and then always manoeuvre to get the enemy at a disadvantage... by a full appreciation of my aircraft as opposed to his. I have studied every type of e/a* and have different rules of procedure for each. This has proved successful in practice for, in my last few flights, not a single bullet has found its mark on my aircraft, and I have more time to look around and keep an eye on the remainder of my flight.

... Each pilot is assessed... Below average: 2 or 3 points. Average: 4 or 5 points. Above average: 6 or 7 points and Exceptional: 8 points.... How I managed it, I don't know, but, in the CO's (80 Squadron) report at the beginning of the year [1941 – shortly before Pattle took over 33] I got 'exceptional' for everything....

In Pattle's two squadrons – No. 80, in which he commanded a flight, and No. 33 of which he became CO, there were five highly experienced and discerning survivors who were also, in their own right, pilots and leaders of signal ability and judgement. Four subsequently enjoyed distinguished Service careers; a fifth left the Royal Air Force after the war, having made a memorable mark, and applied his rare talents to civilian fields.

I put the questions to each: 'How good, *really*, was Pat Pattle? Was he quite outstanding, and apart from other men? Only those of you, who flew and served with him can fairly assess his value.' None of the five knew the others were being questioned.

Here, necessarily abbreviated, are the five answers.

Air Marshal Sir Patrick Dunn, Cookham Dean, Berkshire, 1982

'In 1938, flying was fun for 80 Squadron in Egypt. Pattle, already a flight commander, was struck in contrasting mould: of medium height, spare build, dark hair, striking green-grey eyes. Just a shade older than the others,

*Enemy aircraft.

he shared their enthusiasms – but with reserve, thoughtfulness, and a hint of introspection. He seemed a sound but unremarkable officer.

Sharp sight and quick reaction enabled flying and air combat to come readily. He spotted movement almost before it happened and skilfully exploited the power and pace of his aircraft. Air fighting was his game. In practice combat few could get him in their sights or keep him off their tail. By this expertise, by careful post-flight explanations and unremitting concern for their training, his pilots were inspired and brought on.

Pattle was clearly a remarkable man, with a natural and easy grip upon his Flight; and 80 was a squadron of well-trained regulars of sparkling ésprit, confidence, commitment and humour.

He first brushed with Italian fighters in early August 1940. Grossly outnumbered, the four Gladiators he was leading were all shot down – but not before Pattle had himself destroyed two of the enemy. In the commotion of baling out at 400 feet he felt his parachute spring open in the cockpit. The canopy opened just before he hit the sand. Unmoved, he confided that there was nothing about the opposition which a dozen 80 Squadron Gladiators could not handle.

Four days later this was confirmed by a thirteen-strong formation which took revenge upon twenty-seven CR 42s and Ro 37s, destroying 9 (2 falling to Pattle) and probably accounting for 6 more.

Next day he searched alone in an unarmed Magister for our two missing pilots.

In Greece, his courage and indomitable leadership became legendary. Those close to him believe that, had he survived he might have been the war's top scorer. Indisputably, he was of the bravest; an exceptional fighter leader and brilliant fighter pilot.'

Air Marshal Sir Edward Gordon-Jones, Cambridge, 1982

'Pat Pattle quickly proved himself in war not only as an individual pilot of terrific skill, precision and discretion, but, above all, as a leader. He was a brilliant tactician, and it is worth remembering that, apart from the standard, pre-war Fighter Command attack systems, the tactics of combat had to be developed by those on the spot in the circumstances of the time. His tactical "sense" was equally perceptive on the ground; but, in the air, his flying skills and uncanny reading of a situation, combined with a deadly shooting accuracy, brought him great success. His attributes were an inspiration to all of us.

To understand the depth of his ability, one must try to picture the appalling conditions under which the squadron had to operate, to some extent in the early Desert campaign, but, more particularly, in Greece during the winter of 1940/41. There, the weather was dreadful – one wag observing that every cloud had fir trees growing in it! There were no navigational aids other than a map and a compass, and no devices to help landing in poor visibility. Radio communication between aircraft and aircraft was there, if you were lucky. Wretched airfield and living conditions didn't help and, lastly, there were the odds. The squadron was usually outnumbered in the air 10 to 1.

The historians have suggested that the campaign in Greece had little or no bearing on the outcome of the war. This is at least arguable. What is undeniable is that if Pat Pattle's talents had been displayed in the Western European theatre as they were in the Desert and Greece his contribution would have received much greater recognition. He would not have sought it, nor is the admiration and affection of his colleagues in any way diminished by the lack of acclaim. But his renown would have been enhanced by the wider exposure.'

John Lancaster, Buntingford, Hertfordshire 1982

'In a squadron which contained *nine future squadron commanders* [editor's italics], most of whom were brilliant pilots, Pat Pattle was undoubtedly the best pilot as well as being a brilliant leader.

He had a higher opinion of Italian fighter pilots, as pilots, than he had of German fighter pilots and he would often follow them as they performed perfect aerobatic manoeuvres in front of him. From this he learned that the best avoiding action when being shot at was to fly "ham", skidding and slipping in a manner calculated to provoke the Chief Flying Instructor to ground you for a month. I found Pat's advice invaluable and I believe I owe my life to him and the talks we had in the Mess at night.

Squadron Leader Marmaduke Thomas St John Pattle's history is written only partly in official records and many of these were lost when we left Greece; but in the records and memories of those who flew with him, he was an Ace; many of us would say Britain's No. 1 Fighter Pilot.

No journalists or reporters hung on boasts from Pat's lips. Self-effacing, kind and unselfish to the end, he was a simple officer and gentleman of the Royal Air Force. A South African, and like many of his countrymen who fought with us in the '39–45 war, he was of the highest calibre.'

Air Marshal Sir John Lapsley, Saxmundham, Suffolk, 1982

'Pat Pattle has strong claims to having been the best fighter pilot of the Royal Air Force in World War II.

His life was dedicated to flying and, as a pilot, he had such a remarkable sense of timing, rhythm and judgement of distance, that following him in formation aerobatics became as simple as formating with him straight and level. So precise was his flying, and such was its smooth, effortless ease, that his attacks on enemy aircraft, even in the heat and excitement of battle, were always immaculately judged.

His wonderfully accurate flying was coupled with – or resulted in – his truly outstanding shooting from the fixed front guns of his fighter aircraft. *On one occasion, he shot down an Italian CR 42 at a range of at least 500 yards with the last 10 or so ·303 rounds remaining in one of the wing guns of his Gladiator – the other three guns being out of ammunition* [editor's italics]. In practice shoots, on towed targets, he quite consistently used to score 70 to 80 per cent hits.

Add to his wonderful flying ability and his skill as a shot, a stout and fearless heart, backed by a ruthless determination that forced his attacks right home to the kill, and that was Pat Pattle ... Fighter Pilot Supreme.'

Air Marshal Sir Peter Wykeham, Kingston-upon-Thames, Surrey, 1982

'Pat Pattle was a natural. Some fighter pilots did not last long because they were too kind to their aircraft; others were successful because they caned it half to death, and their victories were accompanied by burst engines, popping rivets, stretched wires, wrinkled wings.

But Pat was a sensitive pilot, who considered his machine, but, somehow, he got more from it than anyone else, and possibly more than it had to give.

He was a reticent man, and we were hardly aware of his South African upbringing. In the Western Desert he looked more at home than we, British, did – trimmer, neater, fitter. His desert khaki seemed as if it belonged to him, while we all appeared to be, more or less, in fancy dress. But he did not give us any of the "wide-open-spaces" treatment. He had a kindly, tolerant nature; accepting us all at our face value. I spent three weeks alone with him on an army support mission in Palestine, and though I got to know him well I still never broke into his background, his early years, or his personal thoughts and beliefs. But perhaps this was normal with young men in the armed forces at that time.

He was so wrapped up in flying, and so dedicated to his job, that we sometimes teased him, calling him Dick Dauntless or the Boy Aviator; but he minded so little that it did not become a habit. He had a strong sense of humour (not just a sense of fun – we all had that) and a kindly disposition.

If the air force had set down on paper the ideal specification for a fighter pilot, it would have been a description of Pat Pattle. So when war came he was fulfilled. Though not bloodthirsty, he was eager to try what he knew to be his outstanding skill.

It therefore came as a nasty shock to him, as it did also to me, when we were both shot down by the Italian Air Force in the first few days of the Desert war. Separated by several miles of desert, and following our different tracks, we walked all night and half the next day until, finally, we were picked up by Long-Range Desert Groups. I was proud that I, who normally rated Knightsbridge to Sloane Square a long walk had made as great a distance in the right direction as our favourite bushman. But Pat took it all quite calmly, even when, two days later, the whole squadron, led by Paddy Dunn, our CO, fell upon the Italians and shot down thirteen of them.

Pat Pattle got a couple that day. He seemed to regard it as just part of the day's work; and, if he was impressed with the drama of this punch and counter-punch, he managed not to show it. Indeed, he took everything as it came to him, easily – not frivolously; he made light of things that could have been heavy going with others.

I think Pat had a germ of greatness in him; but he had a strong idealistic streak which, had he survived the war, would have made it hard for him to compromise – as compromise we must sooner or later. But I only ever saw him doing what came naturally to him.

When we re-equipped with Hurricanes, he began to come into his own, and really became deadly. His victories piled up and, in time, we grew to regard him as indestructible. So, indeed, he might have been but for a fever – and a cloud of Messerschmitts – which descended one afternoon over Athens.'

'High Flight'

Chosen by Wing Commander J.M. Checketts, Christchurch, New Zealand, 1982

Oh! I have slipped the surly bonds of earth
  And danced the skies on laughter-silvered wings;
Sunward I've climbed, and joined the tumbling mirth
  Of sun-split clouds – and done a hundred things
You have not dreamed of – wheeled and soared and swung
  High in the sunlit silence. Hov'ring there,
I've chased the shouting wind along, and flung
  My eager craft through footless halls of air...

Up, up the long delirious, burning blue,
  I've topped the windswept heights with easy grace
Where never lark, or even eagle flew –
  And, while with silent, lifting mind I've trod
The high untrespassed sanctity of space,
  Put out my hand and touched the face of God.

Pilot Officer J.G. Magee, Jnr, No. 412 Squadron,
Royal Canadian Air Force

John Gillespie Magee wrote this poem, and sent it to his parents, shortly before he was killed in a flying accident on 11 December 1941, aged nineteen. Born in Shanghai in 1922 of an American Episcopalian father and an English missionary mother, Magee had had a traditional British public school education. War, however, cut short a promising academic career, for, after winning a top scholarship to Yale, he decided he would return to England and join the Royal Air Force. The United States authorities refused to grant him a visa, so, in time, he went across to Canada and enlisted in the Royal Canadian Air Force as a pilot. He was stationed at Digby in Lincolnshire, with 412, one of the RCAF's Spitfire squadrons, when he died.

The Massacre of 2 Group

D.R. Gibbs, Montserrat, West Indies, 1982

Dennis Gibbs rose, as a wing commander, to command No. 82 Squadron. He began a long and successful spell of operations when the Blenheim crews of Bomber Command's 2 Group were flying their low-level daylight missions against enemy shipping and targets in Germany and occupied Europe. The casualties were shatteringly high. Gibbs, however, was fortunate. He survived to become, in later years, a distinguished figure in Britain's colonial administration and overseas service.

In these reflections, he gives a picture of one of the most dangerous offensive roles of the war.

'We arrived at 82 Squadron's airfield at Bodney in Norfolk on 16 July 1941. On 29th we were ordered to move south to join a detachment of 21

Squadron at Manston, in Kent, rather than fly out to Malta, where our squadron was just completing a tour. Flight Lieutenant Jack Meakin and "A" Flight's flight sergeant, Jock Davidson helped us select an aircraft. Sergeant Stan Pascoe, my air gunner/wireless operator, just had time to fit his warning lights between the turret and the cockpit, and Sergeant Lorrie Cash, the navigator, to collect charts, Verey cartridges and flight plan, before we were off. We landed at war-scarred Manston and Bill Edrich assigned us to "stand-by" the next day.

That evening we listened to the tales of 21's losses, the accuracy of the flak and the ability of the 109s to penetrate the fighter escort. We also met some 242 Squadron (fighter) pilots who provided the close cover for our shipping attacks. They had recently been re-equipped with II C Hurricanes and now, with 4 × 20-mm cannons, could give the flak ship gunners something to make them keep their heads down while the Blenheims went after the merchantmen....

Next morning, 30 July, at dawn, our aircraft was bombed up with 11-second, delayed-action fuses. By noon, our nervousness had given way to the hope that a telephone ring would order a "scramble". It came after lunch. The crews near their aircraft ran to them, the rest piled into the flightvan, jumping off as we passed each aircraft. In minutes six Blenheims and twelve Hurricanes were jockeying for take-off positions in swirls of dust.

We took off in two vee formations of three aircraft and soon dropped below the cliffs heading for France. Hardly had we settled down to cruising than we were advancing the throttles and turning towards six ships dead ahead. At once the sky was dotted with puffs of flak and, seconds later, tracer trails patterned the sky. We were weaving violently when – whang! we were hit, and then again a second time as I saw a Blenheim plunge into the sea, while the Hurricanes were diving on the flak ships.

In the middle of all this, Lorrie, the navigator, reminded me to fuse the bombs. I was now trying desperately to get straight and level over the merchantman. As I succeeded, and pressed the release button four times, I had the feeling that all the bombs would plough into the ship's side. At that moment, a flashing red light on the instrument panel made me whip the aircraft into a turn to port as the sea in front of us was churned into turmoil and Stan fired back. Split seconds later, we saw a Blenheim with an engine on fire; at that instant, an Me 109 flashed by chased by a Hurricane. We continued our turn as we counted eleven and saw the deck of the merchantman explode and debris rise in the sky.

Stan, the gunner, shouted, "Well done, Skip, direct hit; I am sure I peppered the 109." Lorrie switched in to say the Hurricane had downed the 109. Then, as quickly as it had started, it was all over, and four Blenheims closed up, with the Hurricanes in attendance. We had been well and truly blooded in low-level operations.

Almost a fortnight later, we participated in a "Circus" operation with the fighters over northern France while another Blenheim was dropping a replacement leg for Douglas Bader. He had been the CO of 242 and had been shot down the week before leading the Tangmere fighter wing. The 242 boys thought that, with a new leg, Douglas would soon be back! Later we were to learn how nearly right they were.

The object of these sweeps was to entice the German fighters up so that our huge escorts of Spitfires and Hurricanes could have a go at them. We were the bait. In spite of the heavy flak, these high-level operations seemed gentlemanly affairs after the hurly-burly of the low-level shipping strikes.

We did one more "Channel Stop" attack before returning to Bodney. "Flight" Davidson greeted us on arrival and there was a big grin on his face when he saw the three bombs painted on the fuselage denoting three operations. However, it faded when he inspected the hasty patches slapped on at Manston. That night was a mess night to say farewell to "Wingco Attie" and his Malta veterans and to welcome "Wingco" Burt and other newcomers. Nearly all were inexperienced and few lasted long enough to gain the experience to survive. Poor Burt was shot down and the same fate overtook many others on the low-level attacks to Cologne, Stavanger and Schipol and against shipping. The survivors got promoted. Jack Meakin became a squadron leader commanding "A" Flight and I, now a flight lieutenant, was made his deputy.

On 20 August, I was selected to lead a shipping strike against a convoy sheltering along the shores of Heligoland. It was a misty summer's day and Lorrie was navigating the mission. We flew in two shallow vics of three aircraft and dropped to wave-top height as we passed over the now-familiar Cromer lighthouse. There was complete radio silence. We were to be beyond the range of our fighters and survival would hinge on surprise.

After two hours we turned north, running along the northwest German coast. Once we saw some aircraft but they did not see us. At the end of the patrol we saw two barque-type vessels of some 200 tons. This was not the convoy, but better to attack them than take our bombs back. This was a cinch, but it did seem wrong to be attacking unarmed ships.

By the time the hull of one barque seemed large enough to receive our bombs, its masts were high above us. I pulled up and – crash! There were broken perspex and instruments littered about and poor Lorrie, with a terrible head wound, was in my lap. The wind howled in through the broken nose of the aircraft and we were shuddering and climbing too steeply. Using all my strength and throttling back, I got the aircraft level and managed to carry out a gentle turn to port and head for home.

Stan reported a lot of damage in the back and we were trailing wire ropes. The Blenheim was just controllable and, thank God, the engines kept going. The rest of our aircraft were beetling away just above the waves and, as there was no way we could fly down there at their speed, Stan tried unsuccessfully to contact them with the Aldis lamp. A little later he reported that the radio was a casualty, but we still had intercom so I asked him to try and crawl through and give some first-aid to Lorrie, who was badly wounded. The hinged armoured flap behind the cockpit was jammed so he couldn't join us.

It was a little while before I realized that, having got over one danger, we were confronted with another. All the maps and charts had blown away, the blind-flying panel was broken and there was debris over the compass. When I cleared this I was just able to read the compass by screwing up my eyes against the wind but, for the life of me, I could not recall the reciprocal course we had to fly for home. Surely, I thought, England was big enough and we were bound to cross her shores if only we could keep our battered Blenheim flying.

Three hours later, when logic was giving way to fantasy, we sighted some hills and soon afterwards crossed an unfamiliar coastline. I had already decided there was no sense in trying to put down the undercarriage and, with no maps or radio, we were unlikely to find an airfield easily. The all-important thing was to get the unconscious, but still breathing, Lorrie to hospital as quickly as possible. So, when I saw a grass field near a small town, I did a wheels-up landing. It was our Blenheim's last flight.

The Home Guard were with us within minutes and helped us to lift Lorrie out of the shattered nose. Soon he was in an ambulance and Stan and myself were on our way to the sick-bay of RAF Station, Acklington in the north of England. Before the tranquillizers took effect I rang through to our Wingco, Frankie Lascelles, at Bodney. He seemed surprised to hear from me as none of the other crews on the operation had expected us to return!

The next morning we were told the sad news that Lorrie had died. Stan and I were packed off on a week's sick leave. Five days later, while listening to the BBC's 9 o'clock news, I heard there had been an attack on Heligoland and a number of our aircraft were missing. A telephone call to Bodney did nothing to calm my worst fears. On reporting back the next day, I learned the hard news that Frankie and eleven other crew members had failed to return. Two aircraft had turned back on the way out with engine trouble. The rest had been lost.

It was a black day for 82 Squadron. *As a result of these crippling casualties, at the age of nineteen, and within three months of joining the squadron as a pilot officer, I became a Squadron Leader in command of B Flight. Such were the losses in 2 Group.*' [Editor's italics.]

2 Group's losses in 1940 and 1941 were, in fact, horrendous – for highly questionable gain. Squadrons were being turned round in a matter of a few weeks. 82, Dennis Gibbs's squadron, was typical. On 17 May 1940, in an attack on a German panzer column at Gembloux, 12 miles north of Namur, in Belgium, 11 out of 12 Blenheims were lost to flak and fighters after failing to rendezvous with the fighter escort. Three months later, on 13 August, the squadron met similar disaster. In an attack on the airfield at Aalborg, in northern Denmark, a further 11 out of 12 aircraft were lost to enemy opposition. The 12th aircraft turned back on the outward run with a fuel defect.

'Missing'

Reprinted by permission of David Higham Ltd from *Dispersal Point*,
John Pudney (Bodley Head)

Less said the better.
The bill unpaid, the dead letter,
No roses at the end
Of Smith, my friend.

Last words don't matter,
And there are none to flatter.
Words will not fill the post
Of Smith, the ghost.

For Smith, our brother,
Only son of loving mother,
The ocean lifted, stirred,
Leaving no word.

John Pudney

Versatile Albacores
826 (Fleet Air Arm) Squadron – 1940 and 1941

Admiral Sir Frank Hopkins, Kingswear, South Devon, 1982

'Nearly all the various operations we carried out between June and October 1940 were at night. The Albacore's speed of 90 knots made daylight raids too expensive. For example, on our trip to Texel on 21 June we were intercepted by Me 109s – about a dozen of them. We had no fighter escort and there was no cloud cover. Result: 3 Albacores were shot down and most of the rest were damaged. We shot down 2 Me 109s with our single Vickers ·303 machine gun mounted in the rear cockpit; we also set some oil storage tanks on fire.

After the war, I was told by a Dutchman, who was there at the time, that the commanding officer of the Me 109 squadron was sacked for allowing ancient, out-of-date biplanes to waffle in in broad daylight, bomb the oil tanks and lose 2 109s to these ridiculous aeroplanes.

It was not until the 11 September that we were sent out again in daylight. Then it was to attack some ships just off Calais. On the way, we picked up our "escort" of six "fighter" Blenheims, manned by Australians. There was again no cloud cover and long before we entered the target area we could see swarms of 109s taking off inland. If I had been leading those Blenheims I would have got to hell out of it; but they bravely stuck by us.

As a result of this episode, no targets were hit, 4 out of the 6 Blenheims were shot down and, with them, 3 out of 6 Albacores.

All things considered, however, our casualties over the whole period were fairly light and nearly all of them occurred on the daylight raids.

Late in 1940, we embarked in the new aircraft carrier, *Formidable*, and hurried off to the Mediterranean to replace *Illustrious*. There, we had a busy time, culminating in the Battle of Matapan on 28 March 1941. During the engagement we made two torpedo attacks on the new Italian battleship, *Vittorio Veneto*. One was in daylight with six Albacores and the other was at dusk with eight aircraft. Both attacks were most unpleasant and were not something I would wish on my worst enemy.

To attack a modern battleship successfully you needed fifty to a hundred torpedo bombers and a bunch of fighters to suppress the flak. These ships normally mounted 120 AA guns. This was the usual form in the Pacific both in the case of the Japanese (witness the sinking of *Prince of Wales* and *Repulse* off Malaya) and the Americans (e.g. the sinking of the 80,000-ton battleship *Mushashi*, hit by more than twenty torpedoes).

In May 1941, *Formidable* was severely damaged by two bomb hits during the evacuation of Crete and had to be withdrawn to the United States for repairs. Consequently, we disembarked to the Western Desert and operated with the army, mostly against tanks, motor vehicles and coastal shipping.

In December 1941, I was sent to Malta to take command of 830 Squadron. They had just lost five aircraft on one sortie including their CO.'

The Battle of the Atlantic U-boat Warfare: Origin of the Leigh Light

The Rt Hon. Lord Justice Waller (Wartime Flying Officer) London, 1982

'The Leigh Light was a searchlight mounted on the wing of anti-submarine aircraft of Coastal Command and was of major importance in the anti-U-boat campaign in late 1941–43.

Leigh, the inventor, had been a pilot engaged on anti-submarine duties at the end of World War I. He rejoined the RAF in 1939 but was employed solely on personnel staff duties at Coastal Command HQ. He endeavoured to persuade the Air Staff that his idea of a searchlight would be an important aid to the anti-submarine campaign, but for some time without success. However, after Sir Philip Joubert had been appointed C-in-C, the story goes, Leigh found himself in Joubert's office having been sent on some P. staff business. He seized the opportunity to make the case for his searchlight. Joubert's reply was, "There is a Wellington at Thorney Island which is no longer required for degaussing magnetic mines; it has a Ford V8 engine in the fuselage which could be used for providing current for the searchlight; you can use it to carry out trials of your idea."

A searchlight was fitted and preliminary trials were carried out which were sufficiently successful to justify trials with a submarine. Arrangements were made to carry out such trials after dark with a submarine based at Londonderry. This was an H boat, a relic of World War I, and used for training submariners.

I was a pilot in 502, a Whitley anti-submarine squadron of Coastal Command, stationed at Limavady a few miles from Londonderry. I was detailed to be the RAF liaison officer on board the submarine during the trial. The Wellington, with its searchlight was to make several runs over the submarine, homing with radar and had to turn on its searchlight to light up the submarine as it got close. On the last-but-one run the submarine was to be with decks awash and on the last run the submarine was to dive.

I embarked on H 31 at Londonderry just before midnight on 3 May 1941, and was welcomed aboard by the captain. In answer to questions I said that I had never been aboard a submarine before and would be very interested in the experience, including diving.

We set off and about 1.30 a.m. the trials commenced. They were very successful and after the run with decks awash the aircraft signalled cancelling the last run. "Oh good," said the captain, "we can go straight back to Londonderry. Oh, but you wanted to dive, we must dive for you," he said to me. "No, thank you very much," I replied, "so long as it was part of the plan, I was ready and willing, but not just for experience." So we made course for Londonderry without diving.

On the way back I said to the captain who had never flown, that if he would like to fly I could arrange it any day, because there was an aircraft on air test every day. "Me, fly," he replied, "not without a direct order from Their Lordships at the Admiralty."

We docked at Londonderry and I returned to Limavady and made my report.'

Bombing and Rescue Join Hands

Air Chief Marshal Sir Augustus Walker, Brancaster Staithe, Norfolk, 1982

'No. 50 Squadron of 5 Group, Bomber Command, was located at Lindholme, near Doncaster, when I first joined it in September 1940. I assumed command of the squadron in October 1940. It was equipped with Hampden aircraft.

The Station Commander at Lindholme was Group Captain E.F. Waring, a very experienced flying boat pilot. He was thoroughly knowledgeable about seamanship and, as an Air Commodore, became, in due course, the first Director of Air/Sea Rescue at Air Ministry.

To Eddie Waring goes the credit for conceiving an idea which was to improve dramatically the survival chances of aircrew who had to abandon aircraft over the sea. This involved dropping, from a suitably equipped aircraft, a large dinghy to which were attached flotation chambers carrying survival rations and warm clothing. The dinghy was provided with a soluble plug which, when exposed to water, released gas from a cylinder which inflated the dinghy.

The float chambers with rations and clothing were secured on either side of the dinghy by light, iridescent nylon cord, 30 yards long. This arrangement was extended, during trials, to include two equally-spaced cylinders, held on each side of the dinghy with a total of 60 yards of cord.

The cylinders and packed dinghy were housed in the aircraft's bomb bay and were dropped by operating the bomb release gears on a timed sequence.

A smoke float was dropped to enable the crew of the rescue aircraft to establish the position of the dinghy on the sea, and the wind direction over the surface. A loose dinghy will always drift downwind.

Having positioned the float, one of the crew of the rescue aircraft would release the dinghy and float chambers to straddle the craft's line of drift. The floating nylon cord was then available to pull in the chambers and the dinghy which had inflated after making contact with the water. A great deal of experimenting had to be done before the design was perfected and the correct drill established. There was an urgent need for a rescue craft of this type.

When, eventually, we were ready, we prepared a detailed Air Ministry Flying Order for circulation to all Command staffs, Air/Sea Rescue units and flying formations. It was entitled the "Lindholme Rescue Dinghy".

We did not earmark an operational aircraft especially for Air/Sea Rescue duties. All aeroplanes on the operational strength were invariably required for offensive duties. We were, however, usually able to nominate a stand-by aircraft and crew which could, when called, be switched from its primary bombing role to rescue work.

I well remember receiving one "May Day" or SOS call early one morning in June 1941. My crew were alerted and dispatched. Visibility out to sea was good and only a light swell was running. Having reached the search area, which was roughly 30–40 miles due east of the Humber, we began following a creeping line ahead search pattern, straddling the bearing of a signal which had been received. We worked steadily eastwards along this bearing at a height of some 1000 to 1200 feet which was the optimum search height unless the sea was dead calm.

I was just on the point of widening our search and turning on to a reciprocal course when, suddenly, we sighted a dinghy on the water with four or five crew members on board. Elated, I at once flew low over it and then started following precisely the dropping procedure we had evolved. All went exactly to plan.

However, our hopes of a successful, initial rescue, received a nasty setback when, to our dismay and astonishment, we saw the crew in the dinghy paddling strongly in the opposite direction to the wind. I at once instructed my wireless operator, Sergeant Trevor-Roper,* to transmit a flash signal on W/T giving the dinghy's position and requesting that a high-speed launch be despatched from the Air/Sea Rescue unit at Grimsby. Meanwhile, as we waited, frustrated, for the high-speed launch to arrive, I tried desperately, by every means, to direct the crew in the dinghy towards the Lindholme Dinghy; but all to no avail.

Eventually, the launch appeared on the horizon, and having directed it first to the Lindholme Dinghy, which was retrieved, and having seen it transfer the crew of the dinghy safely on board, I set course for base.

As soon as I landed I went to the operations room and asked for a message to be sent to the Air/Sea Rescue unit at Grimsby requiring that the rescued crew be asked if they had ever read the Air Ministry Flying Order on the Lindholme Dinghy. With that I went to bed.

When I woke up I returned to the ops room to hear the answer. It was conclusive. The rescued airmen were the crew of a German Heinkel 111 bomber! I had captured my first, and only, prisoners of war!

There was a curious sequel to this bizarre incident. Years afterwards, I met a friend of my daughter and son-in-law who now live in Sydney, Australia. They had told him to contact me when he reached London.

I was fascinated to find that, through his own enterprise, he had established, in conjunction with the local coast guard authorities, a Land/Air/Sea Rescue service along the west coast of Tasmania. For this he used a small fleet of single and twin-engined aircraft and a powerful sea-going launch.

I asked him what equipment he employed for the sea rescues. "We use the

*Hugh Trevor-Roper – soon to be commissioned and to add the DFC to the DFM for outstanding courage in low-level bombing attacks, including his part in the famous 'Dams' raid when he acted as wireless operator and air gunner to Wing Commander Guy Gibson. His subsequent loss was a heavy blow.

Lindholme Rescue Dinghy," he replied.
More recently, I was the guest of Air Marshal Sir John Curtiss, the C-in-C of the Royal Air Force's Maritime Forces, at his Command reunion. We discussed modern Air/Sea Rescue methods. "What," I asked, "is the equipment you carry in your Nimrod aircraft?"
"The Lindholme Rescue Dinghy," he said.'

Forty years on, it's still in use.... Salute Air Commodore Waring!

Before Pearl Harbor
The New Intake

R.L. 'Dixie' Alexander, Piper City, Illinois, 1982

There now began to arrive in Britain a steadily mounting flow of trained aircrew from the great air training schemes in Canada, the other Commonwealth countries and the United States. Among them was a spirited American contingent who had elected to make our cause theirs long before the Japanese attack on Pearl Harbor on 7 December 1941, had brought the United States to war.

Many of them formed the three famous Eagle Squadrons in Fighter Command before transferring, in September 1942, to their own US Eighth Army Air Force. There, they became the nucleus of the 4th Fighter Group which was to become one of the brilliantly successful fighting units of World War II, led, in the crucial years, first by Chesley 'Pete' Peterson and then by Don Blakeslee, two quite exceptional American officers.

Among their number was Dick Alexander, 'Dixie' to the Royal Air Force, who had enlisted in Windsor, Ontario, in September 1940, to begin his flying training with the RCAF. He sailed for England, as a sergeant pilot, in the armed merchant cruiser, HMS Rampura, *'a sister ship to HMS* Wolfe *and* Empress of Asia, *old British India merchant liners capable of making 20 knots'. The Atlantic crossing, in which the pilots were detailed to 'submarine and aircraft watch', left an enduring mark.*

We had been out about twelve days, zigzagging our way across the ocean, and I was doing a night watch on the port beam with a little grizzled Scots sailor who chewed a small, stubby, unlit pipe. The night was black, but on the swells we could make out the silhouettes of our sister ships, and occasionally see the glow of phosphorous in their wake.

Suddenly, and almost simultaneously, we observed a tiny light in the distance. From our position it was at about three o'clock. We swung off course and headed directly towards the distant light. 'What's going on?' I asked. 'I don't know,' replied my companion, 'but hang on a minute and I'll find out.'

He disappeared into the darkness, and I continued to watch the light until he returned. 'She's an unidentified vessel,' he explained, 'who refuses to recognize our signals. We are going to run her down and make her identify herself.'

I paused. 'And what if she turns out to be *Bismark, Prinz Eugen* or *Scharnhorst*, or one of the other big ones?'

He eyed me carefully, removed the pipe from his mouth and drew himself up to his full height – all 5 feet 2 inches of it – 'Makes no difference, Laddie,' he said firmly, 'she'll hesitate 'fore she tackles a British man-o'-war.'

Oxford Response

His Honour Judge C. Raymond Dean, QC (Wartime Flight Lieutenant)
Boston Spa, Lincolnshire, 1982

'When I was at Oxford in 1941, before joining the Royal Air Force, I was a member of the University air squadron. One of my colleagues was "Dicky" Cecil, a member of the illustrious family of Lord Salisbury. He was a delightful man, but hopeless so far as "bull" and drill were concerned.

Eventually, the Air Ministry sent a team to interview all of us cadets, to see (I dare say) if we were prima facie aircrew material. One of the standard questions was: "Why do you want to be a pilot?" Most of us trotted out some high-sounding rubbish about being inspired by the Battle of Britain, etc.

Not so Dicky. His reply to that question was: "Well, sir, as a matter of fact, I don't want to fight at all, but if I must, I prefer to do it sitting down!"'

Pilot Error

Squadron Leader I.A. Ewen, Waikanae, New Zealand, 1982

'After finishing my flying training, I was posted as a staff pilot, in 1941, to the Royal New Zealand Air Force Station at Woodbourne, Blenheim. Being a newly-commissioned pilot officer, I was detailed, soon after arriving, to ferry a party of ground staff to a drogue-towing airstrip some distance from our base. I had only flown a Vickers Vildebeeste once or twice before.

After taxiing out from the hangar, I was given the green light for take off by the Duty Pilot. Steadily moving the throttle forward, we were soon bowling down the runway and gaining flying speed. Just as I was about to ease the aircraft off the ground, there was a resounding bang; the airscrew quickly stopped turning and we gradually came to a halt, fortunately well short of the end of the runway. Whatever, I wondered could the trouble be?

A ground staff sergeant behind me leant forward. His voice was tactfully quiet. 'Have you switched the petrol on, sir?'

On went the offending fuel cock; the Sergeant jumped out and, with one swing of the propeller, the engine was again turning over nicely.

Pushing the throttle open, we were soon lifting off without the Duty Pilot, or anyone else, having time to inquire into our problem.

What would we have done without the NCOs on the ground staff?'

The End (and the Beginning) of a Legend

Air Vice-Marshal J.E. 'Johnnie' Johnson, *Wing Leader* (Chatto and Windus, 1956; Hamlyn Paperbacks)

Somewhere between 1120 and 1140 hours on 9 August 1941, Douglas Bader baled out of his Spitfire over St Omer, in northern France, and was made a Prisoner of War. He had been leading the Tangmere wing when the three squadrons became embroiled with the Messerschmitts of General Adolf Galland's Jagdgeschwader 26. One other aircraft, that of 'Buck' Casson, a flight commander of 616, the squadron with which Bader had been flying, was also lost.

Two Luftwaffe pilots, Oberfeldwebel Max Meyer of No. 6 Staffel and Leutnant Kosse of No. 5 Staffel, each claimed one Spitfire destroyed (Meyer's at 11.15–11.30 and Kosse's at 11.45) 'in the region of St Omer'. Bader thought he had collided with an Me 109; Galland was sure he had been shot down.

Johnnie Johnson was flying in Bader's section of four that day – the day which, as he remembers it, was now almost over.

It was one of those August evenings which mark the end of high summer. Although it was nine o'clock the sun had disappeared below the Downs and the tall beeches on the other side of the meadow were barely visible in the fading light. It was a quiet time, when all the Spitfires were down and the Beaufighters had not clawed into the air for the night patrols.

Our ground crews were bedding the Spitfires down for the night. Placing chocks under the wheels, tying on the cockpit covers and carrying out last-minute tasks so that they would be ready at dawn. All except the crews of the two missing Spitfires, who stood apart in a restless, disconsolate little group and who occasionally fell silent and strained their eyes to the east, as if, peering hard enough, they would see their two Spitfires swinging in to land.

We, too, were silent when we drove to the mess, for we knew that even if our wing leader was still alive he would have little chance of evading capture with his tin legs. Before this we had rarely thought of his artificial limbs, and it was only when we swam together and saw his stumps and how he thrashed his way out of the deep water with his powerful arms that we remembered his infirmity. At Tangmere we had simply judged him on his ability as a leader and a fighter pilot, and for us the high sky would never be the same. Gone was the confident, eager, often scornful voice. Exhorting us, sometimes cursing us, but always holding us together in the fight. Today marked the end of an era that was rapidly becoming a legend.

The elusive, intangible qualities of leadership can never be taught, for a man either has them or he hasn't. Bader had them in full measure and on every flight had shown us how to apply them. He had taught us the true meaning of courage, spirit, determination, guts – call it what you will. Now that he was gone, it was our task to follow his signposts which pointed the way ahead.

'Our Debt'
31 December 1941

To-night as the old year dies
Look up to the darkened skies
And know that but for those who knew
Where duty lay we could not pray
As now we may:
'God grant them care; and man their due.'

Harold Balfour

PART FOUR

Retaliation

The rough times of 1941 spilled over into 1942 as their sombre aftermath hung dauntingly over the Allied cause. In the Western Desert and the Mediterranean, in the Far Eastern and Pacific theatres, on the Eastern front, and in the West, and in the Atlantic approaches, the first six months of the year were studded with fresh reverses; nor would it rest there. But, as the year wore on, the latent capacity of the Allies to survive, recover and then fight back, began to emerge.

The brilliant United States' air and naval victories in the Coral Sea and at Midway were to turn the Pacific war irrevocably against the Japanese. In Malta, in Egypt, Libya and North Africa, the wheel of fortune settled decisively in favour of the forces of the Allies. In the Western theatre, the war against the U-boats and the escalating air offensive against the enemy in Europe, poured fuel on hopes that better things were ahead.

None underestimated the fighting discipline and qualities of the enemy, but, undeniably, the spirit of retaliation was now abroad.

South Africa in the Desert

Christopher Shores and Hans Ring, *Fighters Over the Desert* (Neville Spearman, 1969)

... It should be realized, though it is a relatively little-known fact, that far and away the largest contribution in this theatre was made by the South African Air Force, which provided half of the light bomber force and a large part of the fighter force. No less than five fighter, one tactical reconnaissance and one strategic reconnaissance, three light bomber and one patrol bomber squadron served with the Western Desert Air Force, and many other squadrons operated in East Africa, freeing British units for other zones....

Retreat of an Army

Lieut.-Colonel J.A.G. Rademan, 'SAAF from Gazala to Alamein,' *Illustrated Story of World War II*, published by Reader's Digest

The unrelenting contribution of the light and medium bomber squadrons to the ebb and flow of the Desert battles of 1942 has sometimes been underplayed. The squadrons' daily round, less glamorous in terms of news

valued than the highly publicized deeds of the Hurricane, Kittyhawk and, later, the Spitfire units, was a continuing and potent threat to Rommel's forces.

None was more active in close support of the 8th Army than the South African Air Force's No. 3 Wing, made up of 12 and 24 Squadrons with their Bostons, and 21 Squadron, soon to be equipped with Baltimores.

Bert Rademan (Lieutenant-Colonel J.A.G. Rademan) was appointed to command 24 Squadron at the moment when the opposing armies were readying themselves for the spring and early summer battles which were to send the Allies reeling back to the Alamein line. His first-hand account sets the scene of the withdrawal.

During April and the beginning of May, it was obvious that both sides were preparing for an offensive. The air forces ... bombed each other's airfields, striving for that vital air superiority.

Lieut.-Colonel H.J. 'Kalfie' Martin, commanding 12 Squadron, and I (both) urged that the Boston squadrons be allowed to contribute to the battle by being used on night operations as well as for the daylight Army-support role. This was agreed to and all flying crews were given a quick night-flying refresher without interference with the routine daylight operations on airfields, supply columns and close support bombing....

The battle continued without a break until the destruction of the British armour during the second week in June. The commanding general ordered the abandonment of the Gazala line.... The 8th Army was withdrawing to the east while the 2nd South African Division, with elements of the Indian Division and a brigade of Guards, were sent to hold Tobruk. The battle was now drawing close to Bir el Baheira....

On 17 June the Boston squadrons were ordered to move to Landing Ground (LG) 07 between Sidi Barrani and Mersa Matruh. The bombers took off for the last time from Bir el Baheira, circled to gain altitude, and, with their faithful RAF and SAAF squadrons escorting, bombed the enemy advance guard before setting course for LG 07.

It was an amazing sight to see from above a whole army in retreat. Thousands of vehicles, radiator to bumper, covered every desert track heading eastwards. What a fantastic target for the enemy air forces! But we had air superiority.

On 20 June, the squadron was briefed for a sad mission – to my mind just a farewell salute to the beleaguered garrison in Tobruk. We flew by way of LG 76 to pick up our escort of Kittyhawks. The target was two enemy armoured columns, which had breached the southeast perimeter of the Tobruk defence system. Only six escorts were available.

What a poignant sight it was as we ran up on target. Beetle-like tanks and armoured cars crawled through the outer defence system. Straight ahead lay ill-fated Tobruk, choking in dense columns of smoke and being ravaged by flames. Demolition had already begun. In the air a massive Stuka party was in progress – little black dots diving almost vertically downwards to release their bombs into the inferno below, their escorts buzzing around busily, watching over them. Our effort seemed so futile. Down went our bombs, scoring three direct hits and causing one large fire, but the ant-like tanks just went on through the gap as if nothing had happened and started fanning out.

The Me 109 escort for the Stuka party 'climbed into us' as we turned away from our target. The badly outnumbered Kitties heroically shepherded their 'big friends' back until met by an RAF 'delousing' squadron of Spitfires.

From LG 07, situated next to the main coastal road, the gallop could be viewed as thousands of vehicles rumbled past night and day. Weary soldiers crammed into their troop carriers, dust-grimed, dirty and hungry, making for the previously prepared defence line at El Alamein....

In fifty days from 17 May to 17 July, our squadron had flown 1001 sorties. No. 12 Squadron, the other Boston squadron, was close to that total. This remarkable achievement was carried out under desert conditions with the army pulling back, and the air force squadrons – bombers and fighters – quitting one airfield after another, each time rapidly re-establishing the ground organization to ensure that there was not a moment's interference with flying operations....

A report by Panzerarmee Afrika HQ declared: 'The heavy losses in personnel and material are attributed to incessant air attacks in strength and to the constantly high rate of ammunition expenditure by the British Artillery.' It also assessed the effect of the day and night bombing on the troops: 'In addition to the extensive material damage caused, the effect on morale was also great. The spirit of the troops was considerably depressed owing to the totally inadequate German fighter cover....'

'I'm Rademan'

Peter Atkins, 24 Squadron SAAF, *Buffoon In Flight, Some Misadventures of an Observer* (Ernst Stanton, Johannesburg, 1978)

A journalist, Peter Atkins was Racing Editor of the Cape Argus *and, subsequently, before his retirement, Racing Editor of the* Star.

Whoever followed Rademan was in for one hell of a job. If he was due to lead, and the target was a relatively easy one, he would wander into the 'ops' tent and declare he did not feel like flying and substitute his name at the top of the stand-by list in favour of one of the other leaders. If the raid looked like being really tough and he was not due to lead, he would have a look at the stand-by list and mutter something about the others 'hogging the raids' and replace the name of the leader with his own....

He was always tidy in his appearance, although towards the end of his tour he allowed his wardrobe to run down until all he possessed was a bush shirt, a shirt, a pair of trousers and a pair of shorts. Due to an unreliable dhobi there were a couple of occasions when his soiled linen was removed before the fresh was returned. This didn't disconcert Colonel Bert; it was not unusual for him to appear in the 'ops' tent in no more than his 'brothel creepers' – the rubber-soled suede shoes worn by just about everyone in the desert – and his identity discs. One morning Colonel Bert, so clad, was sitting in his canvas chair when Lord 'Boom' Trenchard, Marshal of the Royal Air Force, appeared in the tent entrance. We stumbled to our feet – we disliked these unheralded appearances of VIPs – and 'Pop' Barton, the only person present wearing a cap, threw Trenchard a snappy salute. He asked to see the CO and the naked colonel stepped forward with his hand held out in greeting and said, 'I'm Rademan.' Trenchard looked absolutely astonished. Never in his lengthy years in the Service could he have seen a commanding officer so undressed.

ASLEEP IN A SLIT TRENCH

In the Middle East at the time were two American engineers, one from Wright and the other from Pratt and Whitney, whose task it was to sort out any problems we had with the aircraft engines. They made periodic trips to the various squadrons and usually managed to bring some 'goodies' with them.

The afternoon they arrived at Baheira their 'goodies' consisted of two cases of American canned beer, half a dozen bottles of Four Roses bourbon whiskey and an enormous apple pie swiped from an American ship in Alexandria harbour.... The beer – the first the chaps had seen since leaving base – disappeared in no time at all. So did the apple pie. Then we turned to the bourbon. We tried drinking it with the horrible Tobruk water we had in our water bottles, which we carried with us on raids to sustain us if shot down, but it tasted too dreadful for words. So the only alternative was to take our bourbon neat. A considerable while later I decided it was time for bed and set off for my tent. In my befuddled state I got hopelessly lost and finished up by falling headlong into a slit trench. It was a warm night and there didn't seem much purpose in trying to struggle out and have another go at finding that tent. I had a fine night's sleep and was woken up by the 'ops' gong for the dawn raid. I crawled out and ran into one of the senior observers on his way to the 'ops' tent. He exclaimed, 'What a bastard of a night!'

Not realizing that we were at cross purposes I replied, 'Oh, I wouldn't say that. Though I must admit I've got a bit of a hangover.'

He looked at me in amazement and asked, 'Where were you last night?'

'Asleep in the bottom of a slit trench.'

He shook his head and moved on muttering 'asleep' in an incredulous tone. It was only later I learned two marauding Me 110s had found our camp during the night and had had a great time strafing the hell out of it. They didn't do any serious damage but most of our tents were in rags. And I had slept through it all!

BASE WALLAHS

Our next stop was LG 97, some 30 miles east of Alexandria and alongside the main Cairo–Alexandria road. Conditions there were chaotic. The field was too narrow and barely long enough to accommodate the two squadrons of Bostons. There was no water and our water cart had to queue up for twenty-four hours at the nearest point to obtain anything resembling our minimum requirements. Rations were almost non-existent and our opinion of base wallahs, never very high, sank to new depths....

So we set about hitchhiking back. We got to Cairo that evening and an astonishing sight it was. A pall of smoke lay over the city. We thought at first the place had been bombed. Nothing of the sort. It was just the base wallahs burning loads of useless bumph before evacuating to a safer area. As the war was still some 200 miles away to the west this all seemed a trifle premature!

There have been all sorts of explanations for the frightful hammering we had taken in the desert but I have often wondered if the real cause was simply that the fighting part of the army could not carry the appalling weight of the massive staffs sitting on their arses in Cairo busy writing memos to one another. There were, of course, quite a number of dedicated and efficient officers in Cairo but just what the hell the rest of them did is anyone's guess. No wonder they were known contemptuously by the fighting troops as either Groppi's Light Horse or the Gezira Light Horse. Groppi's was Cairo's best-known restaurant and Gezira, a remarkably fine country club. Both, at any one time, were occupied almost entirely by the base wallahs.

In Wavell's day, when he was running a number of campaigns simultaneously, there might have been some excuse for the large Cairo staffs. But surely there was no call for thousands of bods to run the desert campaign from behind – which, judged by our supply problems, they were incapable of doing anyway.

I have never seen a record of what Rommel's rear services comprised, but it is a fair bet that they were slim and streamlined compared to the massive Cairo operation.

Leading Observer

His Honour Judge Cecil Margo, Johannesburg, 1982

The lot of a leading observer in the Desert air war was not to be coveted. The premium on precision was excessive. A small error in calculation could spell catastrophe. With a human and vociferous reporting system always at readiness on the ground, there could be no cover-up. Inaccuracies were exposed for what they were. It was the hell of a responsibility to carry.

Coming from the line of 24 Squadron's commanding officers, Cecil Margo saw the requirements from the pilot's seat in the leading aircraft. As a member of South Africa's judiciary, the impartiality of his assessment is unlikely to be challenged. There is, however, an interest that he would, no doubt, wish to declare. For a while, Captain Peter Atkins, an accepted master of the observer's art, was his navigator. To Atkins, to his unusual skills, and to the success which his navigation brought to the squadron, Cecil Margo dedicates these reflections.

'The main preoccupations of the light bombers in the Western Desert were close army support by day and, at night, attacks on enemy airfields, troop concentrations and road movements. Briefing for these attacks often had to be done in great haste, with corresponding pressure on the leading observer.

The daylight missions were by squadron formations of eighteen aircraft, attacking from medium altitudes. Although escorted by strong fighter cover, which while often attacked by enemy fighters, was but rarely penetrated, the bombers usually encountered fierce resistance from 40-mm and 88-mm flak, the accuracy of which (particularly the 88-mm) at times necessitated fairly drastic evasive action. This was a sore distraction, but it was not the only problem. The haze reduced visibility, and there was the absence of geographical features, both of which hampered navigation and target identification. There was also the dreadful responsibility of having to attack targets in close proximity to our own land forces, with the 'bomb-line' sometimes moving back and forth with the tide of battle.

Because of the absence of a visible horizon and the restricted visibility in haze, the mission leader had to fly on instruments most of the time, and the sole responsibility for navigating the squadron to rendezvous precisely on time with the fighter escort, and then to the target along the predetermined line of attack, fell upon the leading observer. In the run-up to the target, with bomb-bay doors now open, the mission leader's task was to achieve 100 per cent accuracy of height, speed, direction and straight and level flight. This called for precision instrument flying, while the entire responsibility for the accuracy of the approach and bomb-aiming continued to rest with the leading observer. With all bombs gone, the mission leader could take evasive action as violently as eighteen aircraft in tight formation would permit, but yet again the leading observer had to take the squadron out and home along the predetermined lines.

In the night operations, though the aircraft operated singly, the leader's task was to find the target not only for his own attack but also for those following. Here, again, the entire responsibility for navigating, for finding the target, for the run-up and accuracy of the bomb-aiming, devolved upon the observer. Sometimes, because of darkness, interference from enemy fighters, searchlights or ground fire, a second or even a third run became necessary. With it went the hazards of continued exposure to fire and collision with other aircraft in the target area. Again, the withdrawal from the target along the predetermined 'get-out' route, the flight home and the straining in the darkness to pick up the few navigation points – all became the observer's lonely responsibility.

As the air war moved into southern Europe, the squadrons re-equipped with B-26s (Marauders), medium bombers, and reduced the size of the strike formation to twelve aircraft. The daylight close army support operations continued, but the main thrust gradually shifted to semi-strategic targets and the interdiction of the enemy logistics systems, from factories to railway marshalling yards, railheads, road and railway bridges and junctions. As before, the responsibility for getting the mission to the fighter escort rendezvous at the precise time and on the exact heading, adhering to the planned route to the target, turning on, running up and releasing the bomb-load, was that of the leading observer. So, too, was the responsibility for navigating the squadron back to base.

In all operational conditions the leading observer, like his colleagues in the other aircraft, sat or lay prone in his perspex capsule, surveying the earth and sky ahead and below. He could have none of the armour-plated protection of the flight deck. In the Marauder, his little office was equipped with a ·50 calibre Browning firing forward, for use in head-on attacks and offering the same kind of comfort against enemy assault as a miniature lapdog would provide against burglars in modern urban life.'

Unparliamentary Behaviour

The Hon. John L. Waddy, Cremorne, New South Wales, 1982

John Waddy, later to become one of Australia's parliamentarians, amassed an enviable total of victories in the Western Desert. He was with 260 Squadron, one of the Desert Air Force's 'mixed' units, made up of Australians, Britons, New Zealanders, Rhodesians and South Africans, when this incident occurred.

'Date: June 1942; place: Libya (during the great retreat from El Agheila to El Alamein often called the Bardia Handicap).

The enormous tank battle at Knightsbridge on the Gazala–Bir Hacheim line had been fought and lost and the squadron was engaged on dive-bombing and strafing the advancing enemy forces southeast of El Adem. Having completed our bombing we were returning to base at about 3000 feet when two Me 109Fs dived and shot down my number 2 who was on the extreme left of the formation. No one had reported the bandits and the first I knew was when I saw the Kittyhawk, about 100 yards on my left, catch fire and dive away, followed by the two 109s. (The pilot subsequently baled out and was picked up unhurt.) The 109s were following the Kitty down.... This was as perfect a position as one could ever expect and having already shot down some 10 or 11 I was feeling pretty confident.... Or so I thought! I opened fire, but after a few rounds all my ·50 machine guns jammed and I quickly became the hunted....

To understand the sequence of events one needs to picture the terrain.... On my left, a few miles to the north, was the Mediterranean. Not far inland three escarpments, about 150 feet high, ran roughly parallel to the shore. I was flying about 20 feet above the ground with my right wing tucked in as close as safety permitted to the third escarpment with the leading 109 about 300 yards behind me. His number 2 was about the same distance away on top of the escarpment. He could not get at me and number 1 had to be fairly careful of my slipstream. I was, I thought, in the safest possible position under all the circumstances; but a continuous series of bangs and smacks kept me well aware that the number 1's 20-mm cannon firing through the prop of his 109 was working very efficiently....

My consternation was complete when suddenly the escarpment flattened out and there was I, 20 feet up, two Messerschmitts on my tail, no guns and nothing but flat ground ahead. Strangely, I was not afraid but bloody angry. (The fear came later.) Instinct made me do a left-hand climbing turn which evidently took both 109s by surprise.... I then dived back to deck level, having gained a valuable 150 yards. The 109s followed me right back to my airfield. As I went over the top at about 10 feet they fired their last burst.

The squadron had landed and the CO had just stepped down from his aircraft when he saw me coming and a line of bullets kicking up the dust across the landing ground heading in his direction. I can see him now, diving face first into the dust which, because of constant use, had become as fine as powder and inches thick. The Bofors guns opened up but the 109s made off, apparently unscathed, except for a few holes I put in the number 2 at the beginning. I landed and it was found that the Kitty had 124 20-mm and ·30-bullet holes in it. All the perspex down the left side had gone; instruments were shattered; there were no flaps and a flat tyre. I was wet through with perspiration; one earpiece was missing from my helmet; there was a large hole through my flying suit under the arm and another through the suit behind my knee. I was unscratched but my ego had been badly dented....

I was sent for by the CO and given a good dressing down for bringing the 109s back to the aerodrome, though it was never explained to me where I should have led them....

The epilogue: my aircraft was temporarily repaired by squadron fitters before being flown back to the aircraft depot in the Delta for major repairs. It was later completely written off by the ferry pilot who executed an upside-down landing when returning it to the Desert.'

'He was the Best'

Hauptmann Hans-Joachim Marseille was twenty-one when he was killed in the Western Desert. The end came just before midday on Wednesday 30 September 1942, some 7 kilometres south of Sidi Abdel Raman, not far from El Alamein. He died baling out of a new Messerschmitt 109G which had developed a technical fault. As he fell from the smoking cockpit, choked and almost suffocated, he hit the tail unit of his aircraft. His parachute did not open.

Jochen Marseille was undefeated. He was credited in the Luftwaffe's records with 158 victories, 17 of which were said to have come in a single day – on 1 September, a month before he was killed. His opponents never believed his score; but as the years passed and his desert comrades gave their first-hand impressions of his abilities and achievements – and his rare, Bohemian personality – all, in time, came to acknowledge him to be one of the greatest individual exponents of the art of air fighting the combatant countries have ever known.

On 27 October 1982, at a small private gathering in London, after the thanksgiving service for Group Captain Sir Douglas Bader at St Clement Danes in the Strand, the editor put the question to General Adolf Galland: 'How good was Marseille?'

The answer was immediate and direct. 'He was the best.'

Marseille was a squadron commander in JG 27, the Jagdgeschwader commanded by Oberst Eduard Neumann who was himself one of the Luftwaffe's most able operational leaders. Like Galland, Neumann's assessment of his subordinate commander's quality was forthright.

As a fighter pilot Marseille was absolutely supreme.... Above all, he possessed lightning reflexes and could make a quicker judgement in a bigger orbit than anyone else. His senses and reflexes produced what, today, a modern computer would provide. Marseille was unique....

Neumann embellished the picture of the man:

He used to wear a 'bum-freezer' uniform which was so faded and crumpled that it looked rather like a jeans jacket today. His hair was too long and he brought with him [to the Desert] a list of disciplinary punishments as long as your arm. Of the 7 kills he had claimed fighting along the English Channel, 4 had not been confirmed – a large percentage. On top of it all, he was a Berliner.

He was inclined to show off, but this was probably to compensate for some kind of complex; he wanted to feel he was in the circle of important officers. In trying to create an image, he wasn't averse from talking about the many girls he had been to bed with, among them a famous actress. He was tempestuous, temperamental and unruly. Thirty years later, he would have been called a playboy.*

Marseille was heavily decorated – by the Italians as well as by the Germans. After Mussolini had invested him with Italy's highest and most coveted award, and kissed him twice, Jochen commented to a fellow pilot: 'It was terrible. The Duce was unshaven.'

Of those who served with him in the Desert in 1941 and '42, Oberstleutnant Werner Schroer was probably as well placed as anyone to judge him. They had trained together and were close friends. Schroer became one of the Luftwaffe's most experienced squadron and wing leaders, finishing the war in 1945 on the Eastern front having begun it as a 22-year-old leutnant in the Battle of Britain in 1940. In the interim he had been shot down three times. In five, almost uninterrupted, years' operations, he had collected a tally of 114 enemy aircraft destroyed, 102 of them in the Western front, North Africa and the Mediterranean.

A balanced and intelligent man with a perceptive mind, Schroer became, in the post-war years, a senior executive of the Messerschmitt–Bölkow–Blohm company in Germany. In a specially written piece, he has recaptured the memory of the fellow-officer he knew so well in the Desert forty years ago.

Werner Schroer, Ottobrunn, West Germany, 1982

'Marseille's magic sparkled. It lay in his jaunty, cavalier personality, in his very unmilitary behaviour and in his unconventional dress. His headgear was hardly "regulation" – certainly no infantry officer would have been seen wearing a hat styled like his.

His eyesight was exceptional; he seemed to have a third eye in his head. His flying reflected his flamboyance and his dash. He was never one for sticking to the usual disciplines of the air – even when he was a junior officer in the squadron. But there was a genius, a flair about his flying which enabled him to exploit a situation, to shoot from almost any position. When he found himself at a tactical disadvantage he had the opportunism and the speed of mind to extricate himself from it and turn the tables on his opponents.

I saw him using tricks in combat that others would never have

*Eduard Neumann, quoted by Von Günther Stiller in his series *Der Playboy mit dem Ritterkreuz: Bild am Sonntag*, Hamburg, September 1979.

contemplated. He would close the throttle, lower his landing flaps and drop the undercarriage to reduce speed and fox an attacker. By using his flaps and working the throttle he could increase the manoeuvrability of his aircraft and tighten his turns.

His instinct for aiming and shooting was such that he would open fire in a dogfight without actually having the enemy in his sights. He had the confidence in the air which enabled him to feel that in a battle of wits he could outmanoeuvre the Curtisses, the Hurricanes and the Spitfires even if the numbers were against him.

Landings and take offs, however, fell well below usual squadron standards. Once he borrowed a new Macchi 205 from a neighbouring Italian squadron and crashed it on landing. It left a poor impression of his general ability as a pilot.

In the Desert, his sense of fun and his charm were engaging. He had a free and easy attitude to life. He did not allow rules to inhibit him. He listened to the Allies' radio stations which were very much taboo. He said he preferred their music.

Jochen was a practical joker; he was forever playing pranks. He came to see me and my squadron – No. 8 Staffel – one day in his colourful Volkswagen jeep. He called it Otto. After a talk, a cup of sweet coffee and a glass of Italian Doppio Kümmel, he got into his jeep and drove it straight at my tent flattening everything. Then he drove off with a grin stretching across his face.

Against all this there were the serious talks we had in the long evenings when we would all discuss tactics with him – how to break up a defensive circle of Curtiss fighters protecting British ships, how to attack elements supporting the surrounded fortress of Tobruk. He would work out the tactics he thought best and fly them according to his instinct, leaving the rest to try to do the same.

But none of us could emulate his skills and few could achieve victories in the way that they came to him.

I think Jochen felt that things couldn't go on as they were and that sooner or later his luck would run out. He may, in the end, have begun to question his success and his genius. If Fate had not caught up with him near El Alamein, it's a fair bet he would have been killed later attacking the Americans' four-engined bombers.'

Marseille's was a personality which was familiar in any German or Allied fighter squadron in World War II. The colour; the flamboyance and the little affectations; the gaiety and the humour; the infectious friendliness; the mildly arrogant disregard for convention – and yet, despite it, the unwavering loyalty to Service, squadron and cause.... Friend or foe, these features made up the trademark of the units on either side whose members were fighting for their young and boisterous lives.

The fact of it was that in spirit, mannerism, resolve and the unthinking acceptance of war's fortunes, there was little difference between the German squadrons and our own. It gave rise to a remark I remember Douglas Bader once making in my house. 'Next time, old boy, we want to make sure we're both on the same side.'

'The End'

Chosen by Group Captain G.B. Johns

After weeks of damn-all liquor
Comes the beer; then, in a flicker,
Mighty milling crowds upon the Mess descend,
And you say: 'Let's have another,'
But you're told: 'You've had it, brother.'
It's the end, I tell you, fellows, it's the end.

When you've drunk all that you're able,
And you stagger from the table,
And in alcoholic stupor seek your bed;
Man appears, at dawn's first gleam,
Yells: 'Get up, you're on the team.'
It's the end, I tell you, fellows, it's the end.

Then you're up at o-six-thirty,
With the weather blowing dirty,
And by noon dispersal's vanished in the sand;
I.O. natters with a grin:
'It's still twelve kites at thirty min.'
It's the end, I tell you, fellows, it's the end.

When you're jumped by twenty snappers,
And you're dicing like the clappers,
And top cover's disappearing round the bend,
And a voice pipes as you're weaving:
'Bullfrog blue, are you receiving?'
It's the end, I tell you, fellows, it's the end.

When you're shot to rags and tatters,
And your clapped out engine clatters,
And you're heading, willy nilly, for the deck;
And the Spits, your plight discerning,
Cry: 'We're coming, boy, keep turning,'
It's the end, I tell you, fellows, it's the end.

Flight Lieutenant D.H. Clarke, 250 Squadron
(Kittyhawks), Desert Air Force, 1942

Through South African Eyes

Professor Vivian Voss, *The Story of No. 1 Squadron, South African Air Force* (Mercantile Atlas, Cape Town, 1952)

Professor Vivian Voss, the respected South African historian, was a pilot in the Royal Flying Corps in World War I. He served in 48 Squadron with Keith Park. In World War II he became the Intelligence Officer of No. 1 Squadron, South African Air Force, as aggressively successful a fighter unit as any in the Desert Air Force. When, in 1952, he published the squadron's story he had two advantages to lean on – he knew about pilots and flying, and he could write.

PREFACE

Pilots are, in some respects, different from other men. They speak their own language. They spend much of their time far from the earth, moving at incredible speeds, near the gold bar of heaven. There are vistas opened to them [which are] forever shut [to] earthbound mortals. Their friendship is a precious thing. It is not easily acquired by casual visitors. Once upon a time a psychoanalyst came to the squadron. We endured him politely for two weeks. Before he left he remarked despondently to the Padre: 'It's the hardest thing . . . to be accepted into the company of fighter pilots.'

And so the telling of the tale cannot be done by the polished artists who have never had the greatest privilege in the world of living and breathing with the squadron. It must, then, perforce, be told plainly and without artistry, but it is hoped that the deeds of the squadron will shine through and illumine the dull vestment of the tale. . . .

Chosen by Group Captain W.G.G. Duncan Smith

THE SANDS OF THE DESERT

Frequent sandstorms made conditions unpleasant. A bad storm would render a landing ground totally unserviceable while it lasted. Command would keep squadrons informed of conditions at the various LGs. Signals would be received, for example: 'LG 110: sand rising, visibility 200 yards.' This might be followed by another saying that visibility had decreased to 50 yards, or perhaps zero to 5 yards. Legend had it that the Meteorological Section once, in disgust, sent a final signal: 'Birds walking.'

The 'khamseen' sometimes would blow for days on end and prevent all flying. Tent poles would snap, tents collapse and sand would penetrate everything. It was . . . possible to become completely lost . . . trying to walk from a sleeping tent to the mess tent, a matter of 20 to 30 yards. Sandbags placed along the bottom of the side-flaps helped to keep the dust out, but a haze hung inside the tents, which were dark and gloomy even at midday.

Frans le Roux, gazing despondently one day through a chink in the Mess tent at the gale sweeping tons of sand northwards out to sea, remarked sourly: 'And tomorrow, I suppose, the wind will veer and blow it all back again.' (This remark, of course, was duly entered in the squadron line book.)

Late in the afternoon of 2 July 1942, 8 kilometres southwest of Alamein, 1 Squadron fought one of the really decisive air battles of the Desert War. 'Pops' Voss was able to catch the magnitude of the victory.

THE STUKA PARTY

No. 244 Wing Operations, who had their own sources of information, 'phoned [to say] that a Stuka (JU 87) attack was expected on South African troops [positioned] in the bulge of the Alamein line.... No. 1 Squadron, with its Hurricanes, was selected to defend them. The full squadron was to ... stand by ... and wait for the scramble signal.... It was, at all costs, to prevent the Stukas from carrying out their attack. These were critical days. The fate of the Middle East was in the balance and wider disasters threatened.... The troops had been harried and driven back relentlessly, and they were weary. Retreat is never good for morale....

The squadron appreciated that it, of all the squadrons, should have been selected to defend its brothers on the ground. Its old friends, No. 274 Squadron, Royal Air Force, were to provide top cover. No. 1 Squadron was therefore to ignore entirely the fighter cover which the Stukas would have with them and concentrate on destroying the dive-bombers.

The weather was fine and the visibility good. At 1840 hours the order to scramble came. Starter-buttons were pressed and the engines groaned as the props turned over.... Flames ... curled out of the exhausts, to be replaced by heavy bursts of white vapour as the engines, one after the other, roared into life.... The sky was filled with the clouds of desert dust as eleven aircraft swung onto the runway and took off....

Suddenly, the R/T silence was broken with the 'Tally-ho' from Major le Mesurier. Fifteen Stukas, at 6500 feet, were wheeling from east to north and flying in three vics of five, stepped up in line astern. High above, eighteen specks showed their fighter escort. No. 1 Squadron had arrived at the critical moment....

The Stukas were peeling off to bomb as the squadron came in ... from the starboard beam. With throttles open, the Hurricanes roared down on their prey. Major 'Lemmie' was the first to draw blood. He delivered a short quarter-astern attack.... The Stuka blew up completely....

Individual attacks were now taking place all over the sky.... Bombs were jettisoned haphazardly or blew up [with the] aircraft.... Bobby Pryde saw three Stukas burst into flames in the air, three more burning on the ground, and another three crash into the deck. When the Junkers broke formation and scattered, Tony Biden attacked the topmost Stuka closing to within 50 yards.... Flame enveloped the aircraft, which rolled on to its back.... As he turned towards the coast, a 109 flew across his line of flight. Allowing full deflection, Tony pressed the gun-button.... The 109 dived straight into the sea....

Captain Peter Metelerkamp, leader of Yellow Section, dived on the tail of a Stuka, firing a short burst.... The bomber half-rolled and exploded.... He got in a burst at 50 yards range and this [aircraft] too hit the ground.... Peter attacked a third, saw pieces fly off its tail but was then engaged by an Me 109....

Jerks Maclean got in a long quarter-astern burst at a bomber.... Smoke poured from it.... He saw it hit the deck and burn....

Little Harry Gaynor who had never before fired his guns in anger was thrilled to see Major Lemmie's bomber disintegrate.... As the Hurricanes went down on the Stukas, Harry selected his own victim, and was even more thrilled to see it break into a ball of flame....

Ray Connell, Blue Leader, dived on another attacker and got strikes along its mainplane; the port wingtip broke away....

Les Marshall made a full beam attack on a Stuka and blew its hood off.... He saw it burst into flames, turn on its back and crash to the ground.

Moon Collingwood, flying No. 3 in Blue Section, put up the biggest individual score, destroying three Stukas and damaging an Me 109....

As the Hurricanes began to [land] ... news flashed round that the squadron had been dicing.... Groups swarmed round each aircraft and, for a while, there was pandemonium. Excited pilots were describing attacks and, in the way of all pilots, held up their hands, at different angles, to illustrate ... [shooting] positions....

Harry Gaynor was so overcome that he could scarcely speak.... Rapidly, stories were coordinated, and the final (total) emerged: 13 Ju 87s destroyed, 3 damaged, 1 Me 109F destroyed and another damaged.... Wing Intelligence ... were ... astonished and delighted.... Two minutes later the telephone rang.... Wing announced that the army had counted 9 Stukas going down in flames, before the Junkers and Hurricanes had vanished from their sight....

All the pilots paid tribute to the fine protection afforded by 274 Squadron and to the brilliant leadership of Major le Mesurier. He could easily have increased his personal score but, instead, after destroying the first Stuka, he had pulled up and directed the operation, watching his Hurricanes and warning the pilots of the approach of 109s.... Not long after the Hurricanes had landed, Ray Connell arrived back in a 15-cwt truck from Burg El Arab. He was covered ... in oil. A Stuka's shells had pierced his petrol and oil tanks....

Congratulations began to pour in.... A signal was received from Air Vice-Marshal Coningham, AOC, Western Desert: 'Personal: AOC to Major le Mesurier. Magnificent. Congratulations to you, the squadron and the Wing on the grand Stuka party. Your success has greatly heartened the Army.... Messages were [also] received from Field-Marshal Jan Smuts and General Sir Pierre van Ryneveld.... Air Chief Marshal Sir Arthur Tedder [C-in-C, Middle East], visited the squadron to convey his congratulations and good wishes. The other squadrons of the Wing came in to celebrate the victory.

Press officers arrived to get news for British and South African papers. Major 'Lemmie' was invited to Cairo to broadcast the story.... He could not spare the time so Moon Collingwood took his place.... During his broadcast, Moon shot the immortal line: 'We'd have shot down more only we ran out of Stukas.'

The victory was a fillip for morale.... The ground crews had worked heroically on the shagged-out Hurricanes and these positive results greatly heartened them.... Goosen, the Cape coloured mess waiter, was asked what he thought of the party. 'Honest, Sir,' he said, 'I work so hard and our Officers never shoot down nothin'. Now I'm *proud* of them.'

The ground crews worked far into the night to get their aircraft ready for the next day's operations. In the blackout ... the work ... [went on] ... with the aid of pocket torches.

It may be of interest to record a piece of information ... [received] ... from an enemy source ... fifteen months later. Moon Collingwood was then operating with a Royal Air Force Squadron in Italy. An Italian, serving (after

Italy's surrender) with the Allied forces, and who had previously been on Rommel's Air Staff in the Desert, told Moon that on a certain occasion 15 Stukas, with Me 109 escort, had gone out and not a single bomber had returned. A 109, too, was missing. When questioned ... the Italian ... produced a diary and ... [confirmed] ... that the incident had occurred on 3 July 1942....

The squadron may have done even better than it thought.

George Barclay

Edited by Humphrey Wynn, *Angels 22: A Self-Portrait of a Fighter Pilot* (Arrow Books, 1977). First published as *Fighter Pilot* (William Kimber, 1976)

Squadron Leader R.G.A. Barclay, a young officer of exceptional courage, charm and character, was killed near El Alamein on the evening of Friday, 17 July 1942, leading 238 Squadron against the enemy.

Having fought with 249 Squadron in the Battle of Britain, George Barclay had later been shot down over France. He evaded capture and, after some weeks made his way back to England. In April 1942, he was posted to the Middle East. Before he left, he went to see his devoted family at Cromer, in Norfolk, where his father, the Reverend G.A. Barclay, was vicar.

His father described their last reunion:

'Full of joy in his new command, though he had less than a week's notice he allowed himself forty-eight hours at home to say "goodbye". It was Easter Sunday evening. I was celebrating Holy Communion and had just got to the Prayer of Consecration. Charlie* had come home for the half day and was with his mother in the chairs just inside the chancel door and opposite me, when the door beside them gently opened and there was George in all his beauty. He hesitated a moment till he saw his mother and then went and knelt down by her. I was almost overcome by emotion; and again when I administered to Dorothy and her two sons, one in khaki and the other in light blue. It was a marvellous and a God-given farewell.'

Flying with 213 Squadron south of Alamein the same evening that Barclay died was another young officer, Flight Lieutenant Neil Cameron (Marshal of the Royal Air Force Lord Cameron), who had travelled out to Egypt by sea with him. He has added a reflection on their time together on the voyage and, afterwards, in the Desert:

'Barclay ... was balanced and calm when others were getting fed-up and bolshie. At the time, I thought it was just his natural leadership qualities coming out in his considerable interest in the welfare of the troops; but I came to realize that he had a great spiritual strength to support him, which others lacked, when things got difficult. He didn't carry it on his shoulder, but later I heard of his Christian faith....

*The Barclays' eldest son, later killed at Kohima, Assam, while serving with the Norfolk Regiment.

One now knows that when George returned home to his father's vicarage after having made his epic escape from France, his family knelt together to give thanks to God – and in doing so, George didn't forget to remind his family to pray for the Germans as well. . . .

I have often thought about Barclay and wondered what might have become of him if things had been different. At Stowe, his old school, his favourite hymn (I discovered later), set to the glorious melody of *Finlandia* by Sibelius, contained these lines:

"Be still my soul the Lord is on thy side
To guide the future as he has the past."

George Barclay's future ended at El Alamein . . . leaving many of us who survived with the memory and example of a great leader.'

'Graves: El Alamein'

Reprinted by permission of David Higham Ltd from *Dispersal Point*, John Pudney (Bodley Head)

Live and let live.
No matter how it ended,
These lose and, under the sky,
Lie friended.

For foes forgive,
No matter how they hated,
By life so sold and by
Death mated.

John Pudney

Daylight Attack

Donald M. Judd, Winchester, Hampshire, 1982

For the Fleet Air Arm, the Desert provided a testing and varied battleground. Some of its roles were far removed from those which the crews had been trained to expect. In the tradition of the Royal Navy they made light of their experiences, but often they were bizarre and hazardous.

Donald Judd, one of the exponents of the Desert art, survived to practise as a solicitor in the City of London, but there was a moment when he wouldn't have given much for his future in the Law.

'I shall not forget the morning of 10 March 1942, in a hurry. It was on this day that I found myself 100 feet over the Western Desert, in an Albacore, a biplane with a top speed of 100 m.p.h., head-on, at 100 yards' range, to a Messerschmitt 109. It's surprising that I'm still here to tell the tale.

I was in 826 Naval Air Squadron at the time. We had been in the Desert for nearly a year flying from our base at Ma'aten Bagush,a levelled piece of sand, just east of Mersa Matruh. We had chased the Germans up to Benghazi only to be chased back again by Rommel. Now the commanding general of the Afrika Korps was occupying a line, Gazala to Bir Hacheim, poised to attack Tobruk and then, as it turned out, to drive on to El Alamein.

Our forward base was at Bir Amud, an even less level piece of sand on the eastern perimeter of Tobruk. Bir Amud was too close to the enemy to maintain the whole squadron there, so we kept a succession of three aircraft on this landing ground ready to carry out night dive-bombing attacks on enemy airfields and panzer concentrations.

I had been detailed to lead a relief of three aircraft for a tour of duty at Bir Amud. We were each to carry up stores – six 250-lb bombs and flares strung under the wings and the rest of air stores packed into the observer's cockpit. The only briefing I got, ironically as it turned out, was to bypass the Sidi Barrani/Sollum area as German fighters were known to patrol there. We decided to fly at 100 feet in extended line abreast.

It was one of those perfect desert days – a cloudless sky, hot and clear and the desert below stretching out in all its strange beauty. An RAF Wellington was a mile or so away to our left when I spied an aircraft flying across and behind us in the left quarter.

I called to my observer, Lieutenant Macintosh, a South African, down the voice pipe: "Mac, there's an aircraft on the port quarter!" "Yes," he said calmly, "one of ours, I think."

I felt relieved, although he had no more knowledge of the identity of the aeroplane than I had.

"But look," I shouted, "the Wellington is being attacked." Seconds later the bomber crashed and burst into flames.

"That's a Messerschmitt, Mac," I said, "and look, there are two of them."

Everything stopped still and I froze at the sight of the fighters. We were sitting targets for the 109s – a real piece of cake for them.

Then the fun started. Dicky Nathan, on my left, crash-landed and was being shot up. Now it was our turn. The only chance of survival in a slow aircraft being attacked by a fast fighter was to keep turning inside him.

"Mac," I said, "tell me when they come into range – each one in turn." The next moment Mac in his steady voice and without panic started his commentary: "They are turning into us from the port quarter now .... They're straightening up .... Wait for it .... They're coming in .... TURN NOW."

Hard port, rudder stick over to the left and hard back into my stomach. What a sluggish old crate this is, I thought, bombs and all. The Mes now appeared in front of me and had apparently missed. I felt we weren't beaten – yet.

Mac started speaking again. "Starboard quarter .... Attack coming."

I reacted in the same way, turning into the attackers. This time they scored hits in the rear cockpit and through the aircraft to the instrument panel. No time to freeze, as this was followed by a third, port quarter attack .... More hits, but, amazingly, the propeller kept going round and the controls still answered to the touch.

Now came a lull. It seemed a long time, but was probably only seconds. I was looking round the cockpit at the damage. Then something made me look up and there it was coming head-on and closing at maximum speed. All I could do was to jam the stick forward and hope. The bullets went over the top and hit the tailplane.

I was beginning to feel our luck couldn't last – nor, I thought, would my nerves. But then, with amazement and relief, I saw the 109s flying away, a pair in line abreast, making towards their base.... Running short of fuel, I suspected, and very disappointed.

It had not, however, been all one-way shooting. Our rear gunner, Leading Airman Harper, had been firing away from the rear turret as the enemy attacked. He believed he had hit them.

We dropped the bombs in the sea for safety and crash-landed at Sidi Barrani. We suffered no injuries but the Albacore was peppered with bullets. As we climbed out, the relief, the thankfulness – and the elation – were indescribable.

The old crate wasn't so bad after all.'

Naval Pathfinder

David R. Foster, Mission Hills, Rancho Mirage, California, 1982

David Foster, like others among the Navy's flyers in Egypt and Libya, had rather more than his share of the action. As the war progressed, he was to become one of the Fleet Air Arm's stars whose experience stretched from Europe, across the Mediterranean to the Middle East and then, in the final months of the Far Eastern conflict, to the Pacific. In the peace that followed, Foster rose to the presidency of the Colgate organization, worldwide; but life's variety, and the passage of time, have done little to dull the recollection of those taxing desert days.

'Initially, to me, the Desert meant sandfly fever, malaria and paratyphoid – the last two being contracted together, both at the same time!

Years later, it must have meant something quite different to two interns in New York hospital who were logging my medical history. Somewhat surprised at the record, one turned to the other: 'I didn't know we had those diseases in Arizona.'

When pronounced fit, I was appointed in April 1942, to No. 821, a new Albacore squadron. This had been sent to Egypt to reinforce 826 Squadron, 815 having been given an anti-submarine role in the eastern Mediterranean. So now, with two full Albacore squadrons in the Desert, it became their job to act as night "pathfinders" for the Royal Air Force's Wellington bombers. With naval observers to plot a course, as if it were over water, targets were pinpointed and marked. As the last flares were gliding down, and the bombers were on their way back to rearm, the Albacores would dive down on their own and bomb the tanks or the troops concentrated below.

As the Germans advanced, so the Allied armies fell back. We would take off at dusk from one forward landing ground, illuminate and bomb enemy tanks and transports, and then land back on another strip further east, nearer to Alexandria. Two bombing runs a night were common practice.

Sometimes the diet was varied and, instead of marking targets for the Wellingtons, the Albacores were diverted on to illuminating targets at Mersa

Matruh for sea bombardment by heavy cruisers and destroyers. The port of Mersa was being actively used to victual the Afrika Korps for its impending attack.

On 9 July, as my aircraft was one of the few in 821 Squadron equipped with a long-range fuel tank, we were ordered to join 826 Squadron at Wadi Matruh, an airstrip west of Cairo, to take part in "Operation Chocolate", an unusual name for a most unusual operation.

Fifteen Albacores and six Bombay troop-carrying aircraft took off late in the afternoon. The flight was made at 50 feet through the Qattara Depression, below sea level, to avoid detection by enemy radar. We landed at LG 113, a deserted airstrip, *some 200 miles behind the German lines* [editor's italics]. The Bombay aircraft carried aviation spirit for refuelling the Albacores and supplies for army commandos for defending the landing party should it come under enemy attack.

At 0100 hours, after being refuelled, the Albacores took off to attack their designated target – an important enemy convoy of ships approaching Benghazi with vital supplies for Rommel's forces. Having located the ships and dropped our bombs, the Albacores turned east to make the long flight back to Dekheila. Meanwhile, the Bombays, having seen our aircraft safely off on the mission and re-embarked the commandos, took off for Cairo after another successful clandestine operation.

Rommel made his expected move on 25 August. As General Montgomery had anticipated, it was in the south that the Commander of the Afrika Korps aimed his armoured thrust. Bad weather helped him, and with the sandy "khamseen" making it difficult to see your hand in front of your face, bombing missions by day were restricted to the minimum.

Contact with Rommel's advance guard was temporarily lost, but Intelligence was anxious to know when and where the "Desert Fox" would wheel left and make his attempt to encircle and trap the Allied forces. Thus it was that when, on the night of 1 September, my observer, "Scruffy" Cooper and I went into the Operations Room for pre-flight briefing, we could see at once that something was afoot. Six senior army officers, with their Fleet Air Arm counterparts, were peering intently at the Desert map. RAF Operations officers were pointing out special features.

Our job on the mission we were allotted was to find the head of the enemy's column and illuminate it so that the Wellingtons, which had been idle for several nights, could pound away. After finishing the prescribed leg and dropping our first flares, we could see nothing below us – nothing except sand and scrub! Another Albacore, to the west of us, was having no better luck.

Disappointed, "Scruffy" and I elected to take a wide sweep to the south and east and then work back towards the original drop area. After 10 minutes' flying on this new heading we dropped another flare. As it drifted down, illuminating the desert below, a dramatic sight struck us. Moving in line abreast were four columns of tanks, throwing up sand-trails behind them. They were followed by column upon column of armoured cars, troop carriers and thin-skinned supply vehicles. We fired off a number of red Verey cartridges to tell the Wellington crews that we had illuminated a particularly important target. The bombers were soon vectored on to our position.

We headed for home as soon as we had exhausted our flares and bombs, satisfied that none of the attacking aircraft could now miss the target area. The Royal Air Force, with their naval illuminators, had already left many trucks and tanks ablaze.

At our debriefing, we were given more than the usual perfunctory attention. This time the red-tabbed army officers were more relaxed than they had been a few hours before. After bombing up again, we flew a second successful sortie that night against the German armour.

On 4 September, Rommel turned back. He had been unable to breach the tough New Zealand Division's defences. His planned encirclement of the 8th Army, and the subsequent dash for Cairo and the Delta beyond had failed . . . .

The 8th Army launched its attack at Alamein on 23 October. Flying at night over the battlefield one hardly needed flares, so great was the illumination that the artillery on both sides provided. In the holocaust, we led the bombers to the lush supply columns. Petrol and ammunition dumps were left blazing.

By 4 November, after being defeated in the great Alamein battle, the German rear guard was racing down the Desert road, heading west. Our job now was to keep the head of the retreating column illuminated through the night with the Albacores operating in relays, one following another. As the bombers were hammering the head of the column to destroy it or, otherwise, to slow it down, over the escarpment we could see, in the glow of our flares, another race taking place. A column of Allied tanks was running, neck and neck, parallel with the enemy, intent on cutting him off. It was an exhilarating spectacle after all the reverses of earlier months.

With victory at Alamein, the job of the Albacores in the Western Desert was done. On 30 November, 821 Squadron flew into the Naval Air Station at Halfar in Malta. There it took up again the duties it was primarily trained to do – striking and torpedoeing enemy shipping sailing between southern Europe and that part of North Africa which the enemy was still struggling desperately to hold.'

Clive Robertson Caldwell – Exceptional Australian

INTRODUCTION

Christopher Shores and Hans Ring, *Fighters Over the Desert*, (Neville Spearman, 1969)

'Caldwell, the most successful of all the Allied Desert pilots . . .'

Pilots of the Commonwealth Air Forces claiming . . . 9 . . . or more victories in the Western Desert June 1940 – December 1942

Name	Squadrons	*Score in Desert*	*Total in War*
C.R. Caldwell	250, 112	$20\frac{1}{2}$	$28\frac{1}{2}$
A.E. Marshall	73,250	16	$19\frac{1}{2}$
J.L. Waddy	250, 260, 4 SAAF, 92	$15\frac{1}{2}$	$15\frac{1}{2}$
E.M. Mason	80, 274	$15\frac{1}{2}$	$17\frac{1}{2}$
L.C. Wade	33	15	25
J. Dodds	274	14	14
B. Drake	112	14	$24\frac{1}{2}$
P.G. Wykeham-Barnes	80, 274, 73	13	15
A.W. Barr	3 RAAF	$12\frac{1}{2}$	$12\frac{1}{2}$
V.C. Woodward	33	$12\frac{1}{2}$	$21\frac{5}{6}$
J.H. Lapsley	80, 274	11	11
A.C. Bosman	4 SAAF, 2 SAAF	$10\frac{1}{2}$	$10\frac{1}{2}$
R.H.M. Gibbes	3 RAAF	$10\frac{1}{4}$	$10\frac{1}{4}$
R.H. Talbot	274, 1 SAAF	10	10
D.W. Golding	4 SAAF	$9\frac{1}{9}$	$9\frac{1}{9}$
P.H. Dunn	80, 274	9	9
E.L. Joyce	73	9	10
J.H. Wedgewood	92	9	13

Colin Burgess, *Transit* (Qantas Airways Magazine), Kareela, New South Wales, 1983

Clive Robertson Caldwell was born in Sydney on 28 July 1910, the son of J.R. Caldwell, a banker. Fairly undistinguished ... academically ... at school (he *did* win the English prize in his final year at Sydney Grammar) ... he found, instead, the competition ... of sport ... a formidable, but enjoyable, challenge .... He was stroke of the first four in the 'Head of the River' event ... became State junior javelin champion, and represented New South Wales at the Australasian amateur track and field championships of 1930 and 1932. In ... 1931, he [broke] the NSW 440 yards hurdles record ....

In 1938 Caldwell joined the Royal Aero Club of NSW.... But when war was declared in 1939 he found himself with a problem – he was three years over the maximum age for RAAF fighter training. And so he did the logical thing ... he altered his birth certificate to make himself twenty-six .... Thus armed, he walked into the Recruitment Office.

'It was darned near too late for me,' he recalled recently. 'I was three years too old for single-seater training ... so I had to have my birth certificate expertly altered. That got me in. Strangely enough, the corporal at the centre in Erskine Street was ... a keen follower of athletics.... He looked at my certificate, and then ... at me and said "I know you – you're a lot older than it says here!"

Very quickly I replied: 'No, you're getting me confused with my elder brother, B.C. Caldwell; (in fact, we weren't even related!). I'm C.R. Caldwell, and Cliff is three years older than me. We both went to [Sydney] Grammar, and I think you've got the two of us mixed up!'

He still thought there was something fishy going on, and said so. But then I challenged him. It was not, I said, up to him to cross-examine me. I had complied with the request to produce a copy of my birth certificate, and that's what I had done ....

I was accepted.'

'KILLER'

Clive R. Caldwell, Sydney, New South Wales, 1982

'The nom de guerre they stuck me with – "Killer" – derived mainly from what I preached and practised concerning the very worthwhile results to be gained from properly conducted strafing which the desert conditions so favoured. Preferably just two or even a single aircraft sent deep into enemy territory and then home. Sweeps aimed solely at engaging other fighters were really just fighting for fighting's sake; they did not touch the cause itself.

My good friend, Bob Gibbes, with whom I flew many operations both in the Desert and, later, in the Pacific, always inclined to the view that I was a "lucky" shot in the air. As I so often told him and others, "the more I practise at the shadows the luckier I get". *I'm convinced that the most useful contribution I was able to make to our cause was in introducing this simple, do-it-yourself, self-correcting method of learning to shoot – the practice of "shadow shooting"'.* [Editor's italics].

TILTING AT SHADOWS

The outstanding shot in the air, like the exceptional driven game shot (there was an affinity between the two), usually started with an innate, natural advantage. Coordination of the faculties counted for much. But the great shots of World War II – Broadhurst, Stanford-Tuck and Johnson of Britain, Pattle and Malan of South Africa, Beurling of Canada, Marseille of Germany, Gray of New Zealand, Bong of the United States, Caldwell of Australia, among the alumni, all owed something to thought application and repetition. And none more so than Caldwell himself with his dedication to the concept of his invention – 'shadow shooting'.

Late in 1941, Arthur Tedder, then the Air Officer Commanding-in-Chief, Middle East, ordered a directive to be issued to all thirteen fighter squadrons of the Desert Air Force exhorting them to practise what he called the 'Caldwell method of shadow shooting.'

The holder of the copyright has preserved a picture of the process and the circumstances of its derivation:

'This was of great value to me as it was to all who used it properly. Without question, it resulted in the destruction of very many more enemy aircraft in the Middle East than otherwise would have been the case ....

The time when the concept dawned, I was flying rear cover to some aircraft returning eastward at zero feet from a first-light operation.

I was pondering how to improve my deflection shooting when the significance of the clear-cut aircraft shadows I could see racing across the desert floor in the bright, early morning sunshine struck me like a revelation.

If I could hit the racing shadow, then, by the same procedure, I could hit the object that was casting it. If I didn't hit the shadow then, surely, I should be able to observe, by the pattern on the desert surface, my error and take corrective measures.

Making a 30-degree quarter/stern attack, my short burst kicked up earth over the line of flight and behind the shadow. A second attack from the opposite side was an improvement but also missed.

Here, then, was the plain answer to my problem. A simple, do-it-yourself, self-correcting, visual method by which I could instruct myself in fixed gunnery, deflection shooting at aircraft in flight.

Using this method I felt I could become proficient in:

1. Line of flight
2. Range
3. Deflection
4. The curve of pursuit required to hold deflection throughout a 3-second burst.

This way I could build a repertoire of shots most likely to strike home.'

POSTSCRIPT

On the important subject of armament, one comforting point about flying US built fighters like Kittys, Mustangs and others (which we certainly didn't enjoy in Spitfires) was that no enemy aircraft was better armed. The lethal density of six ·50-cal machine guns concentrated into a 4-foot circle at 300 yards was truly impressive. I missed these beauties sorely (but, naturally, could hardly say so to my colleagues) when, subsequently, I flew with the various Spitfire Wings in the UK and Pacific.'

Axis Verdict

With Montgomery's victory at Alamein and the pursuit of Rommel westwards through Cyrenaica into Tripolitania, 1942 closed with the Allies well established along the North African shores of the eastern Mediterranean. Much rigorous fighting remained before the enemy could finally be dispatched, but the great Desert battles were over and with operation 'Torch' (see page 170) successfully launched, the Allied air forces were now dominating the southern sky.

In their extensively researched work, *Fighters over the Western Desert* (Neville Spearman, 1969), the air historians, Christopher Shores and Hans Ring, have collected the retrospective impressions of leading Axis participants in the Desert war. Further, their scoreboard of Luftwaffe claims for 1941–42, when set against the Commonwealth air forces counterpart for 1940–42 (see page 163), offers scope for argument and thought.

Pilots of the Luftwaffe claiming ... 16 ... or more victories in the Western Desert, April 1941–December 1942:

Name	*Unit*	*Score in Desert*	*Total in War*
Hans-Joachim Marseille	I/JG 27	**151**	158
Werner Schroer	I and III/JG 27	**61**	114
Hans-Arnold Stahlschmidt	II/JG 27	**59**	59
Gustav Rödel	II/JG 27	**52**	98
Gerhard Homuth	I/JG 27	**46**	63
Otto Schulz	II/JG 27	**42**	51
Gunther Steinhausen	I/JG 27	**40**	40
Friedrich Körner	I/JG 27	**36**	36
Karl-Heinz Bendert	II/JG 27	**36**	54
Rudolf Sinner	I and II/JG 27	**32**	39
Karl-Wolfgang Redlich	I/JG 27	**26**	43
Ferdinand Vögl	II/JG 27	**25**	33
Ludwig Franzisket	I/JG 27	**24**	43
Karl v. Lieres u. Wilkau	I/JG 27	**24**	31
Horst Reuter	II/JG	**20**	21
Erwin Sawallisch	II/JG 27	**17**	38
Franz Stiegler	II/JG 27	**17**	28
Jurgen Harder	III/JG 53	**16**	64

NEUMANN

Eduard Neumann, Kommodore of JG 27 in Africa:

'My fighter unit, I/JG 27, was the first of the German fighters to move to North Africa in April 1941. This Gruppe was activated during the first weeks of the war and had participated in the Battle of France, and later in the Battle of Britain where it had suffered considerably but had gained much experience ....

'During the course of 1942 the Geschwader (two more Gruppen having arrived in Africa in the meantime) was equipped with the Bf 109F and later with single specimens of the Bf 109G. These types had a lot of advantages ... but the ascendancy of performance over enemy types shrunk more and more. The British received a lot of aircraft of American origin in 1942, and between June and October of that year their operations were very reserved for good and right reasons. During these months of regeneration for the Western Desert Air Force, the German fighters were worn out escorting the obsolete Stukas at a time when the Ju 87 should have been withdrawn from service. The Ju 87 was too slow and therefore very difficult to cover, and these missions were very costly to JG 27. The initiative had passed over to the British in mid-1942.

'The North African theatre was considered by German pilots as not so hard as the Channel front, but much more difficult than the Russian front. On one hand this was the consequence of the excellent fighting spirit of the British pilots and their good aircraft, and on the other hand to the specific conditions of the Desert. The food was defective because the supply over the sea did not function due to the failure of the transports. The climate injured the health of all pilots with a desert time of more than six months – these were the majority. Life on the ground was rendered more difficult by the activity of the extremely effective English commando troops. We could expect on any night that submarines would land sabotage troops, and during periods of bright moonlight, bombs fell. Altogether these efforts were of considerable effect....

'We really began to feel the British bombing attacks during the second half of 1942. It was just the time when the German fighters were already weakened by losses and sickness, and they were also not skilled at fighting bomber formations. In my opinion this explains why the results against the bombers were apparently so small....

'... We were on friendly and often cordial terms with our Italian comrades so far as was allowed by the Italian higher staff. The Italians were always willing to cooperate closely with us and when we flew together during Stuka escort missions for example, the Italian pilots provided valuable support. That the Italian aircraft were never good enough ... and that the ... political authorities in Italy seemed not to approve of too good relations between the German and Italian troops at the front, is another thing.

'It may be a little difficult for most people to understand today that the British flyers always enjoyed our respect.... This is more conceivable if one knows that in all German pilots' messes in peace time the old veterans of World War I always spoke of the British pilots, of air combat with them, and of the British fairness, in the most positive way. Many German pilots could speak English better than any other foreign language and were thus quickly on good terms with captured pilots. We endeavoured always to be hospitable in order to make their passage into captivity a little less bitter. We were ... impressed to see how the captured pilots tried to conceal their regret and feign indifference and frivolity. For example, one time a very young Australian Flying Officer was rescued by the chief of our Desert Rescue Staffel immediately after landing in mid-desert. When he got out of the Fi 156 he said to us with a bright smile: "What a good service!"

'Another time one of our older reserve officers asked a captured British pilot brusquely what the hell he wanted in the Desert. (The English had bombed the camp early in the morning.) The British pilot replied laconically: "Exactly the same as you!" ...

'... We had no hate against our enemies. We were sorry that we must fight against Englishmen, but I can only add that the air war in North Africa was extremely fair, as far as a war can be.'

FRANZISKET

Professor Doctor Ludwig Franzisket, Staffelkapitän in I/JG 27 during 1941–42:

'Our activity ... over North Africa had positive and negative aspects for us. Absolutely positive was the unique good comradeship amongst the pilots and ground crews. The psychological reason for this was probably the circumstances of our being in the Desert where the conditions were equally hard for all ranks. For me, as an inexperienced young officer who had never had to exercise leadership (as I was at that time), it was not difficult to win the respect and obedience of corporals and men, who were mostly older than myself. The pilots dwelt in tents like the ground crews, ate the same food, but their fighting was more dangerous and could be seen by all. Nobody envied the officers and pilots, and there was no resentment about anything.

'Another positive factor was the knowledge that we were fighting a very fair enemy in the air. We knew this from news received of our missing comrades from captured British pilots. Therefore we treated all British pilots very nobly. We respected and esteemed each other, but the British pilots stayed only a short while with our Geschwader before we had to send them back to Fliegerkorps headquarters ....

'Negative points were the enormous technical difficulties and the lack of supply. An important but not decisive psychological factor was the news of the immense super-victories of the German pilots in Russia. But we all had the feeling that these victories were gained much more easily than ours. I never envied the German Jagdgeschwaders on the Channel front; I have a very unpleasant memory of my own tour of several months over England, and therefore I had the highest respect for the results of the pilots on the Western front.'

KÖRNER

Oberst Friedrich Körner, Oberleutnant in I/JG 27 in 1942 until he became a POW:

'I was posted to I/JG 27 on 4 July 1941 and joined the 2nd Staffel where I stayed until I was shot down on 4 July 1942 exactly one year after I had arrived in Africa ....

'Our top aces were very self-willed characters who had been able to develop their own methods of attack which could not be copied by the average pilots. The average pilots in Africa were very well trained and full of fighting spirits (an old slogan was "Der Geist machts!" – it is the spirit that counts). I believe our tactics of breaking formations into pairs at the start of a combat was the key to our success. It should be mentioned that all top aces had outstanding eyesight ....

'My Kommandeur Neumann (later Kommodore) was on good terms with the Italian fighter pilots. For we other pilots the relations with the Italian air force were more neutral. Some joint missions to Tobruk and an evening with beer in our bar at Martuba were not enough to produce more friendly terms. Over and above this, there was the language difficulty. I doubt if, officially, there existed a real interest in better terms. For us the Italian air force was a real disappointment .... Their equipment was very bad, their aircraft obsolete. Some outstanding Italian pilots with dash and fighting spirit (aerobatic flying in low-level flight) could not wipe out our impression ....

'I cannot understand why the Spitfire was introduced so late to the Desert. Perhaps there was a priority for the defence of the motherland .... The first news of the appearance of the Spitfires caused a certain anxiety and tension, especially among some of the older pilots who had experienced the Spitfires over the Channel in 1940. Fortunately some Spitfires were shot down in the first days of their appearance and the spell was broken ... After a few days we lost our fear of the Spitfire ....

'After twenty-five years I cannot conceal that my personal adventures in Africa seem very far away. The knowledge and remembrance of all details never had the importance to me that it did at the time, or as it would have had if the war had ended otherwise. The terrible events of the later war years have produced a certain aversion in me to all that happened in World War II . . . .'

SINNER

Rudolf Sinner, Technical Officer of II/JG 27 during 1942:

'I fought as a young officer and fighter pilot for over eighteen months in Africa . . . . This was a great adventure, the culmination of a boy's dreams. Now . . . in later years, I confess that this was an unforgettable part of my life, a time of good and bad experiences, a time I would not want to have missed . . . .

'The air war in Africa favoured the "experts". . . . The wing men, "Kaczmareks" and "No. 2s", had only a small chance to gain any victories.

'The law of natural selection brought the more talented pilots quickly to the position of Rotten or Schwarm leader, and from that moment they had the opportunity to build up their personal scores . . . . However, it cannot be said from this that the pilots who were not prominent were below average, poor flyers or lacking in fighting spirit. Such generalizations are not possible, and likewise one cannot compare the German and British pilots so far as their claims are concerned. The difference between the aircraft and tactics employed were too great . . . .

'If I compare the war theatres, I . . . [would] . . . say that the fight on the Channel front was somewhat harder . . . than [that] in the Desert. With the exception of the Murmansk region, the fighting in Russia in 1941–42 was much easier because of the technical inferiority of the Russians, allowing us to claim large numbers of aircraft. But this was compensated for in other areas by the chances of survival after a forced-landing in enemy territory, whereas in Russia the pilots were killed. I personally had no trouble with the African climate, and preferred it to the Russian winter . . . .'

BISLERI

Dr Franco Bordoni Bisleri, Tenente in 3° Stormo in 1941–42, and later top-scoring fighter pilot in the Regia Aeronautica:

'My first visit to Libya was in August 1940 . . . . I returned on 29 January 1941 when I served with 18° Gruppo (3° Stormo) until 14 August 1941. My final period in the area was from 20 July 1942 to 21 November 1942, again with 3° Stormo. I arrived as a Sottotenente, and was promoted in May 1941 to Tenente, the rank I held until just before the Armistice in 1943.

'In 1941 the CR 42 was a good aircraft, easy to fly, strong and manoeuvrable, but it lacked sufficient speed and armament. It was possible to fight Gladiators, Blenheims and Wellingtons in this aircraft, but against other aircraft it was outclassed. In 1942 3° Stormo received the Macchi 202, and the situation became very different, as with this new aircraft it was possible to successfully combat Hurricanes and P40s and it was easier to intercept bombers.

'The fighter pilots I flew against were generally very skilful, and the bomber crews were also good, their tactics being quite varied.... The night bombers bothered us a lot, particularly the Wellingtons. I tried repeatedly to intercept these in a CR 42 during moonlit nights, but without radar or ground control ... it was a most unrewarding job.... On the night of 3 August 1941 I nearly collided with a Wellington over Benghazi.

'Two things particularly stay in my mind, one, an attack at night on our tents near Abu Hagag camp, near Fuka, made by slow aircraft (Swordfish or Albacore) with anti-personnel bombs. 18° Gruppo had seven dead (six pilots) and a lot of wounded, five badly. The second thing was the number of Allied aircraft flying in the Desert from mid-September 1942; they were impossible to count, and I still do not know how my comrades and I managed to survive.

'Our allies, the German fighter pilots were very brave and had a wonderful aircraft* which they knew how to employ the right way, particularly making the best use of its good armament.'

In the Vanguard with 'Torch'

Group Captain P.H. Hugo, Victoria West, Cape Province, 1982

Piet Hugo, born in Pampoenpoort, Cape Province, was another in that small, but significantly successful group of South Africans who left their native land in the 1930s to come to Britain and join the Royal Air Force. With Pattle and Malan, 'Zulu' Morris, Rabagliati and others, Hugo found advancement and friends. A group captain, he left the Service in 1950 to farm in East Africa, only to find, with others, that hard work, investment and the good that employment brings, offer no safeguards against the hazards and disruption of political change.

Starting again on a farm in Cape Province, the spirit that spurred others in wartime wrestles afresh to reverse the turn of fortune's wheel.

'At dawn on 8 November 1942, the "opening day" of the Allies' North African landing, I took the first Spitfire wing, No. 332, commanded by Group Captain Charles Appleton, from Gibraltar into Maison Blanche, east of Algiers. Operation "Torch" had begun.

For two days we were busy defending the harbour of Algiers and covering the army moving rapidly east. On 11 November, a landing was made at the small port of Bougie but Allied shipping sustained heavy casualties as it was difficult for us to defend it from Maison Blanche. It was therefore decided to move one squadron forward from Algiers to Djidjelli, a small airfield east of Bougie. Once there, I was ordered, on the 12th, to escort a formation of American C 47s carrying paratroops to capture the harbour of Bone and then to land back at Djidjelli to continue the defence of Bougie.

With No. 154 Squadron (Squadron Leader "Gus" Carlson), I escorted the closely packed gaggle of C 47s flying at about 2000 feet on an easterly course directly for Bone; this took us over some very rugged mountains. Halfway there, while flying to the starboard of the transports, I suddenly saw a Dornier 217 diving steeply to attack the formation from the port side. The heavily armed, fast 217 could do great damage to the slow, unarmed C 47s. So calling Flight Lieutenant "Shag" Eckford to follow, I turned to cross over

*Me 109 (E,F and G).

the formation and to intercept. The Dornier immediately steepened its dive, intending to get under the transports and pull up into them. I was thus forced to dive through the C 47s to head off the enemy, while Eckford went round after him.

My passage through the formation clearly caused considerable consternation; there were some very close shaves, and a lot of highly unorthodox evasive action by the normally staid transports. But I got through and positioned myself between them and the Dornier, which at once turned away.

This gave Eckford an opportunity to attack and immediately afterwards I opened fire from starboard. There were strikes all along the fuselage and into the wing root; the starboard engine slowed, was feathered and then stopped. The Dornier rolled quickly over and dived down into a narrow, deep gorge leading up into the mountains. The pilot levelled off over the bed of the gorge and started "hedge-hopping" up it, twisting and turning to follow its course.

I called Eckford to stop attacking as I could see that having once got in, the Dornier couldn't get out again flying on a single engine; nor could it maintain height above the rising level of the gorge. The aircraft got slower and slower, there was a spurt of dust as the left wing-tip touched the surface – a pall of dust, debris and bits of wreckage were thrown up as the enemy ploughed to a stop.

I reduced speed, put down my flaps and entered the gorge higher up. As I flew slowly down it I could see no sign of life in the tangled wreckage.

The next day I moved No. 81 Squadron (Squadron Leader "Razz" Berry) to Les Salines, an airfield just east of Bone; there I went to discuss airfield defence with the commander of the US paratroops. I asked whether they had had any casualties the previous day.

"None," said the CO, 'but some of the boys got roughed up a bit when some goddamned British flyer took his pursuit ship right through our formation."

Silence, I remembered, can sometimes be a virtue.'

Island Climacteric – 1942

Winston Churchill, *World War Two, Volume IV: The Hinge of Fate* (Cassell, London; Houghton Mifflin, New York)

During March and April all the heat was turned on Malta, and remorseless air attacks by day and night wore the island down and pressed it to the last gasp.

The World's First Skyjack

This is one of the strangest stories to emerge from the great battle for Malta in the spring and summer of 1942. It is told in two parts, the first by Lieutenant-Colonel Ted Strever, a South African, from Klerksdorp, the aggressive captain of a torpedo-carrying Beaufort of No. 217 Squadron, based on the island. The second, the sequel to it, is related by a fair-haired, placid, but granite-hard New Zealander from Auckland, Flight Lieutenant Harry Coldbeck, an accomplished photo-reconnaissance pilot, who made his name during the peak of the struggle.

Strever, with others in the squadron, had been detailed, on 28 July to attack an important Italian supply ship of 12,000 tons, escorted by destroyers, sailing south down the Greek coast, bound for North Africa. He had with him in his crew an Englishman from Liverpool, Pilot Officer Bill Dunsmore, and two rugged New Zealand sergeants, Ray Brown, the radio operator, and John Wilkinson, the gunner.

At 1220 hours, the target was located and Strever pressed his attack, scoring a direct torpedo hit on the merchantman. But his aircraft, and the Beaufort immediately behind him, were both hit by gunfire from the destroyers. The second aircraft sank immediately, but Strever and his crew, after ditching in the sea, struggled from the wreckage into their dinghy. Later they were picked up by an Italian Cant seaplane and taken to a small air base on the Greek island of Levkas where they were given dry clothes and other comforts. Strever continues the story.

Lieutenant-Colonel Ted Strever, 'The World's First Skyjack', *Illustrated Story of World War II,* published by Reader's Digest

The hospitality stunned us....

The four of us were given a huge meal of steak, tomatoes and potatoes. Malta, by comparison, was besieged and there were desperate shortages. Food was rationed and the troops lived on a rough, combat diet of bully-beef.

Wilkinson joked: 'Why don't we come here more often?'

But the laughter was hollow. We were captives ... and this hospitality would not last for long.. ..

After breakfast on 29 July, we were loaded into another Cant seaplane. Our destination – Taranto and life behind barbed wire.

I was surprised to see that the Italian crew consisted of only two pilots, an engineer, a radio operator and a young armed corporal to guard us. I discussed it quietly with Brown and Wilkinson and decided that we should make an effort to take over the plane in flight ... but as I spoke to Dunsmore we were ordered to be quiet.

The plane headed to the west into the brilliant sky over the Ionian Sea. The four of us were in the rear of the plane with the radio operator. The engineer and the two pilots were up front. The armed corporal stood between them.

He was young, he was relaxed, and he was enjoying his first flight....

At 10.20 a.m., 2500 feet above the shimmering Ionian Sea and 30 minutes from Taranto, the world's first skyjack swung spontaneously into action.

Wilkinson, who was nearest to the Italian radio operator, tapped him on the shoulder and pointed out of the window behind him. The man looked around and up, and Wilkie let him have a haymaker right on the jaw.

As I saw the punch landing, I jumped up and landed on the radio operator. Wilkinson was the only one who could get to the guard and it was vital to get the radio man out of the way. Brown helped me to pull him towards the back of the plane. As we did so, Wilkie lunged at the guard, who was just turning in surprise from the view that had transfixed him for so long.

Wilkie was tough, thickset ... a really hard man. He actually pulled the holster and gun right off the guard's belt and threw it to me. I snatched it from the holster and pointed it straight at him.

The youngster reeled, clapped his hands to his belt and cried out: 'Mi pistola, mi pistola.' I grabbed him with my free hand and dragged him behind me to Dunsmore and Brown.

At that moment Wilkinson cried: 'Look out!'

The first pilot had turned around in his seat. Only his eyes were visible ... and the short muzzle of an automatic pistol.

For seconds that seemed like hours a deadly stalemate froze the nine men in the seaplane. And the second pilot, taking over the controls, turned the Cant's nose down towards the waves. Death stared at us all.

Our reckless plan was suddenly saved in a way which was totally unexpected.

The engineer was obviously slow at realizing what was going on, and was fortunately facing away from us. He looked up ... and found himself glancing at the pilot's gun. I don't know what went through his head, but he dropped to his knees and began backing towards us, not taking his gaze off the gun.

I looked at Wilkinson. As one man we planted our boots in the seat of his pants and heaved ... he shot forward and fell in a confusion of arms and legs on the first pilot. Wilkie and I pounced forward and overpowered them both. The pilot's gun fell to the floor. The plane was ours!

We tied up our five prisoners but still had plenty of problems. None of us had flown a Cant. We needed help.

The engineer was hauled forward and the gun thrust at him. I made it clear with explicit actions that he must switch over to the reserve fuel tanks ... or we would kill him.

The Italians all began shouting together: 'No gasoline to Africa. No gasoline to Africa.'

I told them: 'Africa – No! Malta. Malta.'

The Italians shouted with fear: 'Spitfires.'

Well, there was no doubt that would be bad enough when we got there – if we did. But how would we? There were no maps aboard. None of us had navigated from that far north before.

There was only one way, we must fly on ... until we reached the very south of Sicily, and then strike off for Malta.

It was early afternoon when we finally reached Cape Passero, the southern-most point of Sicily. I set the Cant at 220 degrees ... and headed towards Malta.

Three times in twenty-four hours we had dodged death, but we knew that the greatest danger to our amazing escape was still to come....

I peered at the island looming up ahead.... I had no time to congratulate myself on a brilliant piece of makeshift navigation before being shocked by a violent shout from Ray Brown: 'My God, here come, the Spits.'

In one, unique moment we knew both the fear of facing the Spitfires ... and the comfort of knowing that these men were really on our side.... I hurled the first pilot back into his seat and ordered him in hurried sign-language to throttle back and land in the sea. As he eased the engines, rapidly slowing the Cant, the first fusillade of fire from the leading Spit spurted into the sea just 30 feet in front of us....

Dunsmore had now whipped off his vest and was trailing it out of the top hatch, a hopeful sign of surrender.... The little plane suddenly stuttered. It dropped. Then it floated silently to a halt on a sloppy sea. It had run out of petrol. And we were just 2 miles off Malta....

A launch came out and we were towed to St Paul's Bay. What a reception we had! ... We learned soon just how unfortunate our Italian captives had been. Through an interpreter they explained that they had been given the task of taking our crew to Taranto only because they had been going home on leave. Now they were on the way to several years in a prisoner-of-war camp.

The four of us did our best to return the hospitality we had received in Levkas the day before. But it was the Italians who once again came to the rescue ... they produced some wine they were taking home for their leave and shared it with us.

Harry Coldbeck has added the epilogue.

H.G. Coldbeck, Remuera, Auckland, 1982

When they were asked, the crew of the Cant agreed to show the Royal Air Force how to handle the aircraft on the understanding that it would only be used for mercy/life-saving missions. This was readily accepted as Malta had no float plane available for air-sea rescue at this very difficult time. Moreover, with the imminent arrival of the August convoy from Gibraltar, or what was to remain of it, aerial activity was sure to intensify.

Once in service, however, the Cant, manned now by an RAF crew, was involved in a humiliating experience. A search was being made for an Allied crew in a dinghy, but there were difficulties with the aircraft's radio. It wasn't receiving R/T messages which added to the problems of the escorting fighters. At this moment, a German Dornier 24 appeared looking for one of its own crews. Spotting the dinghy and thinking the airmen were theirs it landed alongside and whisked the crew away – presumably to captivity.

About this time, I was handed a package in some secrecy together with a note from Air Headquarters, Malta. I was instructed to drop this parcel at Catania (in Sicily) the next time I was returning from a photographic mission further north. I was told that the package, which was addressed to the Regia Aeronautica, contained family letters, lottery tickets and other personal items belonging to the crew of the captured Cant seaplane.

Catania struck me as an odd choice for such an operation with the whole of Sicily and much of Italy to choose from. Catania, and the adjacent airfield at Gerbini, were strongly defended and accurate flak could always be expected below 25,000 feet.

To make the parcel more conspicuous on the ground, I got the ground crew at Luqa to cut up some serrated-edged fabric which, in the past, had been used for Gladiators, Swordfish or other fabric-covered aircraft. With this we made nice long streamers on each of which the address: 'Regia Aeronautica, Catania,' was chalked in red. These were secured firmly to the package.

On 10 August, I set off in Spitfire No. BP 915 on a mission to Messina and

Naples. I kept Catania in view as I passed flying northwards.

After taking the photographs I turned south for Catania with some apprehension. There was no sign of enemy activity. After descending to between 3000 and 4000 feet, I reduced speed as I approached the airfield. I opened the hood and, over the aerodrome, heaved the package over the side with streamers flowing; I felt very self-conscious.

Having got rid of it, I shut the hood, opened the throttle to maximum boost, put the nose down and scooted off out to sea as fast as the Spitfire could go, and so back to Malta. There was no visible reaction from the enemy throughout.

Three months later, on 10 November, I was shot down off Augusta. After taking pictures of Taranto and Messina, I had to drop down to 400 feet to get below cloud to photograph Augusta. Travelling across the harbour, and the three cruisers lying there, at some 400 knots through the flak, I was beginning to head out to sea when I lost control and my aircraft did a violent bunt.* My last conscious recollection was of seeing the windscreen crumpling up in front of me and thinking to myself: 'This is it.'

I came to falling freely through the air. I found the rip-cord at the last moment as I fell, and the parachute opened; I was in the sea before I could do much. It was now dark on the water and, after what seemed like an eternity, a small minesweeper came out with lights ablaze. Soon after, a sleek naval launch came alongside, with a Roman Catholic padre and sickbay attendant on board, and I was transferred. At the harbour wharf an ambulance waited with doors open and I was unshipped on a stretcher.

Eventually, I was laid before an assembly of Italian officers and asked the usual questions to which I gave the usual, formal replies. I was treated very well during my few days sojourn at the base before being sent north. It was during this time that I got the chance to ask an English-speaking guard whether he had heard of the incident of the Cant seaplane. He answered perfunctorily that he had.

Warming to it, I said I had heard that some of the personal effects of the Cant's crew had been returned by air and wondered, in fact, whether they had been received by their next-of-kin.

Laconically, the Italian again confirmed that they had.

Six Feet from the Brink

It is at least arguable whether Malta could have held out without President Roosevelt agreeing to Premier Churchill's request for the use of the United States Navy's massive aircraft carrier, *Wasp*, as a reinforcing medium for the island. With a flight deck of some 800 feet, she was capable of accommodating 48 Spitfires in addition to her own complement of aircraft. This represented a reinforcement potential quite outside the compass of any of the Royal Navy's carriers. There was no other practical way of sending fighter aircraft to the beleaguered fortress.

Wasp made two runs down the western Mediterranean in April and May 1942. The second, completed in company with the British carrier, *Eagle*

*An outside loop – a loop flown round the outside of a circle.

(capability: 16 Spitfires), enabled the great air battles with the Luftwaffe on 9 and 10 May to be fought and decisively won, thus arresting the slide to eventual starvation and surrender.

Acting as Landing Signal Officer in *Wasp* on each of the two operations was David McCampbell (Captain David McCampbell, USN), who was later to excel in the Pacific theatre as one of the US Navy's most gifted and decorated flyers. (Douglas Fairbanks Jr was also aboard *Wasp* in the role of Liaison Officer.)

When, on the second run, the Spitfires were flown off some 50 miles north of Algiers and around 600 miles from Malta, no aircraft of this type had ever before landed on the deck of a carrier without the aid of arrester gear. Pilot Officer Smith, a Canadian, was the first to do it. McCampbell tells how.

Captain David McCampbell, USN (Retd), Lake Worth, Florida, 1982

'Smith had lost his 90-gallon belly tank on take off. There was, therefore, no way that he was going to have enough fuel to reach Malta; nor could he make it back to Gibraltar. He had two choices – climbing up and baling out and being picked up by a destroyer, or going for a landing back aboard ship with no tailhook to check him.

Fortunately I had given all the Spitfire pilots a briefing before take off to acquaint them with the operations aboard ship. One of the things I told them was that during landing operations if anyone saw me jump into the net alongside my platform he would know the plane coming in to land was in trouble and it was the signal for the pilot to go round again and make a new approach. When Pilot Officer Smith decided to make his attempt at a landing without the tailhook, all our planes were in the air so we could give him the whole length of the flight deck.

On the first approach he was much too high and too fast and when I found I couldn't bring him down or slow him down enough for a landing, I simply jumped into the net.

He got the news real fast and went round for a second approach. As I got him to slow down and make his approach a little lower, I decided to give him the 'cut' signal. He landed safely with his wheels just 6 feet short of the forward part of the flight deck.

That night, in the Wardroom, we presented him with a pair of Navy wings.'

Conversation Piece

Lord David Douglas-Hamilton, 'The Douglas-Hamilton Papers', Malta, 1942

Squadron Leader Lord David Douglas-Hamilton, having flown 603 (City of Edinburgh) Squadron off Wasp *into Malta, thereafter led the unit with bravado and easy control and with a paternal regard for those who served him. He played his own performances right down with an aristocratic modesty seasoned with liberal dashes of laughter; yet it was he who made the openings that others might score the points.*

'... The Me 109s still came on ... and we were approaching each other head-on at great speed. I resolved not to give way before he did, and he evidently made the same resolution. We were going straight at each other, and as soon as I got my sights on him I opened fire, and kept firing. He opened fire a second afterwards.

It all happened in a flash, but when he seemed about fifty yards away I gave a violent "yank" on the stick and broke away to the right. As I did so, his port wing broke off in the middle, and he shot past under me. I turned and looked back; his aeroplane did about five flick rolls to the left and broke up. Then a parachute opened.

Next day I went to the hospital to see "my" prisoner. He was a small, muscular, blond-haired, blue-eyed young man of twenty-one, with a squat Germanic face. We conversed in German, as he knew no English and when I told him I had shot him down he surprised me somewhat by seizing my hand and congratulating me. His left arm had been broken by a cannon shell and was in plaster. He did not know what had caused his aeroplane to do flick rolls, and was under the impression that we had collided. He was rather disappointed that he had not hit my machine at all.... He was a Sudeten German who had Czech relatives, and said he was not a Nazi, but just became a fighter pilot for the fun of it! He told me that when he came down his parachute caught in the roof of a house and left him dangling just out of reach of a crowd of angry Maltese. He was rescued by some British soldiers.

The next time I shot at a 109 I was full of confidence, and fired a very long burst at him with my sights dead on. I was sure he would blow up and fall into pieces, but nothing seemed to happen, and he flew away apparently quite unharmed. Most disappointing!'

'Gracious' Message

Winston Churchill, *World War Two, Volume IV: The Hinge of Fate* (Cassell, London; Houghton Mifflin, New York)

It may be well here to complete the story of the *Wasp*. On 9 May she successfully delivered another all-important flight of Spitfires to struggling Malta. I made her a signal, 'Who said a wasp couldn't sting twice?' The *Wasp* thanked me for my 'gracious' message. Alas, poor *Wasp*! She left the dangerous Mediterranean for the Pacific, and on 15 September was sunk by Japanese torpedoes. Happily her gallant crew were saved. They had been a link in our chain of causation.

Hazards of Air-Sea Rescue

W.G. Jackson, Thursby, Carlisle, 1982

High Speed Launch 107
Royal Air Force, Mediterranean Command

Skipper — Flight Lieutenant Price

1st Coxswain	Sergeant 'Taff' Lewis
2nd Coxswain	Corporal 'Beeze' Beezer
Motorboat Crew	
Leading Aircraftman	'Joe' Maitland, Gunner
Leading Aircraftman	'Reg' Hargreaves
Leading Aircraftman	'Ginger' Strickland
Leading Aircraftman	'Bunty' Harris
Aircraftman 1st Class	'Bill' Jackson, Wireless Operator/Mechanic
Leading Aircraftman	Lance 'Yorky' Jefferson, Fitter Marine
Leading Aircraftman	'Geordie' Wilkinson, Fitter Marine

The raucous blare of the crash alarm klaxon came as no surprise to the ten-man crew of the Royal Air Force's High Speed Launch 107, moored alongside the jetty at the Marine Base at Kalafrana at the southern end of the island. We had been on top-line readiness since dawn. It was May 1942, and, in the crescendo of the fighting, Malta's air-sea-rescue service was engaged in exceptional activity....

Before the sound of the alarm had died away, the three 500 HP engines had burst into life, and all made ready, so that the moment the Skipper, Flight Lieutenant Price, set foot on board, 107 could put to sea....

Before his stocky figure could get into the wheelhouse, coxswain, Sergeant Taff Lewis had given the order to cast off and at once the three throttles were eased steadily forward as the sleek, 64-foot wooden craft headed out into the bay and the open sea beyond Delimara Point. The Skipper had already plotted the course that would take us to the last known position of the pilot in trouble.

In the tiny wireless cabin at the stern of the wheelhouse, I had established contact with Operations' Headquarters, deep underground in Valletta, and was settling down, earphones clamped over my ears, ready to deal with the traffic that would be coming my way.

While Reg, Bunty and Ginger were hurrying to stow away the mooring lines and check that all on deck was secure, Joe, the fourth deckhand, had moved aft of the engine room and was now manning his beloved Lewis guns. In truth, the crew doubted whether the guns were likely to be any more effective than the ones on HSL 129 when she was mercilessly attacked by six Me 109s and her skipper and one gunner killed outright.

Below, Yorky, one of the two engineers, settled into his wooden seat between the three bellowing engines to begin his watch in the engine room.

As Taff Lewis kept the bows heading past Delimara Point, Corporal Beezer had switched on the electric log which would record the mileage covered. This would become vital for the square search which would have to be made if the downed pilot was not within sight of the pinpoint which the Skipper had been given.

At 38 knots it was going to take almost another hour to reach the search area. Meanwhile, those not on watch could not relax. All eyes would be scanning sea and air for any sign of enemy activity. This time we did not have the usual escort of Spitfires. They had other work to do.

One of the problems faced by an HSL rescue crew was to reconcile the pilot's position, when he put out his Mayday call, with the place he actually landed in the sea. If he had been separated, other pilots would not have given a'fix'. And if he baled out his parachute could have been carried for miles on a strong wind before it settled down on the water....

With bodies beginning now to ache from the pounding 107 was getting as she bounced from wave top to wave top, we had just reached the search area when Ginger, at his favourite lookout position well forrard, suddenly pointed to something he had spotted in the sea off the port bow.

There, moving about on the waves was an object feared by all mariners – a drifting mine. Normally, when we came across these horrible things, we logged the position and then signalled it to HQ for others better equipped to deal with. This time, however, the gunner, Joe, had different ideas. With the Skipper's consent, the coxswain throttled back, keeping just enough speed to maintain direction and keep the bows in the sea.

Secure in his 'dustbin' turret, Joe lined up the two Lewis guns on the distant, deadly target. Water splashed up all round it as he fired successive bursts.

With speed cut right down, 107 was rolling and pitching. It made an unsteady gun platform, and after Joe had emptied a full pan of ammunition the mine was still drifting. As Joe fed in a new pan of ammunition and began to line up his guns there came another shout from Ginger. He had the reputation for having the sharpest eyes of any of the HSL crews. Instead of just shouting and pointing, he ran along the deck until he was alongside the Skipper, sitting up on his perch. The noise from the engines drowned his words from the rest of the crew.

Swinging round to where Ginger was pointing, the Skipper put his binoculars to his eyes, at the same time signalling to Joe to hold his fire.

It only took seconds for the rest of the crew to identify the tiny speck moving low in the waves. There, floating in his mae west life jacket was the pilot we were looking for.

But the elation we felt at the discovery was tempered by the proximity of the floating menace which had been the target for 107's guns. Using the throttle with special care, Taff steered the bows between mine and man while the deck crew rolled the crash ladder over the starboard side. Clambering down until we hung on only inches above the water, two of us were ready to grab the pilot's arms as 107 was eased forward. As we pulled him close enough to let him start scrambling up the rope ladder, the rest of the crew became only too well aware that the danger was by no means over.

With the pilot safely aboard and 107 acting like a magnet, the deadly flotsam began to drift closer, its lethal spikes now clearly visible. Reg hurried forrard towards the bows with a boathook. The suspense was unnerving as he carefully lowered the hook-end towards the black object drifting in the water.

For the moment the metal end of the boathook wavered horribly close to one of the spikes; but, taking an even firmer grip, Reg, with great composure, gently touched the metal casing of the mine and gradually applying his full weight on the long pole, pushed it away. Taff, who had been anxiously watching Reg through the forrard window, now eased the centre engine throttle forward to send 107 slowly gathering way ahead.

Everything else was forgotten as the whole crew watched Reg, boathook in hand, walk slowly back along the deck. The relief, as the mine floated away behind the square stern was intense.

The pilot was led down to the Wardroom to have his wet clothing stripped and to be warmed up by a tot of rum. Shaken by his awful experience, he was silent for some moments. Then he turned to Reg. 'Thank heavens,' he said, 'you were a rotten shot.'

Malta 'Made' Them

Adrian Warburton, an Englishman, and George Beurling, a Canadian, were two utterly opposite characters. Yet they had two features in common. Both were non-conformists in the Service sense. Each was 'made' by Malta. It is improbable that either would have sparkled elsewhere with the same brilliance. The chances would not have occurred. Indeed, Beurling failed in Western Europe where the operations didn't suit his style.

When Warburton disappeared without trace, flying across the Alps in 1944 to Italy from the UK, his reputation was barely known outside the Mediterranean where he was god. On the other hand, Beurling, by the war's end, was transatlantic news.

What were these two enigmas like? For the answers, go to those who served or flew with them.

WARBURTON
'... FEW, IF ANY, EQUALS'

Group Captain E.A. Whiteley, *The Royal Air Force Quarterly*, Spring 1978

Adrian Warburton, DSO and Bar, DFC and two Bars, and American DFC, has been variously named 'King of the Mediterranean' ... and the 'Lawrence of Malta'. When I last saw him in October 1943, he had completed over three hundred operational sorties – mainly strategic PR* trips as opposed to tactical operations. He had destroyed at least 10 enemy aircraft before he and his original two aircrew colleagues were rested in October 1941. By the time 'Warby' disappeared ... his operational record must have had few, if any equals....

... Shortly after our arrival in Malta, I decided to train two navigators (Warburton and Devine) as pilots. Both had flown Ansons – but Warby not recently. Warby's first solo almost ended in disaster. He eventually landed across wind with the top strands of the fence round his tail wheel – watched by an irate wing commander from HQ. I was firmly reprimanded.

A few days later I sent Warby off again. Unfortunately the Bob Martins† had one vice; in a cross wind they tended to swing into wind and then keep swinging. A pilot undergoing training in Malta had to be accompanied by a crew to provide warning (via W/T) of enemy aircraft. So on Warby's early

*Photo reconnaisance.
†Glenn Martin 167 – Martin Maryland.

flights Sgts Bastard and Moren lived through circular take offs and zigzag landings. I was tempted to give up but Bastard and Moren were always prepared – after high speed taxiing practice – to go off with Warby 'just once more'. Warby was greatly indebted to those two.

Paddy Moren, as wireless operator air gunner and Frank Bastard as navigator ... continued to fly with Warby.... Both survived the war. Moren (Later Lt. Moren, RN pilot) has contributed the following:

'I flew over a hundred operations and participated in shooting down 10 enemy aircraft with Warburton. Bader, Gibson, Tuck and Johnnie Johnson rightly became famous, but little mention has been made of Adrian – not even in the official Royal Air Force *History of the Second World War*. When the siege of Malta was at its height and things were at their worst, Warburton's exploits made him a living legend. His enthusiasm for any task and his ability to complete the impossible were unparalleled. His crew, whenever they were flying with him, were made to feel 10 feet tall.

'On one occasion we were asked to photograph the Tripoli/Benghazi road – or a section of it.... "Warby" was briefed to tackle it in a series of five different trips. It was typical that we did the whole job on one operational trip, notwithstanding the fact that we were chased by enemy fighters.... His persistence and skill were backed by his sheer unorthodoxy....

'He was the first photo-reconnaissance pilot to mix photographic duties with aggression. The 10 enemy aircraft shot down by himself and his crew, as well as 6 destroyed on the ground, is evidence of a team working as a unit.

'The photographic-reconnaissance pilot had to have the flying accuracy of a bomber pilot, the alertness and manoeuvrability of a fighter pilot plus technical skill and persistent purpose added to the endurance to bring his photogaphs back to base....

'Warby was totally unorthodox, a complete individualist with both courage and flair.... A man who knew no fear. He was ... a truly oriented perfectionist who was dedicated to the task that he had to perform.'

'Titch' Whiteley, the Australian, was Adrian Warburton's commanding officer in Malta, first of 431 Flight and then of 69 Squadron.

Roy Nash in the London *Star*, 18 March 1958
(Reproduced by permission of Associated Newspapers, Ltd)

Warby's cap was almost as much of a legend as Warby himself. It was incredibly ancient, unbelievably shabby.... Yet he could dress up when he wanted. No RAF officer looked smarter than Warby at the mess dances and parties with which Malta tried to forget the bombs. Sometimes he was accompanied by a pretty, fair-haired English girl, a cabaret dancer who had been stranded in Malta at the start of the war and had stayed on as a plotter in the RAF underground HQ.

Was Warby ever afraid? Frank Bastard who flew with him on many operations as navigator thinks he 'failed to realize danger'.

'There were definite stages in a man's reaction to flying in the war,' he said. 'First, you didn't fully realize what was happening to you. Then suddenly you were afraid. I have known chaps fall down by their beds at night crying and wondering whether they could get through their next flight.

'This, too, passed and then you didn't care a damn – the most dangerous stage of all because then a man got careless.

'Finally you rose above the whole thing. It became just a job –like going to the office each morning. I think Warby had reached that stage when I met him, but, being Warby, he had probably skipped a couple of stages on the way.'

BEURLING – ONE GENIUS

Laddie Lucas, *Five Up: A Chronicle of Five Lives* (Sidgwick & Jackson, 1978)

George Beurling was untidy, with a shock of fair, tousled hair above penetrating blue eyes. He smiled a lot and the smile came straight out of those striking eyes. His sallow complexion was in keeping with his part-Scandinavian ancestry. He was highly strung, brash and outspoken. He hadn't, I judged, had much education, but he was a practical man and had certainly made a deep study of aircraft and flying. He was fascinated by them; this was what he lived for.

He was something of a rebel, yes; but I suspected that his rebelliousness came from some mistaken feeling of inferiority. I judged that what Beurling most needed was not to be smacked down but to be encouraged. His ego mattered very much to him and, from what he told me of his treatment in England, a deliberate attempt had been made to assassinate it. I was, therefore, all the keener to bring to full fruition his manifest skills and flair....

I promised Beurling I would give him my trust and he would get his chance. I said he had individual flying and fighting attributes far above the average on the island, and it was right in this theatre that he should have full rein. I added a caveat. If he abused this trust then he would be put on the next aircraft to the Middle East.

When I said all this those startlingly blue eyes peered incredulously at me as if to say that, after all his past experience of human relations in the Service, he didn't believe it. He was soon to find that a basis for confidence and mutual trust did exist. He never once let me down....

He had an instinctive 'feel' for an aircraft. He quickly got to know its characteristics and extremes – and the importance of doing so. He wasn't a wild pilot who went in for all sorts of hair-raising manoeuvres, throwing his aircraft all over the sky. Not at all. George Beurling was one of the most accurate pilots I ever saw. A pair of sensitive hands gave his flying a smoothness unusual in a wartime fighter pilot.... This acute sensitivity told him that a Spitfire was only a fine gun platform if it was flown precisely. He therefore set out to make himself the master of the aeroplane.... He never let it fly him....

Two other attributes enabled him to capitalize his skill. First, his eyesight was quite exceptional. This gave him the edge on the rest.... Allied to his eyesight was a precise and finely trained judgement of distance. I never saw Beurling shoot haphazardly at an aircraft which was too far away.... He only fired when he thought he could destroy. Two hundred and fifty yards was the distance from which he liked best to fire. A couple of short, hard bursts from there and that was usually it. He picked his targets off cleanly and decisively, swinging his sight smoothly through them as a first-class shot strokes driven partridges out of the sky. It was a fluent and calculated exercise.... For Beurling the confirmed kill was the thing.

His desire to exterminate was first made manifest in a curious way.

One morning, we were on readiness at Takali, sitting in our dispersal hut in the southeast corner of the airfield. The remains of a slice of bully-beef which had been left over from breakfast lay on the floor. Flies by the dozen were settling upon it....

Beurling pulled up a chair. He sat there, bent over this moving mass of activity, his eyes riveted on it, preparing for the kill. Every few minutes he would slowly lift his foot, taking particular care not to frighten the multitude, pause and – thump! Down would go his flying boot to crush another hundred or so flies to death. Those bright eyes sparkled with delight at the extent of the destruction. Each time he stamped his foot to swell the total destroyed, a satisfied transatlantic voice would be heard to mutter 'the goddam screwballs!'....

So George Beurling became 'Screwball' to 249, to Malta and to the world. It was an endearing appellation. It suited him exactly. What's more he liked it. It helped his ego. It made him feel he was now regarded as an established member of the team. He felt the gaze beginning to be focused on him. At last he was a figure in his own right....'

Beurling was not invincible. On 14 October 1942, while he was lacing into one Me 109, another, taking him unawares, drilled his aircraft with cannon shells from behind. Screwball, injured quite severely by shrapnel, baled out low down. He landed in the sea and got into his dinghy. Malta's air-sea rescue service quickly came to his aid. L.G. Head, a member of the crew of HSL 128, has the incident fixed in his memory.

L.G. Head, Stokenchurch, Buckinghamshire, 1982

'... When we picked him out of the water he was most concerned for a few moments because he was unable to locate a small bible that he had been given by his mother. He told us that he would not fly without this bible and was most relieved when we found it and handed it to him....

Before we got him ashore, he was most adamant that he was going to fly and fight again within a few hours, but it was obvious to us that the wound in his heel in particular would put him out of action for some time* ....'

The Mediterranean fighting was over for Beurling, but not the danger. With Arthur Donaldson (Group Captain A.H. Donaldson), and others, he left Malta on 1 November in a B-24 (Liberator) to return to the United Kingdom via Gibraltar.

Donaldson had been Beurling's wing commander flying at Takali in the

*The assessment was correct. Beurling was in hospital for a fortnight and did not fly again in Malta. His total of victories was now 28 – to rise to 31⅓ by the end of the war. (Ed.)

closing months of the battle. The youngest of the three brothers, Arthur had, by his leadership, added fresh lustre to the family's name before being wounded in the fighting. Besides collecting a bar to his DFC, he had also gained a DSO to add to the two that Baldy and Teddy had, between them, won in Norway and France. (The Donaldsons' ultimate tally of decorations – 1 CB, 1 CBE, 3 DSOs, 2 DFCs, 4 AFCs and 1 Legion of Merit, USA – is quite without family precedent.)

The B-24 reached Gibraltar in the middle of a violent thunderstorm. Short of fuel, the pilot had no chance of diverting elsewhere. Disaster now attended his landing attempt. Donaldson recorded his impression of the dreadful accident shortly before he died in 1980.

Group Captain A.H. Donaldson, The Donaldson Archives, London, 1979

The nineteen passengers were all in the bomb-bay ... sitting on parachutes or anything comfortable they could find. Next to me was a beautiful Maltese girl called Bella; she had a son of about one year with her and she was going home to join her husband, a Fleet Air Arm pilot.

We were told we had arrived over Gibraltar and, since we were not strapped in, we just sat there waiting for the landing. I can remember the sequence of events vividly. The first intimation we had that all was not well was when the aircraft touched down with a bump and then there was a burst of all four engines; we realized, although we could see nothing, that the Liberator had attempted to take off and go round again before trying another landing.

People who were watching from the aerodrome later told me that the pilot overshot the runway, landing about two-thirds of the way down it. It was obvious that he could not pull up the huge aircraft in the remaining length of ... runway....

He had then opened up his engines at full throttle as he saw the end of the runway coming up fast, and, beyond it, a fairly high barbed-wire fence. To clear the fence he had to pull the stick back hard. The Liberator rose up sharply, cleared the fence, but as it did not have sufficient forward speed it stalled and flopped into the sea about 100 yards off the runway. The force of hitting the water broke the aircraft's back and it sank immediately.

... I can remember a loud noise as we hit the sea; then the bomb-bay started to take in water rapidly. In no time, we hit the sea bed. Naturally we were all thrown forward, not being strapped in, and everything was chaos. I can remember, too, seeing that, above me, there was a small crack which looked lighter; this turned out to be where the aircraft had broken its back.

I was lucky since I was not hit on the head and was, therefore, fully conscious. I simply fought my way up through the broken hull to the surface. The scene of the accident was amazing; there seemed to be literally hundreds of men swimming out to the position of the wreck and soon someone got hold of me and pulled me towards the shore.

I can remember seeing one friend of mine, a Spitfire pilot named Hetherington, who had just been awarded the DFC for his part in the air war, near me.

He seemed quite all right and certainly conscious though I noticed a trickle of blood coming from his mouth. I heard him shout to his rescuers 'I'm OK. Save some of the other passengers who need help.' Seconds later he died, presumably from internal injuries.

I never saw Bella and her child again. I got away with an injury to my right arm caused, I think, from being brushed against one of the hot engines and getting burnt.... Only a scar reminds me of how lucky I was.

I gather that the crew were safe because they were strapped in, but out of the nineteen passengers only three were saved, one of them being the famous Screwball Beurling.*

Donaldson, after recovering from his traumas, commanded operational stations at home and finished the war in charge of day operations at 11 Group headquarters at Uxbridge. As a station commander, responsible for the administration and discipline of those serving under him, he never forgot an incident which occurred while he was instructing at the Royal Air Force's famous Central Flying School at Upavon in the early days of the war. Group Captain James Robb (Air Chief Marshal Sir James Robb) became the Commandant during Arthur's time there. It was his opening shot which made the impact.

Group Captain A.H. Donaldson, The Donaldson Archives, London, 1979

'His appearance was unusual.... He was almost completely bald and he wore glasses. Under that strange head he had a fine brain and superb powers of leadership.... James Robb was a great leader...

The first thing he did when he arrived to take over was to assemble all the officers together and address them.... He spoke for only one or two minutes.... I can remember his principal words off by heart because they had such a tremendous effect upon me.

"Gentlemen," he said, "I am the new Commandant of CFS. You may wonder what sort of man I am. I will tell you. I treat all officers as grown-ups; I give them as much rope as possible. If they hang themselves, then it is their own fault.

"Good morning, gentlemen."

I always tried to model myself on Sir James's way of commanding a station.'

Sinking of *Prince of Wales* and *Repulse* A Japanese Account

Edited by Nobuo Ando, *Kaigun Chuukou Shiwa Shuu* (A Recollection of Haruki Iki) published by Haruki Iki, 1980)

At 0714 hours on 10 December 1941, twenty-six Mitsubishi G4M bombers of Kanoyal Kokutai, led by Lieut-Commander Shichizo Miyauchi, took off from the base near Saigon to find and attack the British fleet sailing south off

*This was Beurling's fifth crash. There were others. The tenth and last, which killed him, came three years after the war, on 20 May 1948, when a Canadian-built Norseman, in which he and another pilot were flying, burst into flames as it landed at Urbe airfield, near Rome. George Beurling was then twenty-six. (Ed.)

the east coast of Malaya. I was leading the 3rd squadron of nine aircraft.

We flew due south for some 600 miles, but after searching we could not find the warships. At 1058 we began to turn back towards the north. We then received an uncoded message to confirm that Ensign Hoashi had found the fleet. After radioing back to base that we had received the message we then turned on to a heading of 280 degrees. Forty-eight minutes later we sighted the British ships through broken cloud. Each squadron used the available cloud cover.

The fleet consisted of three destroyers, leading, with *Prince of Wales* some 2000 metres aft of them and *Repulse* approximately 2500 metres aft of *Prince of Wales*. The ships were heading on a course of 160 degrees at an estimated speed of some 20 knots.

The leading aircraft attacked the first battleship from starboard, five other bombers from the 1st and 2nd squadrons following suit immediately afterwards. We observed five distinct fountains of water which indicated a torpedo hit in each case.

The remaining five aircraft of the 1st squadron attacked *Repulse* from starboard while six of the 2nd squadron made their attacks from the port side. Our squadron had at first intended to attack *Prince of Wales* but, by the time we reached her she had already received several further hits, so we switched and made *Repulse* the target. The battleship was turning to starboard at the time and making some 20 knots. Her anti-aircraft fire was very heavy.

My section of three aircraft attacked with torpedoes from the port side while W.O. Yuji Yahagi, the observer, maintained a running commentary, giving distance and height over the intercom.

After the attack, NAPI/c Tamotsu Maekawa shouted: 'Squadron Leader, a hit.' And then, again, 'another hit'. The next moment I saw the second aircraft burst into flames right in front of me and dive into the sea 300 metres from *Repulse*. The third aircraft also dived into the sea 50 metres to the left of the first.

The second and third sections fired torpedoes into *Repulse* from the port side as she turned hard to starboard. My squadron's attacks finished at 1202. In the face of heavy anti-aircraft fire, our gunners responded by raking the deck of the battleship. We then withdrew outside the range of AA fire. As we did so we saw the British man-of-war turn turtle and disappear, leaving a whirlpool on the surface of the sea. At this moment, our aircraft was filled with the 'Banzai' call.

I formed up with four other G4Ms and returned to base, my aircraft being the first to land. At Headquarters, we were welcomed by the staff who, to express their joy, tossed me in the air.

Our Kanoya unit confirmed 7 hits on *Repulse*, 2 on the port side and 5 on the starboard. My aircraft, No K-301, received 17 hits according to the ground crew's report.

'Three Lasts'

T.W. Watson, 232 Squadron (RAF), Mississauga, Ontario, 1973. By courtesy of the Directorate of History, National Defence HQ, Ottawa

I can't think of many firsts that I have had in my life but there are three lasts

– the last plane out of Singapore, the last plane out of P2 (southern Sumatra) and the last plane out of Java. All in all, I was lucky....

They Bombed Tokyo

(Reproduced with permission of the *Los Angeles Times*, 17 April 1982)
Chosen by Group Captain Sir Douglas Bader, 30 April 1982

A nice sunshiny day with overcast of anti-aircraft fire
Sgt Eldred Scott, Gunner on Tokyo raid, April 18 1942

by JERRY BELCHER, *Times* staff writer

Forty years ago Sunday, 80 American airmen won the first victory of the last war the United States ever won.

The spring of '42 was the worst of times for this nation and its embattled British, Russian and Chinese allies. In the Pacific the situation was especially bleak – everywhere the Japanese Empire was triumphant.

Pearl Harbor was still in shambles from the 7 December sneak attack by carrier-based Japanese aircraft. Guam, Wake, Manila and Singapore had fallen. The Dutch East Indies and the Solomon Islands had been invaded. On 9 April, after a gallant fight against overwhelming odds, weary American and Filipino troops on Bataan surrendered to the Japanese. Corregidor, the last American foothold in the Philippines would hold out less than a month more.

American morale was at its lowest ebb.

Then on 18 April 1942, sixteen US Army Air Corps B-25 medium bombers struck at the heart of the Japanese Empire. Bombs fell on Tokyo, Yokohama, Nagoya, Osaka and Kobe.

Although the raids caused negligible physical damage (each twin-engined B-25 carried only a ton of bombs), it badly damaged Japanese morale. It was the first successful attack on the home islands of Japan in the history of that ancient and proud nation. An attack that the Japanese people had been told was impossible.

And, military historians have said, the Tokyo raid goaded the Japanese High Command into the face-saving 5–6 June Battle of Midway in which the Imperial Navy was dealt a blow from which it never recovered. It was the turning point of the Pacific war.

This morning in St Petersburg, Fla, during a private ceremony at the otherwise public commemoration of the 40th anniversary of the raid, a stocky, flagpole-straight, 85-year-old man will lift his brandy-filled silver goblet.

'To those who have gone,' retired Lieutenant-General James H. (Jimmy) Doolittle will say.

The thirty-one others in the room – younger than the old man they still call 'The Boss', but all more than sixty now – will raise their silver goblets: 'To those who have gone.'

Nearby, eighteen identical silver goblets will stand upright on a table – one for each of the survivors of the mission who could not make it to the reunion this year. Another thirty-one goblets will be arrayed, but with their bowls turned downward, each bearing the name of a Tokyo raider now dead. There will also be a single unopened bottle of fine brandy.

... The Hornet steamed out of San Francisco Bay at 10.18 a.m. 2 April.

On 13 April, she and her escort rendezvoused with the carrier *Enterprise* and her escort vessels. Originally the plan was for a take off late on 19 April, with Doolittle dropping incendiaries on Tokyo to guide the other planes for a night raid.

But at 3 a.m. on the 18th the *Enterprise*'s radar picked up two enemy patrol boats, part of an armada of picket vessels stationed 600 to 700 miles off the Japanese mainland....

Messages flashed.

Soon afterward, another Japanese patrol boat was spotted by a Hornet look-out. Then another. Halsey flashed this message from the *Enterprise* to Admiral Marc Mitscher on the Hornet:

'LAUNCH PLANES. TO COLONEL DOOLITTLE AND GALLANT COMMAND GOOD LUCK AND GOD BLESS YOU.'

Doolittle was first off at 8.20 a.m., easily airborne after a run of only 427 feet. Stiff headwinds helped lift the overloaded bomber. At about the same moment the cruiser *Nashville* was sinking one of the patrol boats with shells – but not before it had relayed a radio warning to another Japanese vessel....

Robert (Doc) White ... retired from his medical practice and living in Palm Springs ... the group's flight surgeon who doubled as top turret gunner ... remembered the moment:

'Then [one of the cruisers] let go a broadside. It was one of the most exciting things I've ever seen. Then it suddenly occurred to me – you know if he's shooting at somebody, somebody's liable to be shooting back pretty soon. I got scared then and I guess I stayed scared the next six days....'

... David Pohl ... a retired business executive living in Manhattan Beach, was the youngest of the eighty men who clambered into their B-25s that morning.

He recalled that he was not particularly anxious to fly the mission. But neither was he quaking in his flying boots. 'When you're twenty years old, going into combat for the first time, you consider yourself immortal.'

He even turned down an offer of $100 from a sailor who wanted to take his place on the raid. Pohl, a sergeant, was a turret gunner (twin ·50 calibre machine guns) on Captain Edward J. York's B-25, the eighth to take off.

'The hairiest time of the entire mission probably was the 5 minutes when they were wheeling our plane into line,' Pohl remembered, 'then the actual take off was scary.... The Hornet was taking green water over the flight deck.

'I just got myself into the turret – and kept the turret facing aft. I didn't want to watch the take off. But we made it with feet to spare which surprised everyone.'

It surprised everyone in Tokyo, too, when Doolittle's B-25 roared in over their roof-tops, pulled up to 100 to 200 feet and then dropped his bombs at 1.30 p.m.

After Doolittle made the initial run he encountered intense anti-aircraft fire, but the rest of the B-25s, save one, managed to hit either their primary or secondary targets. . . .

The Tokyo raiders were supposed to land at airfields in China. But because of the premature take off (they had planned to launch at about 400 miles and strike by night but lifted off by day at more than 600 miles) and because of headwinds, foul weather and other problems none made it. . . . The survivors went through some incredible ordeals and adventures evading the pursuing Japanese troops. Finally, with aid from Chinese guerillas and civilians, they made their way out of China. Lieutenant Ted Lawson, suffering from head, back and leg injuries, was carried to a Chinese village after crash-landing on a beach. Doc White, who had saved his medical gear, made connections with Lawson and several other injured airmen some time later.

On 4 May, because of infection, White was forced to amputate Lawson's injured leg. As Lawson wrote later. . . .

'I watched a buddy saw off my leg.'

Everyone got medals. But ask survivors if they were heroes and these are the answers you get. . . .

David Pohl: 'Who's a hero? All I did was sit in a turret and watch the scenery. . . . Never fired a shot at another aircraft.'

Ted Lawson: 'Good lord! No! Other people make heroes. . . . We just had a job to do. . . . I guess we did give the country a shot in the arm at a time when everything was going against us.'

Doc White, shrugging off the suggestion, thought the Chinese fisherman who took him and his crew in was . . . the hero:

'I'm sure he had never seen a white man before. I'd like to think that if I were called out in the middle of the night and met four giants – we were 2 feet taller than he was – in strange uniforms, speaking a strange language and obviously in trouble, I like to think I would have the courage and the humanity to ask them into my house. That's what that little man did. . . .'

Doolittle's raiders will continue to hold their reunions. When the survivors are down to two – and actuarial projections place that time shortly before the turn of the century – they will open that bottle of fine brandy (Hennessy, 1896), pour it into their silver goblets and raise them in this toast: 'To those who have gone.'

Friendly Message

Forty years before, myriad messages had greeted Doolittle's safe return to the United States after the Tokyo raid. A telegram from his old friend and fellow racer of pre-war days, Roscoe Turner, made its mark:

'Dear Jimmy,
You son of a bitch.
Roscoe.'

Japanese Torpedo Bombers at Midway

Hidetoshi Kanasawa, *Kuubo Raigekitai* (Konnichi-no-wadai, 1975)

We sailed from our homeland on 27 May 1942, and two days later I was sent to the armoury to draw seventy revolvers for the crews. These weapons were not for self-defence but for committing suicide should any of the crews be forced down and taken prisoner. Japanese soldiers were forbidden to surrender.

After taking part in the attack on Midway and being wounded in the foot, I then witnessed the bombing of the carriers *Akagi, Kaga* and *Hiryu* by ... [carrier-borne] Dauntlesses ... [of the United States' Navy]. Thereafter I was engaged in the strike against the attacking enemy carriers. The decision to launch the attack was taken when it was reported that three US carriers had found the ... [Imperial Navy's] ... 3rd attack force....

The bell on the aircraft lift told us that, one by one, the ten Nakajina B-5N torpedo bombers, each carrying a torpedo, were being sent up to the flight deck. Among them was Lieut. Tomonaga's aircraft with its three red bands on the yellow tail, making it look very conspicuous. This machine had been damaged in the left wing fuel tank on the previous attack on Midway. It could now only carry sufficient fuel to fly to the target, but Lieut. Tomonaga refused the offer of another aeroplane; he said that fuel for the single journey was sufficient.

While the engines were being run up, the crews gathered together to brief and brace themselves for the next attack. Lieut. Tomonaga just said: 'Don't be worried. Remember death can come any time. We are all expendable. The difficulties will come later. Then is the moment for you to do your duty. When the captain addressed us he ended his words with 'goodbye', not with 'good luck'. Lieut. Tomonaga, the leader of the attacking force, told us: 'You guys must stick to my tail until you hear the tallyho.'

Before the sign was given for take off, the leading aircraft had already begun the process, fearing another attack by the enemy. Altogether, ten torpedo bombers and six escort fighters were airborne. None of us had encountered the US fleet at sea before. What would it be like, we wondered?

Suddenly it was sighted. We could see the outline of ships on the horizon, but no smoke was coming from any of them. None appeared to be damaged.

As we closed the range to 10,000 metres, a heavy black smokescreen was being laid down. We flew straight through it. There didn't appear to be any fighters about.

Delaying the time of the usual tallyho signal, Lieut. Tomonaga looked left and right to confirm that the other aircraft and crews were still with him. Then he grinned and waggled the wings of his aircraft to tell us to spread out. As we fanned out we also started to dive steeply to increase speed. I couldn't identify the type of carrier in front of me.

As there were still no fighters about, we attacked from the port bow; I have always thought this the best position from which to attack. But our B-5Ns flew straight at the carrier from every direction. Twice there were big splashes. These must have been our own aircraft, but there was no time to look round.

Now the enemy carrier was looming up.... 'Get ready ... fire!'

My aircraft zoomed up after the dive; I wasn't particularly conscious of the flak any more. Just in front of me the yellow tail unit of the leading machine was engulfed in flames and it broke away from the fuselage. Ablaze, the aircraft splashed flames over the flight deck of the carrier before exploding and disintegrating totally. My eyes were blinded by a flood of tears.

We zigzagged away from the target, eluding the opposing fire. On the return flight, when the three of us in my aeroplane got within sight of our fleet, we could see that a heavy pall of smoke hung over each carrier. And I thought the *Hiryu* would be safe!

After waggling our wings to identify that we were friendly, the engine began to run rough and cough. The pilot put his head out of the side of the cockpit to see what was happening; the windscreen was covered with oil and dirt after the attack. The airscrew began to slow right down and I could now smell burning. Were we going to force land? I must have blacked out completely. The rest was a blank.

When I regained consciousness my first sensation was of being soaked, then of being aware of pain and the bright sunlight in my eyes. I realized then that, drenched to the skin, I had been picked up by a destroyer....

Editor's note: Hedetoshi Kanasawa joined the Imperial Navy in 1937 as a flying cadet. He was then attached to *Hiryu* in November 1940, and served in the carrier as an NCO. After *Hiryu* was sunk, Kanasawa then served with other units until the end of the war, being promoted in due course to Ensign.

The Meaning Of Midway

Winston Churchill, *World War Two, Volume IV: The Hinge of Fate*
(Cassell, London; Houghton Mifflin, New York)

... Reflection on Japanese leadership at this time is instructive. Twice within a month their sea and air forces had been deployed in battle with aggressive skill and determination .... The men of Midway, Admirals Yamamoto Nagumo and Kondo, were those who planned and carried out the bold and tremendous operations which, in four months, destroyed the Allied Fleets in the Far East and drove the British Eastern Fleet out of the Indian Ocean. Yamamoto withdrew at Midway because, as the entire course of the war had shown, a fleet without air cover and several thousand miles from its base could not risk remaining within range of a force accompanied by carriers with air groups largely intact ....

One other lesson stands out. The American Intelligence system succeeded in penetrating the enemy's most closely guarded secrets well in advance of events. Thus Admiral Nimitz, albeit the weaker, was twice able to concentrate all the forces he had in sufficient strength at the right time and place. When the hour struck this proved decisive. The importance of secrecy and the consequences of leakage of information in war are here proclaimed.

This memorable American victory was of cardinal importance not only to the United States but to the whole Allied cause.... At one stroke the dominant position of Japan in the Pacific was reversed. The glaring ascendancy of the enemy, which had frustrated our combined endeavours throughout the Far East for six months, was gone for ever.... No longer did we think in terms of where the Japanese might strike the next blow, but where we could best strike at him to win back the vast territories that he had overrun in his headlong rush....

Guadalcanal Adventure

J.J. Southerland, Lieutenant, US Navy, 26 August, 1942
(Reproduced with permission of US Marine Corps Historical Centre)

Lieutenant James Southerland II, peacetime product of Annapolis, led eight VF–5 Grumman F4F Wildcat fighters off the US Navy's carrier, Saratoga, *at 1215 hours on 7 August 1942, to cover the transports engaged in the landing of US marines on Guadalcanal in the Solomons.*

Against the transports, anchored off shore, Rear Admiral Sadayoshi Yamada, commanding the 5th Air Attack Force (25th Air Flotilla), had ordered a strike of 27 Mitsubishi G4M1 Type 1 medium bombers, supported by 18 Zero fighters. After the ensuing clashes, in which his aircraft was repeatedly hit, 'Pug' Southerland eventually found himself locked in deadly combat with the Japanese 'ace' Saburo Sakai, flying a Zero fighter. By this time, Southerland's tough Grumman had been holed and damaged. It was also out of ammunition.

With Sakai's final thrust, pressed home to close range, the Wildcat exploded and Southerland baled out over Japanese-held territory from a very low altitude. He recounted the story of the next couple of weeks in a plain, factual, but tense report he gave on 26 August as he was beginning a period of convalescence after recovering from the first impact of his deranging experience.

'... My first grateful thought was to thank God that I was alive. A hasty survey revealed the following damage: one large hole in my right foot which was the most painful of my injuries, three holes in the calf of my right leg, one in my left knee, one in my left thigh, three in the upper part of my left arm, one glancing shot through my right eyebrow which bled considerably but was not serious, and one small piece of shrapnel in my scalp.

Also noted: two flash burns from the explosion, one on an exposed portion of the right forearm and one on my left wrist between glove and sleeve. Minor abrasions on my left leg from the fall through the trees were of little consequence. My right shoe was full of oil, blood and dirt. I took it off, removed the dirty, oily black sock, stuffed it into the bullet hole to stop the bleeding and tied my shoe on tightly.

I wanted to get out of this area quickly as I knew I was in Jap territory and thought they might have seen me bale out and would attempt to make me a prisoner. Tried to make the coast by keeping to the hills but realized after about an hour's struggle, with frequent rest periods, that I would have to

Versatile Fleet Air Arm. The FAA, with the Royal Air Force's Malta-based strike aircraft, wrought havoc in 1942 on Axis shipping crossing the Med. to Rommel *(top right).* Three years later, in January 1945, 849 Squadron, led by David Foster and operating with others from the carriers *Victorious* and *Indefatigable,* obliterated the oil refineries at Palembang in eastern Sumatra *(bottom right).* Bobby Bradshaw *(below),* most talented of all the Navy's flyers, sparkled in the Western Desert and elsewhere (p. 158 *et seq.,* 342 *et seq.* and 376 *et seq.*)

Malta climacteric, spring and summer 1942. Devastated, but resolute, Valletta *(facing page, (bottom)*. Takali airfield *(right)*, pictured from 15,000 feet by Harry Coldbeck *(bottom right)*, New Zealand's outstanding reconnaissance pilot (here with Norman McQueen, right). The redoubtable David Douglas-Hamilton *(bottom left)*, CO of 603 (City of Edinburgh) Squadron and five pilots, who helped to put 249 Squadron at the top of the British and Commonwealth league, giving Intelligence the news *(facing page, top)* – left to right: Laurie Verrall (New Zealand), Les Watts (UK), Chuck Ramsay and Frank Jones (Canada) and, sitting, Raoul Daddo-Langlois (UK)

Mediterranean memories, 1942. Pilot Officer Smith (Canada) *(left),* who landed a Spitfire on USN carrier *Wasp* without tailhooks (p. 17) and the Regia Aeronautica on 'ops': SM 79 starting a bombing run over Malta *(facing page, top)*; Cant-Z 501 seaplane on patrol *(below)* and ground crews working on an Italian JU 87 at Trapani 'Milo', Sicily *(facing page, bottom)*

Same play, different theatre, Luftwaffe *v.* the Soviet Air Force, Eastern front. 1943, Johannes Wiese, successor to Steinhoff as Kommodore of Jagdgeschwader 77 in 1944, and credited with 131 victories, 'feted' on the Russian Front after claiming his 100th 'kill' in his Me 109G6

Fighter pilot F. Khimich, *(top)*, hero of the Soviet Union, congratulated by squadron comrades after a 'success' near Kursk; and 'chocks away' – Soviet ground crew seeing off an IL-2 Stormovik, workhorse of the Russian Air Force *(bottom)*

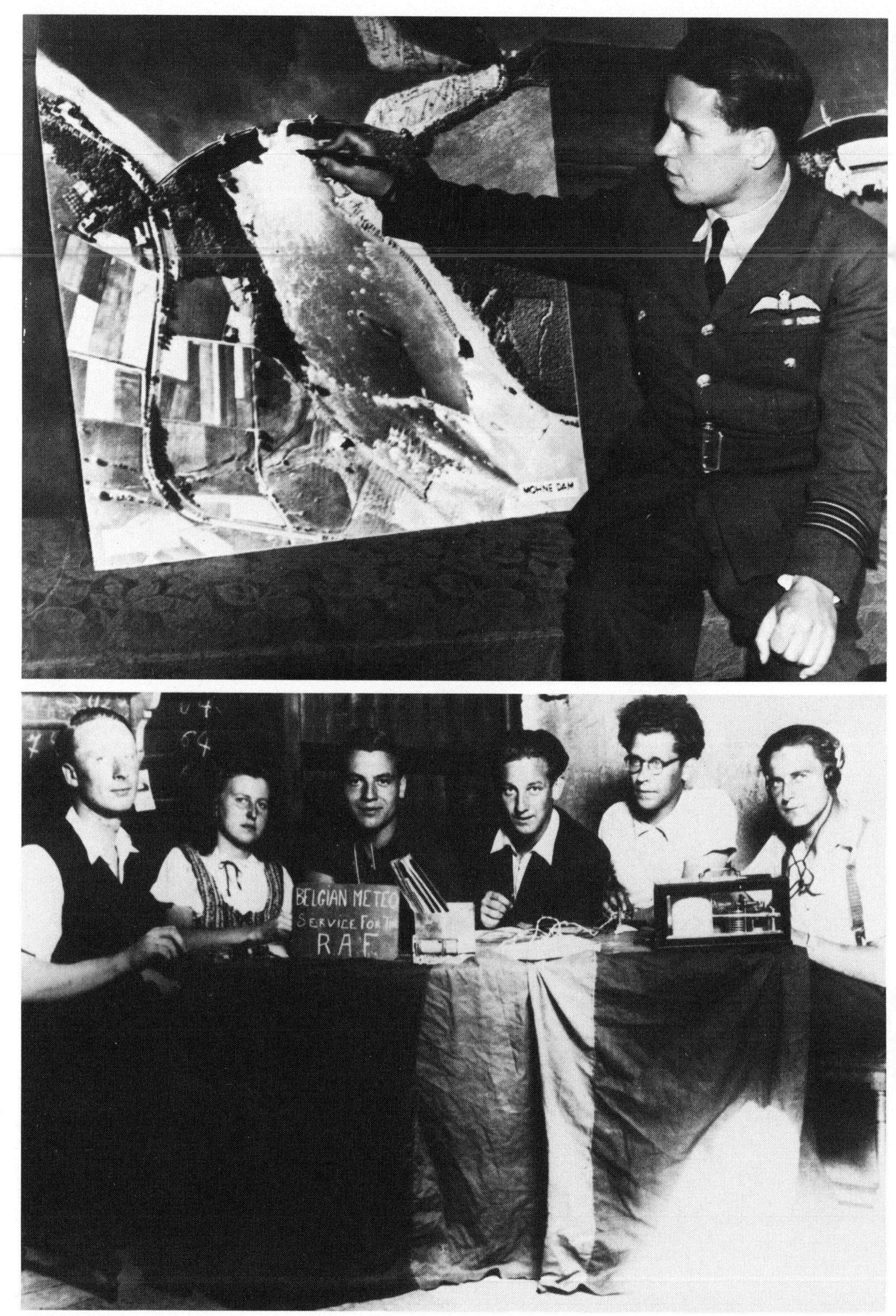

Operations extraordinary, Guy Gibson, VC, *(top)* pointing to a photo of the breached Mohne Dam after the great raid of 16/17 May 1943. And six of the brave team of Belgian agents who, constantly in danger, sent secret WT 'met.' reports daily to England between 1942 and *(bottom)*. The picture was taken in 1943 (p. 314 *et seq.*)

penetrate the jungle if I wanted to make any headway. I sat down and started sliding downhill (this was easier on the foot), while lining up a couple of high trees above the jungle growth which would give me the direction to the water in case I got lost. Suddenly I stopped and crept behind a bush to my right. In a high tree, to the left of a lofty dead tree I was using for a bearing, were two Japs on a wooden platform apparently designed as a lookout post. Fortunately, I saw them first.

Continued toward the jungle, keeping as well concealed as possible and crawled into a discouragingly thick and seemingly impenetrable undergrowth. There were masses of thick, thorny vines, spiders' webs with big, formidable-looking spiders, and unseen but imagined wild life of all sorts, mostly poisonous snakes and boa constrictors. The imagination really goes to work in a jungle ....

Through the grove I could see the water. Just to my right, on the edge of this grove, was what appeared to be a deserted Jap camp. I scouted it for a few minutes, concealed in the tall grass and, satisfied that it was unoccupied, proceeded to investigate and see if I could find anything I needed. The Japs had cleared out completely and all I could find was an old flashlight with no batteries but a large reflector I thought I might use for signalling....

Progress was slow as I kept concealed as well as possible .... I at last made the beach. To my right was an open point of land. Our planes were passing overhead frequently returning to the carriers, so I decided to get out on this point and see if I could attract their attention.

On reaching this destination, I stripped and waded into the salt water. The coral was rough on my feet but I was glad for the opportunity to cleanse all my wounds. Thoroughly washed, I tore my underwear into bandages and made a compress to stop the bleeding in my right foot. Used my collar insignia to pin the bandage on as the strips were too short to tie. The rest of the holes had stopped bleeding. I dressed hastily, polished my reflector with sand and stone as well as possible and attempted to attract the attention of returning planes by reflection of sunlight in the pilot's eyes. I had landed at about 1400 and it was now, I guessed, about 1730 as the sun was very low.

After a few disappointments my system apparently worked for one SBD, soon followed by several others, started circling warily above me descending slowly. I discarded the flashlight and started waving my yellow life jacket which seemed to be the best means of identification. I then signalled SOS in semaphore, though one plane dropped a smoke bomb, flew low over me rocking his wings, and then headed off with the rest for the carrier.

I was concerned about being so exposed and concealed myself between a large log and a pile of sand while awaiting the arrival of a rescue plane or boat. Passed a very cold and windy night on the beach, plus about 3 hours after dawn the following morning, before I realized that our forces were probably too troubled with other major difficulties to be concerned with dangerous rescue work.... Decided to make my way along the beach toward Kukuum, about 25 miles to the eastward, which I hoped would be in the possession of our marines.

The going was tougher today as my foot had swelled considerably and was much more painful. I soon came to what appeared to be a deserted missionary's home .... There were pigs and chickens in the yard. I investigated fully, in search mainly of fresh water. This I found in some tanks to the rear. . . . I drank just enough to quench my thirst as there was a danger that the tanks were contaminated. Further search of the grounds disclosed a garage, containing an old truck, and a tractor nearby. Both tanks were bone dry and no fuel was in the vicinity, so I had to abandon hope of motorized transportation.

Soon thereafter I spotted three native boys headed toward me. . . . They were obviously friendly. I had to assume so anyhow, as they were a formidable-looking trio dressed only in sarongs, with large cane knives at their waists, beaded ornaments on ears and arms and bracelets made out of boar tusks. I was unarmed and definitely harmless. I pointed toward Kukuum and said "Japanese." One answered "No Japanese."

Later I came to a beautiful fresh water stream running clear and rapidly to the bay. It was about 3 feet deep. I waded in, drinking and washing in the cool water, and really hated to proceed on my way.

About noon I decided it was time to eat. Having selected a good coconut along the way, I now broke it open, loosening the fibre by pounding on a tree stump and ripping it off with my hands. Broke open two of the holes at the top of the shell with my Eversharp pencil and really enjoyed the cool and refreshing coconut milk. . . .

Continued slowly and soon saw three wild horses coming toward me. Here seemed another opportunity for transportation of a sort. My foot was troubling me to such an extent that any idea to give it a rest seemed good. The horses, however, [would] . . . only get so close then bolt away. This idea too was soon abandoned.

I passed several deserted native villages consisting of a number of thatched huts. In one of these I found an old high bowed war canoe of the type dramatized in the movies. I cleared a path through the brush and rocks and after about 1½ hours' labour, with the help of some logs I used as rollers, the war canoe was launched. I made a paddle out of the bottom of a palm frond and climbed aboard. She leaked like a sieve, but I figured this was natural until the wood swelled to close the seams, so I started bailing out with my shoe. This system was doomed to failure as the canoe leaked faster than I could bail. After about a five-minute cruise I beached the rescue ship, just in time, and continued once more on foot.

. . . . Passed more deserted villages, one of which had a little thatched Catholic chapel and school house. . . . Then I spotted two more natives. They were coming from behind and also responded favourably to the raised right hand in greeting. We shook hands. One, who spoke very good "pigeon English" said: "Me very sorry" and pointed to my foot. Next he said "Come with me." Noticing my hesitation he said: "Father told me to be good to white man."

. . . I was worn out and figured I had little to lose so went with them. We returned first to the Catholic chapel. I told them I was a Catholic which pleased them greatly. The younger one told me in pigeon English that his Christian name was Joseph, the older one's name was Jonas, and that they had gone to the missionary school for thirteen years. Catholic was a magic word. They brought me some more oranges and fussed over me trying to make me comfortable.

Joseph fixed me a pole to use as a crutch and we worked back toward their village. I stopped at a stream for water and Joseph made a perfect drinking cup from a plant leaf. They then took me to a fresh water pond. Jonas got some soap from the village and I bathed myself thoroughly, getting plenty of soap in the gunshot wounds to act as a disinfectant. Joseph washed my clothes and brought me a pair of white shorts (waist size about 46) to wear while my clothes dried.... I lay down gratefully while they went to get me food and water.

They soon returned with food "the white man liked", sweet potatoes, coconut, pumpkin, papaya, small green onions and some green oranges. Potatoes and pumpkin were boiled. I tried all but the onions, with little relish, except for the oranges, as I was not very hungry.

Joseph and Jonas stood watch, sleeping on thatch laid over the dirt floor of the hut. My rest was a fitful one, as I heard tremendous volumes of gunfire which I later learned was a night cruiser battle. My benefactors had promised to take me by canoe to Kukuum the next morning as some of their scouts had informed them that this town was now in possession of the white men. However, their minds were changed the next morning since the gunfire of the previous night had scared them. Most of our ships had slipped out of the harbour and they, as well as I, were worried that the Japs might have retaken the island.

Sunday was sideshow day. Many natives came in to see the white man and scrupulously shook hands under Joseph's careful coaching. All but one seemed quite friendly. He was an old man, about seventy-five to eighty years of age, tall, straight, and very active but thin as a rail. He looked to me like the medicine man of the outfit. I don't think he approved of the younger set's solicitous care of the white man, for he stood apart and looked with disgust at me for about 10 minutes, then walked out muttering something which sounded very much to me like: "Used to eat white man, now don't know what to do." I was glad to see him go and hoped his influence was small in the community.

Joseph had sent one boy into the hills to bring back all the medicine the natives could find. He appeared with three bottles, two small ones labelled APC capsules and Aspirin, and one large one with no label and about half full of a brownish liquid. I smelled its contents, however, and noted the odour of plain old household Lysol. I then had Joseph boil some water and soaked my foot in hot Lysol solution. There was considerable swelling and redness in the area of the wound so I figured some disinfectant was badly needed. Joseph also indicated the APC bottle and kept saying "Quinine". Since the mosquitoes and flies had been having a field day with me, and this was known to be malaria country, I decided to take some tablets.... I laid one on the tip of my tongue. The taste of quinine was unmistakable so I took two of them.

All day Sunday I tried to persuade them to take me to Kukuum but they were still afraid.... In the evening we had a feast. Joseph and Jonas went in search of food and returned with a wild chicken and two eggs. I ate heartily of a delicious meal.

My two protectors stood guard in the hut again Sunday night. The only light was the glow of their smouldering pulp wood.... All night long I could hear the crackling of the lizards and kept wondering when one would drop on me. I kept hoping there would be no gunfire to change the scheduled trip in the morning.

The boys came early, appearing in the door of the hut about an hour before sunrise. I was anxious to be on my way so we launched the canoe about half an hour later and I took leave of some real friends. Jonas and two other boys came with me.

We headed toward Kukuum hugging the shoreline. I tried to persuade them to get farther out but they were afraid of Jap planes and no amount of reasoning could convince them that no plane would waste ammunition on a native canoe.... We passed many Jap landing boats, beached and damaged, which I later learned were there as a result of a foiled landing attempt early Sunday morning.... Soon I saw men running toward the beach with rifles ready. They wore overalls and helmets similar to those of the Japs so I didn't know who we were running into. The natives also questioned their identity.... I waved to them. They waved back, which convinced the natives we were old friends, so they paddled rapidly to the beach.

What a relief it was to see our own marines! I asked them to take good care of the natives so they handed them food, candy and cigarettes. I was laid in a blanket while they went for a truck to take me to Brigade Headquarters' first-aid station.... That night, whilst resting in a medical tent in the woods near the airfield, the Japs decided to make their first attack from the hills in which they were hiding. Machine gunfire was terrific. Tracers were skimming close when they decided to evacuate the patients. Four corps men, who deserve some recognition for devotion to duty, carried me on a stretcher, and we spent from 11.30 that night until dawn the next morning shouting "Hallelujah" (the password), jumping in ditches, behind trees, and running like mad toward the beach to avoid gunfire.

After repeated transfers from ship to hospital and back again, I came aboard the *Argonne* on Monday 24th, seventeen days after baling out, to wait for a ride home.

Maybe I'll get another crack at them. If so, I hope to have an altitude advantage to start with. Also, if we only have 8 fighters to tackle 27 bombers and about 20 Zeros, I hope they keep us together. We would have accomplished much more this time if those two points had been observed. Anyhow, I'm told the bombers didn't get a hit which is a cause of much satisfaction.'

'Pug' Southerland returned to operations and, as a full commander, was in action again in March 1945, off Okinawa, leading Air Group 83 in the Western Pacific in the US Navy's fleet carrier, Essex. *Commander Southerland was eventually killed on 12 October 1949, on a night take off while serving as air group commander in the battle carrier,* Franklin D. Roosevelt.

To the Far East – and Problems

D.R. Gibbs, Montserrat, West Indies, 1982

82 Squadron was posted overseas early in 1942. It was certain that our

destination would be the Far East as the rapid Japanese advance towards India had to be stopped. The air and sea parties of the squadron did not meet up again until March when the Japs were on India's borders and all our twenty brand-new Blenheims, handed over to other squadrons operating in this theatre, had been shot down or destroyed on the ground.

There were rumours that we would be re-equipped with Mosquitoes which we all very much wanted following the morale-boosting display which Geoffrey de Haviland had given us at Bodney a few months before. He had feathered one engine on take off and then done a steep turn around the airfield into his dead engine – a certain recipe for suicide in a Blenheim! However, the Mosquitoes were not to be. Instead we were downgraded to a new US single-engined dive-bomber, the Vultee Vengeance. The pluses were that it carried twice the Blenheim's bomb load and had four forward-firing machine guns. Being a single-engined aircraft squadron, we lost our wing commander to Delhi and I found myself the Squadron Commander, a post I held for the next twenty-seven months until we finally did get our Mossies. My first job was the unhappy one of reducing each crew from three to two.

The Vengeances were shipped out so quickly that flight testing had not been completed. Vultee had, therefore, seconded a group of technicians under Test-Pilot Spinney Leach and Engineer Blackie to help us solve the inevitable teething troubles.

The first Vengeances arrived over our Madras airfield in August 1942, followed by white puffs of smoke just like flak. This indicated serious piston-ring failures, due to the engines not having been inhibited before shipping. In the end, all twenty aircraft had to be grounded for new engines. Jock Davidson, our newly-commissioned squadron engineer, had, in addition, a long list of modifications necessary to make the aircraft operational. Back at Chittagong we were finally made operational just before the 1943 monsoon started in earnest.

This was a very different war from the one the squadron had fought in Europe. There was little flak due to the Japs' logistic problems. The RAF Beaufighters and USAAF P-47s and P-38s had attacked the Japanese airfields within range to such good effect that all enemy aircraft had been withdrawn to Indo-China only to be brought forward under cover of darkness when an attack was planned. We seldom saw any enemy aircraft and if we were spotted nothing could follow us in a vertical dive with dive brakes out. Once down in the jungle valleys, we were hard to see and easily able to get away. All this was very different from 2 Group in England, from the openness of the North Sea and the flak and fighters the Germans threw at us.

However, we lost some experienced crews. One aircraft apparently received a direct hit over Akyab as it blew up in the dive. Soon after this casualty, an armourer broke the weld on a fastening lug of a trunnion band, an essential part of a device which threw the bombs clear of the prop. In retrospect, this loss over Akyab was clearly due to the bomb falling into the propeller. Sandy, an Australian flying officer, found out the reason why some aircraft never pulled out from a dive. He was himself doing a loop over the airfield when, to his dismay, the stick broke as he pulled out of the dive. Luckily, he had the height and was a good enough pilot to be able to bring the aircraft under control and land it on trimming tabs and the little bit of control column that remained.

Others were not so lucky. However, generally speaking, we saw the same faces week after week in the dispersal hut. When we did lose an aircraft we grieved for some well-known friends and comrades. It was all so unlike 2 Group where there was seldom time for friendships to mature.

Once these technical failures were overcome the Vengeance became a most reliable aircraft. 82 Squadron never lost a single aircraft through engine failure. In fact, the squadron not only consistently topped the Command's monthly list of sorties flown, it was also at the head of the aircraft serviceability state.The accuracy of our attacks made us very popular with the army who felt they could safely move up close to an objective while we were bombing it.

When the enemy attacked Imphal and Kohima we were sent up to Khumbirgram to give close support to our troops in these areas. We then took part in the counter-attack which, a year later, enabled Burma to be recaptured. We were back in Arakan engaged on close support sorties, when the 1944 monsoon broke and we had hastily to abandon our paddyfield strip. We flew down to Kolar in southern India and this time the Mossies were real. There was even an old Blenheim for twin-engine familiarization!

Indian Nightmare

The primitiveness of the early operations from India is described by Ralph Fellows, observer of 215 Squadron.

'We climbed aboard our Wellington 10 ... and began stowing our 'chutes and equipment. As always, my job as observer (or, if you prefer it, navigator bomb aimer) required me to lug the cumbersome green bag of charts, maps and other things into the aircraft.

It was 1 May 1942 and our target was to be Magwe, a Japanese-held aerodrome in the Irrawaddy Valley. Originally, the powers-that-be had wanted us to attack Rangoon, now in enemy hands, on this first raid from India. *However, we had been able to satisfy them that this target would be at least two hundred miles outside our range* [editor's italics]. Five other aircraft from the squadron were detailed for operations that night.

We had been in India for three weeks and, at Pandaveswa, our base, 200 miles northwest of Calcutta, for ten days. Before that, in the autumn and winter of 1941/42, we had been operating with 214 Squadron in 3 Group of Bomber Command from Stradishall in East Anglia. We were familiar with operations against Germany and the ports of enemy-occupied Europe. Ours were the first "heavy" bombers to arrive in India.

We were a fairly typical crew. My pilot, Jim, was an American who, under age, had crossed into Canada and joined the RCAF in late 1940. He was still very young – not yet twenty. Our second pilot was "Moose", a large Canadian from Montreal. The other four of us were English. At twenty-five, I was the oldest.

The Jap advance had soon brought targets within our range. It had moved with alarming speed. The towns and aerodromes of the Irrawaddy Valley in Burma had fallen quickly and eventually Akyab had been overrun.

As I went up the ladder into the nose of the aircraft, I said to myself, as I had so often before on the missions to Germany, "Please God, take care of us this flight." I felt we really needed His protection.

I had made it a habit not to distract His attention for air tests and little items of that nature. He needed that time for others without any diversion from me. But whenever I climbed up the little wooden ladder to enter the aircraft before an "Op", I said very quietly and to myself, "Please God take care of us this trip." And as it worked each time, I always remembered to say, "Thank you, God," when we were safely back.

Compared with Bomber Command's operations from England, things in India were primitive.

The map sheet we had for the Pandaveswar area gave coverage to the north of the aerodrome for only some 20 miles. We had no map sheets for the area north of that.

We had our astro sextants in the aircraft but, as yet, there were no astro tables available to us. Without these, we could not perform the necessary calculations to convert the sextant reading of a star's altitude into a position line.

We could forget such sophistications as a radio "fix" or "bearing". The only radio help we could expect would be Calcutta civil radio station which would broadcast a signal at five minutes to the hour for a duration of five minutes.

Any "met." or weather information available to us was at least twenty-four hours old – but we could have the "met." and the winds from the same date the previous year!

There was no such luxury as a flare path. Instead there were set out for take off and landing about ten hurricane lanterns. A small searchlight was brought in and shone vertically upwards to help guide us home.

We were airborne at 2035 and as I sat there in the darkness I wondered just what I could do. In England it had been very different. There, we had no radar help such as Gee or Rebecca/Eureka, which was to come much later in the war, but at least we had something to go on. Usually there was a good "pin point" in the Orfordness area as we left the English coast. Crossing the North Sea, if the night was clear, we could drop a flame float into the sea and the rear gunner could provide a reliable drift reading. Stars and a sextant would offer some guidance when above cloud while the wireless operator usually obtained a number of radio bearings. Navigation over Germany depended upon the night and the target but on the return leg home radio beams and flashing occults helped.

Here we were in India, however, climbing away from our base of sorts in thick cloud and with virtually nothing by which to navigate.

I had asked George, our wireless operator, to try to pick up the Calcutta beacon at 2255. He interrupted my thoughts by reporting he had tuned in an indistinct signal at about the right time and he had obtained a bearing. It must have been Calcutta because, to my knowledge, it was the only signal available. I plotted it and it made sense but I felt I could not place any great reliance on it. In any event we had only been in the air for 20 minutes.

For the next three hours we ploughed ahead using the winds we had been given at briefing. Whether they were twenty-four or forty-eight hours old, or even the winds from last year, I hadn't a clue.

George had tuned in to try and pick up Calcutta at each hour but without success. My navigation, if you could call it that, had consisted of plotting our Dead Reckoning position on a chart using the winds we had been given. I had had no opportunity to check them.

We levelled out at 12,000 feet for the very good reason we could climb no higher. The performance of the aircraft with its full fuel and bomb load didn't allow it. Most of the time we were in cloud but on several occasions and for a few fleeting moments we encountered breaks. It was then we could see, very dimly, stars above us or a blackness below. This latter must have been the ground because what looked like the occasional village fire flickered and disappeared.

It was in one of these upward breaks in the cloud that we saw ahead of us dark, heavy stuff – thunder clouds, darker and heavier than we had ever seen before. They went up far higher than our altitude and seemed to stretch out across our path as far as we could see on each side of it.

In a few minutes we were enveloped by them. Suddenly we became, it seemed, a small fragile shell with thick, dark heavy layers pressing in on us from all sides. It was useless to try to turn to port or starboard for we had seen the immense width of the cloud formation. We could not climb because we had reached our altitude limit. We did the only thing we could and flew on into it.

The buffeting began quite quickly and was severe. We had experienced thunder-cloud flying before but this was different. I stood up front behind the pilots' positions, hung on and watched in a kind of hypnotic fright. Jim and Moose were working together at the controls but seemed to be helpless. We lost and gained height at the will of the storm and without regard to the position of the controls or the engine settings.

There were a few brief intervals of relative calm and during one of these alarmed chatter came over the intercom system from the gunners. They reported blue flashes and blue lights ringing their turrets. It all seemed very electrical and they were afraid to touch the metal of their guns or turrets. Subsequently we were to be told this was St Elmo's fire and quite harmless. At the time it happened, however, we were all very scared. We brought the gunners out of their turrets and into the main cabin area with us. It did not seem possible but conditions worsened. Lightning seemed to be flashing all over the sky. Suddenly there came a blinding flash more vivid than before, which lit up the interior of the aircraft as if it were daytime. Both engines cut out for a few seconds. We fell out of the sky at an alarming angle and then came a wild surge of power as both engines roared back into life.

Sammy, in moving from his position in the rear turret had disconnected his intercom. He must have reconnected at a spare socket in the main cabin area for suddenly we heard his Lancashire voice:

"This is just bloody stupid," he said. "We should get out of this."

At that, we began to turn and Jim asked for a course back to base. As we retraced our steps the wild bouncing of the aircraft gradually eased until at last we found ourselves in lighter cloud. True, we were still in cloud but at least it was normal cloud and we could cope with it.

Strangely, I think it was at this moment that I experienced my wildest panic. It dawned on me that my five companions were, no doubt, breathing sighs of relief that we had given it an honest effort, had been beaten back by the elements and were now returning home.

Only I knew – so I believed – that we were hopelessly lost. We had been flying for 4½ hours using winds I had been unable to check, had been thrown all over the sky and on several occasions considerably off our compass course by the worst storm we had ever experienced. To add to our problems, the influence of the electrical storm had rendered our compass suspect.

For me, one of the most frightening times in flying always came when we were unsure of our position and losing height to descend out of cloud. So easily – and vividly – imagination could conjure up the possibility of flying into high ground whilst still in cloud.

It was at this point that Jim, Moose and I discussed our petrol situation. We were getting quite low on our main tanks according to the fuel gauges but we still had our emergency nacelle tanks.

As we were still some fifteen minutes from my ETA* at Pandaveswar we flew on holding our height at about 1200 feet, just below the cloud base. The gunners had gone back to their turrets some while before and now, with the exception of George, busy on his set, all of us scanned the ground and the horizon. We hoped, desperately, to locate the light of our base searchlight. We saw nothing.

I was standing up front ready to try and map read as we should shortly be flying on to my map sheet. The time was about 0530 and the sky was getting lighter by the minute.

Suddenly the Wellington was thrown into a violent banking turn and we came round as tightly as a Wellington could. Before us, and quite close it seemed, was a hill which stretched up well above our 1200 feet. If dawn had not been quite so near or if Jim had not been quite so alert we would have buried our nose in it.

There and then we decided to circle until dawn had broken completely. We would continue our flight south when we could see the ground ahead of us.

For about the tenth time on this one trip it became panic stations all over again. Jim and Moose, between them, determined the problem quite quickly. The nacelle tanks were empty. They could not have been filled before take off and since their last use. We were down to the dregs of our main tanks.

We found a dried up river bank nearby and from our height of 1200 feet I let the bomb load go. A few seconds later there came a sharp, loud crack. Earth flew and we felt a small shock wave.

Jim now picked out a landing spot. As he turned into wind and began letting down he said, quite suddenly: "That field over there is longer than the rest. The surface looks firm. I think I can get her in there, wheels down." Little did we realize then what a momentous decision he had made.

Wheels were selected and flaps came down. We just hoped petrol would suffice and we would not have an engine cut out.

Jim reduced his speed as close to the point of stall as he dared. He had a very limited distance to play with. We banged down and there followed a

*ETA: estimated time of arrival.

series of alarming brakings. Each time the tail of the aircraft came up until we seemed in danger of standing on our nose. With a sideways swing, bringing us round at right angles to our landing direction and parallel to the far "paddy bund", we came to a dusty halt.

We tumbled out with a great deal of laughter and chatter and with voices pitched unnaturally high. We were down and down safely. We had been in the air for 8 hours and 25 minutes.

As I stood there on the flat, brown and dusty earth of the paddyfield I said to myself – and I said it with fervour – "Thank you, God."'

The crew's troubles were by no means over. From the remoteness and isolation of the landing site, arrangements had to be made to contact the airfield at Pandaveswar, to transport aviation fuel and clear a sufficiently long strip for take off.

The pilot and the navigator, learning that the nearest large town, Bhagalpur, on the banks of the Ganges, was some 30 miles away, set off on foot with a young guide. In the heat of an Indian sun, a helping hand was offered by the driver of an open ox-cart. With bells jingling, and after some ten hours' travelling over scarcely passable tracks, the pair reached the railway halt at Kaithiya. Police and friendly officialdom were alerted in Bhagalpur and in due time all the resources of Service, constabulary and local administration were marshalled.

Back at the aircraft, however, things looked ominous. Crowds of Indian villagers, appearing seemingly from nowhere, were pressing closer and closer to the Wellington. Unable to force them back, the remaining crew members fired coloured Verey cartridges over their heads. The effect was instantaneous and the onlookers were sent scattering.

Meanwhile local labour, rounded up by the police, worked through the heat of several cloudless days to prepare a path for take off. Eventually on 12 May, nearly a fortnight after the mission to Magwe had begun, the ground party from Pandaveswar reached the aircraft. By then, the navigator had drawn some consolation from hearing that had the crew continued to fly south at the time the fuel ran out they would certainly have seen their base.

As they prepared for take off, there was a final and unnerving blow. A 250-lb bomb was found to have 'hung up' in the bomb rack in the belly of the aircraft. It had defied release when the remainder of the load was jettisoned.

Ralph Fellows completes the tortuous story:

'The take off over a rough, curving and abbreviated strip, was quite obviously worthy of my "please God" ritual and I felt justified in employing it, which I did.

All went off without incident. We were soon back on my map sheet and, shortly after that, over base. Landing was no trouble although we did not push our luck by taxiing away to a dispersal point. Armourers were alerted to the problem of the bomb and dealt with it. Only then, when all was safely concluded, did I express my thanks to the Almighty and finally release Him.

Of the six aircraft which took off on the night of 1 May 1942, four returned to base early, being in the air no more than one or two hours. Two of us went on. The other crew was never heard of again.'

Opinion

D.R. Gibbs, Montserrat, West Indies, 1982

'The truth is that except for those flying the Hump, the Far East air war was nothing like as dangerous as the operations I experienced in Europe in 1941.'

Channel Calamity

On the face of it, it appeared to be a triumph for the German Navy and the Luftwaffe. For the British air and sea forces it seemed to suggest humiliation. Coming within nine months of the Fleet Air Arm's telling contribution to the sinking of *Bismarck*, it had momentarily, all the effects of a punch to the solar plexus. That was the short-term picture. In the long term, results were rather different....

For months, the German battle cruisers, *Scharnhorst* and *Gneisenau* had been bottled up in Brest, on the French Atlantic coast, hammered constantly by bombers of the Royal Air Force. It had even been put about that, as a result of these attacks, the ships were unfit to put to sea.

On 11–12 February 1942, however, they gave their dramatic answer in one of the epic operations of the war. Lord Kilbracken, who, as Lieutenant-Commander John Godley, RNVR, had a distinguished wartime record both as a Swordfish pilot and as a squadron commander in the Fleet Air Arm, has written, from the air standpoint, probably the most objective and authoritative summary of the affair.

After explaining that 825 Squadron, led by Eugene Esmonde, an officer of immense courage and dash, who had already won glory in the attack on *Bismarck*, was relatively inexperienced operationally, he points to the question marks.

Lord Kilbracken, *Bring Back My Stringbag* (Peter Davies, 1979; Pan Books, 1980)

Why they sent 825 is a question that has never been explained – and it would take some explanation. There is no attempt to do so in the report of the Board of Inquiry. Shore-based operations against enemy shipping in the Channel were not a normal function of naval aircraft. The RAF had many squadrons of fast modern bombers standing by for the operation to which Esmonde's little striking force would make an insignificant addition. Bomber Command had 242 aircraft ready to take part; there were also fifty Whitleys of which it would be reported, though they were far faster and less vulnerable than the Swordfish, that they were 'a type unsuitable for day bombing'. Coastal Command had three dozen torpedo-dropping Beaufighters briefed and ready. To support these, thirty-four squadrons of fighters were standing by, comprising over 500 Spitfires and Hurricanes.

The entire debacle, perhaps the sorriest of the war, went wrong from the beginning. For reasons hard to understand, it had been thought more likely that the battle cruisers would sail from Brest in daylight and one solitary aircraft – a Hudson of Coastal Command – had the job of watching the harbour from dusk till 2300 when another would replace him. It was a pitch dark night and the Hudson's ASV just happened to pack up at 1920. It was 2238 before the next took over. A careful watch was then kept up till morning and nobody realized that *Scharnhorst* and *Gneisenau*, accompanied by the heavy cruiser *Prinz Eugen* and a powerful escort of destroyers and E-boats, had slipped out at 2120 during those 198 minutes when Brest wasn't covered at all.

The RAF flew two other single-aircraft patrols during the night, and a dawn sweep by two Spitfires, but the enemy armada was not sighted until 1042, when two Spits who weren't even looking for it but chasing a couple of Messerschmitts just happened to fly over it by mistake. The fleet was then approaching the Straits of Dover, having covered 300 miles undetected. For unexplained reasons the pilots were in no circumstances permitted to use their wireless and had to return to base, where they landed at 1109, to report the sighting for which half the RAF was waiting.

The events which followed pass all comprehension. Alone of all the aircraft, Esmonde's squadron, as a naval unit, came under the command of Vice-Admiral (Dover), who received the news at 1130 and ordered the six Stringbags to take off for their attack one hour later. This was supposed to be coordinated with an attack by four Beaufighters, the first little wave of all the air force aircraft that were waiting in their hundreds. But the Beaufighters couldn't get airborne till 1340, seventy minutes after Esmonde's though they had been standing by for a week. It was known that extremely heavy air opposition would be encountered and it was therefore arranged on the telephone between VA (Dover) and Fighter Command that five squadrons of Spitfires and Hurricanes would accompany 825. However, only one of these reached Manston by 1230 and its ten Spitfires soon lost contact with the much slower Stringbags. So Esmonde's Swordfish, which took off precisely on time in very bad visibility, flew alone and virtually unprotected towards their immensely powerful enemy, now 10 miles north of Calais....

... They approached in two flights of three. Before reaching the powerful destroyer screen they were engaged by the enemy's most modern fighters in strength. Esmonde's own aircraft was the first to be badly damaged. His port mainplanes were 'shot to shreds' (according to a survivor) but the Stringbag kept flying as Esmonde headed at low level over the destroyers towards the capital ships they encircled. Both the pilots with him, Kingsmill and Rose, were also hit but flew on, though Rose was badly wounded, his air gunner killed, his petrol tank shattered by cannon fire. Esmonde's aircraft crashed into the sea when hit again some 3000 yards from the battle cruisers. Kingsmill and his observer, Samples, were wounded, the aircraft further damaged. But they closed within range of their target, *Scharnhorst*, and Kingsmill could aim and drop his torpedo before being forced to ditch. Rose did much the same: he pressed home his torpedo attack, also on *Scharnhorst*, and was able to turn back over the destroyers before his engine succumbed.

Of the second flight, led by Lieutenant Thompson, nothing can be reported. Flying astern of Esmonde's, they were never seen again by any of those who survived.

So ended the most incomprehensible, most badly planned, most gallantly led operation in the history of naval aviation. No hits were scored on the enemy fleet. Of the eighteen officers and men taking part, thirteen were killed and four seriously wounded. Lee, who was Rose's observer, alone emerged unscathed.

Esmonde was awarded a posthumous VC. The four surviving officers, all RNVR sub-lieutenants, received DSOs, the surviving gunner a CGM, the next highest decorations for gallantry. They had been picked up by allied torpedo boats after over an hour in their dinghies.

During the rest of the day, the enemy flotilla was attacked by the RAF as well as by torpedo-boats and a handful of ancient destroyers, but to no avail whatever. Of the 242 modern bombers sent out with full fighter escort, 188 'failed to locate the ships or were unable to attack them owing to low cloud'. So states the Board of Inquiry's report. No hits were scored by the twenty-eight fast torpedo-bombers (Beaufighters) sent out additionally, three of which were lost. The only redeeming feature was that both battle cruisers were quite badly damaged by mines laid ahead of them by aircraft of Bomber Command.

The Board found that no blame could be attached to anyone for the whole shameful disaster.

Writing almost a decade after the event, Churchill contrived to put a different complexion on what had, at the time been a bitter-tasting pill:

Winston Churchill, *World War Two, Volume IV: The Hinge of Fate* (Cassell, London; Houghton Mifflin, New York)

'... Very soon however we found out, by our Secret Service, that both the *Scharnhorst* and the *Gneisenau* had fallen victims to our air-laid mines. It was six months before the *Scharnhorst* was capable of service, and the *Gneisenau* never appeared again in the war. This, however, could not be made public and national wrath was vehement.'

On the other hand, Adolf Galland, who was privy to the German High Command's intentions and plans well in advance of the Channel sortie, has given his country's verdict.

Adolf Galland, *Die Ersten und die Letzten* (Franz Schneekluth, Germany, 1943; Methuen, London, 1955 and in paperback by Collins, 1975)

'In spite of all protests, including those of our former enemy, it must be said that ... objectively ... the breakthrough ... by the ... battle cruisers ... under the umbrella of the German fighters constitutes, in planning and execution, a great and impressive military victory....

It confirmed once more that naval forces in coastal waters can stand up to attack and defence only if they are covered by superior air forces. The superiority had been undeniably achieved by the German Air Force with little more than two hundred fighters and a few bomber wings....'

Australian Bomber Captain – Pilot Officer R.H. Middleton VC

Peter Firkins, City Beach, Western Australia, 1982

Rawdon Middleton, from Brogan Gate in New South Wales, was a 26-year-old pilot officer of 149 Squadron when, on 28 November 1942, he and his crew of a Stirling bomber were briefed at Lakenheath, Suffolk, for an attack on the Fiat works at Turin, in northern Italy.

It was the crew's twenty-ninth mission – one short of the requisite thirty needed to complete their operational tour. For Sergeant S.T. Martin, the front gunner, it was his thirty-third operation, 'three over the quota having volunteered to continue with Middleton until he had finished.' The Australian captain encouraged that kind of loyalty.

In a hazardous flight across the Alps, 'with the four engines labouring to squeeze every extra foot of altitude.... Middleton used excess fuel to climb to the necessary 12,000 feet as he fought his way past the great mountain peaks which at times towered above as they flew down the passes ... in the darkness of the winter's night.' Over Turin, he dived the aircraft 'to 2000 feet positively to identify the target'.

There was violent enemy reaction over the aiming point. The Stirling was hit hard and three of the crew, including the captain and the second pilot, Flight Seregeant L.A. Hyder, were injured.

Peter Firkins, Australian author, and narrator of this story, who himself completed a tour of operations as an air gunner in Lancasters of No. 460 Squadron, Royal Australian Air Force, continues his account of this courageous feat of arms.

'A shell splinter struck Middleton's face destroying his right eye and exposing the bone over the left, also wounding him in the body and legs.... The second pilot ... was wounded in the head and both legs and bled profusely.... And so they turned to cross the Alps a second time....

...By now Middleton must have known he had little chance of survival but he kept going for the sake of his crew. The possibilities of ... parachuting or landing in France were discussed but Middleton was determined to reach the English coast so that his crew would at least have the chance to survive and fight again. "I'll make the English coast. I'll get you home," he said several times over the intercom.

For four hours Middleton's wounds worsened; his agony was almost unbearable and his strength ebbed.... As the French coast was approached the aircraft, flying at 6000 feet, was again ... hit by intense flak.... Twenty minutes later the friendly shores of England were sighted with five minutes' fuel remaining.... The captain then decided he would avoid risking civilian casualties by flying across the English coast and possibly crashing into a built-up area; so, gathering whatever reserves of his extraordinary strength remained, and with the fuel gauges showing empty, he gave the order to bale out as he flew parallel with the English coast. He then headed out to sea again with two of his crew, Mackie, the front gunner, and Sergeant J.E. Jeffrey, the flight engineer, who refused to leave him.... Each ...

parachuted into the sea but neither survived the night.... however, five of the crew had survived.

It was approximately 0300 hours on 29 November when Middleton crashed into the sea. On 1 February 1943, his body was washed ashore at Shakespeare Bay, Dover.'

Firkins concludes his narrative by quoting from 'a remarkable letter which a group of old aircrew friends' wrote subsequently to Middleton's mother.

'Greatly as his country has honoured your son, he deserved no less. He could not possibly have done more. We, who knew him, also honour him, and we would like to tell you why.

He was so quiet and so unostentatious. He was so quiet, in fact, that he was not easy to get to know. His crew did the talking when they were interrogated after an op. He spoke only to give his judgement when that was needed or asked. Yet underneath that quiet was a strength of character and a gentleness that only the strong possess. How strong he was we suppose you always knew. We only suspected it, or, at the best, half knew until – well, until he won the Victoria Cross.

He would hardly be noticed in a crowded room.... Yet there was always a constant turning to him for confirmation of a point. "Didn't we, Ron?" or "Wasn't it, Ron?" ... was so often heard. He was popular, though he never sought popularity. His crew thought highly of him as a captain, a pilot – and a man. We who fly know what the high estimation of a captain means.

Victoria Crosses are only rarely awarded. You know why. They only go to the bravest among the brave. His exploit caused a tremendous wave of emotion throughout the country. We knew he would not have wished it, but such things could not have been in his power to prevent....

We, in the RAF, are accustomed to hearing of bravery, so much so that our appreciation of it tends to be blunted. But this was just outstanding, "unsurpassed in the annals of the RAF", the official citation itself said. So it was....

We have some appreciation of his fortitude in bringing that machine back, wounded though he was, but other men have done that. What made his courage "unsurpassed" was that when he flew his machine out to sea he knew exactly what would happen.... Of him it soberly might be written, "For greater love hath no man than this...."

We do not offer sympathy. But at the risk of intrusion we would like to say that we humbly share your pride. He was not wasted. With the inspiration of such an act behind him, what other man, whether in the fighting services or not, would dare to do less than his best? The British Commonwealth is safe while it sends out men like Ron Middleton.'

'Short Prayer'

Chosen by John Rennison, Cheltenham, Gloucestershire

Almighty and all present Power,
  Short is the prayer I make to Thee;

I do not ask, in battle hour,
For any shield to cover me.
The vast unalterable way
From which the stars do not depart
May not be turned aside to stay
The bullet flying to my heart.
I ask no help to strike my foe,
I seek no petty victory here;
The enemy I hate, I know
To Thee, O God, is also dear.
But this I pray, be at my side
When death is drawing through the sky.
Almighty God, who also died,
Teach me the way that I should die.

Hugh Rowell Brodie

Flight Sergeant Hugh Rowell Brodie, Royal Australian Air Force, a former schoolmaster from Victoria, was killed in action over Germany on 2 June 1942, aged thirty, flying with No. 460 Squadron. So far as is known, this short verse which was first printed in *War Service Record 1939–45*, published by the Victoria Education Department, and subsequently appeared in a booklet produced in New Zealand by the Methodist Church Group, was Brodie's only poem to be published. It is reproduced from AD REM by kind permission of The Butterfly Company Ltd.

Cologne

Group Captain Leonard Cheshire VC, *Bomber Pilot* (Hutchinson, 1943; White Lion Publishers, 1973)

'What about the bombs? Have we still got them?'

'Certainly.'

'Well, we'd better go find Cologne.'

Taffy looked back over my shoulder, shouted out 'Jesus!' and dashed off down the fuselage. What he went to do I don't know. I only know there was, for a while, a confusion of cries and noise and violent movement, and then Taffy came back and disappeared into the front turret. The shells were still as fierce as ever, but now that there had been a diversion it was not quite so bad. Someone flopped down beside me. I looked up. He was squatting on the step, his head down below his knees and his arms covering his face. I leant across and pulled him gently back. Pray God I may never see such a sight again. Instead of a face, a black, crusted mask streaked with blood, and instead of eyes, two vivid scarlet pools.

'I'm going blind, sir. I'm going blind!'

I didn't say anything: I could not have if I had wanted to. He was still speaking, but too softly for me to hear what it was. I leaned right across so as better to hear. The plane gave a lurch, and I fell almost on top of him. He cried out and once more buried his face below his knees. Because I could not stand it, I sat forward over the instruments and tried to think of something else, but it was not much good. Then suddenly he struggled to his knees and said:

'I haven't let you down, have I, sir? I haven't let you down, have I? I must get back to the wireless. I've got to get back. You want a fix, don't you, sir? Will you put the light on, please, so that I can see?'

So it was Davy. Davy: his very first trip. Someone came forward and very gently picked him up. Then came Desmond. He sat down beside me and held out his hand. I took it in both of mine and looked deep into his smiling blue eyes.

'Everything's under control.'

'God bless you, Desmond.' Never have I said anything with such feeling. 'What about Davy? Is he going to die?'

'He's OK. Revs is looking after him.'

'Thank God. Tell me the worst. What's the damage?'

'Pretty bad.'

'Will she hold?'

'I don't know. About evens I should say. The whole of the port fuselage is torn: there's only the starboard holding.'

'How about the controls?'

'I don't know. They look all right, but it's difficult to tell. Shall I go and look at them more carefully?'

'No, it makes no odds. We're going to make a break for it however bad they are. If they're damaged, I think I'd rather not know. If this ack-ack doesn't stop soon I shall lose control of myself, Desmond. I can stand all the rest, but this I can't. They've got us stone cold. We can't turn, and we can't dive, and we can't alter speed, and it's only their bad shooting that will – '

A staccato crack, and Desmond covered his face.

'Desmond, Desmond! Are you all right?'

'Yes, sure.'

'You're bleeding.'

'That's nothing. A bit of perspex, probably: the splinter missed me. Taffy's signalling. Have you still got the bombs or something?'

'Yes. Which way?'

'Right. Hard right. Go on, much further.'

'Tell him to shut up. What the hell does he think this is? A Spitfire?'

We went on like this for some time. Turning all the time, very gently, but none the less turning, and always to the right. Then at last the bombs went. I felt the kick as they left the aircraft. Desmond stood up and went back to Revs and Davy.

'Where was it?'

'Cologne.'

Yes, there was the Rhine right beneath us. I recognized the wide curve just south of the town. What a long way we must have flown back into Germany, and all through my carelessness. Then Taffy came back. He jerked his thumbs and laughed, but his face looked a little drawn.

'What luck?'

'Wizard. In the middle of the yard.'

'Fair enough. Go and work me out a course for home, will you? I'm flying 310: more or less right.'

He looked me full between the eyes. 'Shall we make it, sir?'

'Why, of course, Taffy.' Never have I seen such a genuine look of relief over anyone's face. I thought: 'He's a trusting soul, Taffy. I hope no one ever takes advantage of it, because if they do he'll fall with an awful bump.'

Stuka Victory

Hans-Ulrich Rudel, *Stuka Pilot* (Euphorion Books and Transworld Publishers)

Oberst Hans-Ulrich Rudel was one of the Luftwaffe's best known and most decorated officers. He made his name as an anti-tank, Stuka (Ju 87) pilot, flying for much of the war on the Eastern front. In more than 2500 operations, he was credited with the destruction of upwards of 500 Soviet tanks as well as the Russian battleship Marat: *an astonishing record.*

Before Rudel died last year in Austria, aged sixty-six, I asked him to propose some piece of writing for inclusion in this collection. He responded at once and ventured to suggest this passage from his book Stuka Pilot. *It clearly made for him a proud memory.*

. . . The Russian fleet is based on Kronstadt, an island in the Gulf of Finland, the largest war harbour in the USSR. Approximately 12½ miles from Kronstadt lies the harbour of Leningrad and south of it the ports of Oranienbaum and Peterhof. . . . Our chief concern is the two battleships, *Marat* and *Oktobrescaja Revolutia*. Both are ships of about 23,000 tons. In addition, there are four or five cruisers, among them the *Maxim Gorki* and the *Kirov*, as well as a number of destroyers . . . .

There is no question of using normal bomber aircraft, any more than normal bombs, for this operation, especially as intense flak must be reckoned with. . . . We are awaiting the arrival of 2000-pounder bombs fitted with a special detonator for our purpose. With normal detonators the bomb would burst ineffectively on the armoured main deck and though the explosion would be sure to rip off some parts of the upper structure it would not result in the sinking of the ship. We cannot expect to succeed and finish off these two leviathans except by the use of a delayed action bomb which must first pierce the upper decks before exploding deep down in the hull of the vessel. . . .

. . . Our 2000-pounders arrive. The next morning reconnaissance reports that the *Marat* is lying in Kronstadt harbour. . . . Now the day has come for me to prove my ability. I get the necessary information about the wind, etc., from the reconnaissance men. Then I am deaf to all around me; I am longing to be off. If I reach the target, I am determined to hit it. I must hit it! We take off with our minds full of the attack; beneath us, the 2000-pounders which are to do the job today.

Brilliant blue sky, without a rack of cloud. The same even over the sea. We are already attacked by Russian fighters above the narrow coastal strip; but they cannot deflect us from our objective. . . . We are flying at 9000 feet; the flak is deadly. About 10 miles ahead we see Kronstadt. . . . At an angle ahead of me I can already make out the *Marat* berthed in the harbour. The guns boom, the shells scream up at us, bursting in flashes of vivid colours; the flak forms small fleecy clouds that frolic round us; if it was not in such deadly earnest one might use the phrase: an aerial carnival. I look down on the *Marat*. Behind her lies the cruiser *Kirov*. Or is it the *Maxim Gorki*? These ships have not yet joined in the general bombardment.... They do not open up on us until we are diving to the attack. Never has our flight through the defence seemed so slow or so uncomfortable....

...Our diving angle must be between 70 and 80 degrees. I have already picked up the *Marat* in my sights. We race down towards her; slowly she grows to a gigantic size. All their AA guns are directed at us. Now nothing matters but our target, our objective; if we achieve our task it will save our brothers in arms on the ground much bloodshed....

...The ship is centred plumb in the middle of my sights. My Ju 87 keeps perfectly steady as I dive; she does not swerve an inch. I have the feeling that to miss is now impossible. I see the *Marat* large as life in front of me. Sailors are running across the deck, carrying ammunition. Now I press the bomb release switch on my stick and pull with all my strength. Can I still manage to pull out? I doubt it, for I am diving without brakes and the height at which I have released my bomb is not more than 900 feet.

The CO said when briefing us that the 2000-pounder must not be dropped from lower than 3000 feet as the fragmentation effect of this bomb reaches 3000 feet and to drop it at a lower altitude is to endanger one's aircraft. But now I have forgotten that!... I tug at my stick ... exerting all my strength. My acceleration is too great. I see nothing, my sight is blurred in a momentary blackout.... My head has not yet cleared when I hear Scharnovski's* voice:

'She is blowing up, sir!'

Now I look out. We are skimming the water at a level of 10 or 12 feet and I bank round a little. Yonder lies the *Marat* below a cloud of smoke rising to 1200 feet; apparently the magazine has exploded.

'Congratulations, sir.'

Scharnovski is the first. Now there is a babel of congratulations from all the other aircraft over the R/T. From all sides I catch the words: 'Good show!' I am conscious of a pleasant glow of exhilaration such as one feels after a successful athletic feat. Then I fancy that I am looking into the eyes of thousands of grateful infantrymen....

...The whole neighbourhood is full of AA guns; the air is peppered with shrapnel. But it is a comfort to know that this weight of iron is not meant exclusively for me! I am now crossing the coastline. The narrow strip is very unpleasant. It would be impossible to gain height because I could not climb fast enough to reach a safe altitude. So I stay down. Past machine guns and flak. Panic-stricken Russians hurl themselves flat on the ground....

Frustrated Ally

Major General Count I.G. du Monceau de Bergendal, Royal Belgian Air Force, Brussels, 1982

By now, men from the Allied nations, who had made their way out of

*Rudel's gunner. (Ed.)

Europe, were moving from the training establishments in the United Kingdom into the operational squadrons – but not always according to plan.

'I should first mention that I had graduated from the Belgian Military Academy as a cavalry officer in December 1938. Horses becoming obsolescent, and being replaced by the combustion horse power of motorized squadrons, I immediately transferred to the Air Service where horse power was abundant and provided thrills. My family which was – and still is – deeply attached to the quadrupeds, frowned at my wayward behaviour; but I ignored the censure.

Having been born in England during World War I, I was not considered an alien by immigation authorities and after a day or so was dispatched to RAF Station, St Athan, and told to wait.... Wait I did until I found myself at RAF Station, Shawbury, near Shrewsbury, in Shropshire. It was a bomber SFTS,* flying Airspeed "Oxfords". I was delighted to be flying again. I was even more delighted when the school, with unerring vision, decided at the end of the course that I was unfit to fly bombers. I was thus posted to No. 56 (Fighter) OTU† at Sutton Bridge to fly Hurricanes.

My flight commander at Sutton Bridge, Flight Lieutenant Harry Tait, was relieved to hear me speak some English. He had feared having to go back to the "la plume de ma tante est sur la table" routine of his school days. Under his guidance I experienced no trouble flying the Hurricane.

We were given pep talks on the fighter pilot's role in the war. A Flight Lieutenant Hallowes (DFC and Bar), professing scorn, disregard and disdain for the Me 109 called it, all in all, a "dud".

If the Me 109 was such a "dud", how come, I wondered, that RAF planes were shot down during the Battle of Britain? It didn't make sense. Such pep talks left me cold and more than somewhat distrustful....

After roughly 20 hours on Hurricanes, we were asked to give our choices of stations or squadrons for posting. I knew nothing about Fighter Command or its groups, but one thing I knew perfectly well, I had no experience whatsoever as a fighter pilot, and no pep-talk could give me any. So my "preference" was for any squadron in Great Britain – excluding the Shetland Islands.

I wasn't trying to be funny, but somewhere in the upper reaches of command some chairborne warrior must have thought I was.

"Ah! Ha! This lad rules out the Shetlands! Very funny! I'll give him the next best thing."

I was posted to 253 Squadron at RAF Station, Skaebrae, in the Orkneys.

Flight Lieutenant Tait sympathized and suggested that I signal the squadron and ask for seven days' leave before reporting for duty.

The reaction was immediate and understanding: "Ten days' leave granted".

Such humanity warmed my heart. I hoped the chairborne warrior (from now on dubbed "the Stinker") would be annoyed if told of 253 Squadron's splendid behaviour.

*Service Flying Training School.
†Operational Training Unit.

I trained down to London with a bunch of robust and exuberant Australians, where, in spite of (or because of) a couple of night "blitzes" we had a wonderful time....

The train journey to the north of Scotland was mirthless. The scenery became more and more desolate and no one on the train seemed to enjoy the voyage. I reached Inverness in a state of hopeless resignation. Within a few hours I boarded an Avro "Anson" piloted by a very young pilot, named Beddow, who seemed to live in a trance which I could well understand considering the surroundings.

The only other passenger in the Anson was a parrot in an expensive-looking gilt cage. I tried hard, but the bird refused to join in any conversation. I never discovered why it was travelling to Skaebrae. Actually, it never made it. An hour after take off it lay dead at the bottom of the cage, frozen to death. Very sad. Probably another victim of the Stinker.

We landed in time for tea with a bunch of cheery officers seemingly in excellent health.

From what I saw of RAF Skaebrae, all the required conditions existed for a severe epidemic of deep neurosis; yet I could not detect the slightest indication of breakdowns among these lads. They looked surprisingly normal....

My unpacking was interrupted by one of these cheery young men who waved a piece of paper and burst into a laugh:

"You're posted!"

How well I knew.... This was probably the first sign of neurotic imbalance. I tried to be soothing. Of no avail.... The boy got really excited:

"You're posted!" he almost screamed. "Posted to North Weald, near London, you lucky bastard!"

It took me some time to digest the news. I grasped the piece of paper. There it was. Black on white. It was an official signal all right. Or was it another piece of British humour perpetrated by the Stinker?

I slipped on my tunic, took the boy by the arm and asked him to lead me to the Station Adjutant. This officer very obligingly asked for confirmation and got it within an hour. I was truly flabbergasted....

In the officers' mess there was a double celebration; one for my arrival, one for my departure. Most of the chaps there looked at me in awe. To pull a thing like that one had to know the right people! I kept a low profile and retired early. I had to get out of this place before the powers-that-be changed their minds....

The southbound express was like "Ole Man River".... It just kept rolling along – day and night.

To reach North Weald, I had to leave the train somewhere near Stevenage and board some "bummeltrain" which would deposit me within a short distance of the base. Stepping down from the express, I saw my name chalked up on a blackboard; an RTO desired conversation....

The officer asked for identification and perused my papers at length. Having been satisfied he gave me a friendly smile. "You're not going to 56 Squadron," he said. "This message says you have to report forthwith to 609 Squadron at Biggin Hill."

I closed my eyes to avoid a swimming sensation. The Stinker was at large again ... prowling freely....

What on earth was I expected to do at Biggin Hill? The only silver lining I could see in such a predicament was that it would certainly be exciting.

In due time, I alighted from a train at a place called Bromley. There ensued what our American friends call a complete SNAFU of telephone calls, missing transports and wrong railroad stations. After an hour and a half's waiting at Bromley Station I felt the malevolent doings of the Stinker and expected some RTO to appear with a signal posting me to Iceland, Cairo or Singapore.

At that stage of the game I was ready for absolutely anything.... Little did I know!

A passenger car reported and I loaded my things in the boot. The driver sniggered and said in a pointed manner that when Bromley was mentioned in connection with Biggin Hill it always meant Bromley South, not Bromley.

He drove me straight to 609 Squadron dispersal where I suffered one of the most severe shocks of my life.

609 (West Riding) Squadron, Auxiliary Air Force, was flying Spitfires!'

The First of the Fortresses (B-17s)

Lieut.-General Baron Michel Donnet, Royal Belgian Air Force, *Flight to Freedom* (Ian Allan, 1974)

The 17 August 1942 was a historic date when the first of the US 8th Army Air Force's mighty B-17 formations went into action. They carried no less than thirteen heavy-calibre ·5-in. machine guns and a crew of ten men. Their first target was ground installations near Rouen and the formation was led by Major Tibbett who, three years later, was to drop the atomic bomb on Hiroshima.

The briefing had called for the overall formation speed to be cut down so that they could remain in more compact formation and be easier for us to protect. We were to rendezvous at 25,000 feet over the English coast, and all three Mark IX-equipped Spitfire squadrons would provide the close escort right round the bombers....

The first leg went precisely as planned.... The weather was clear and we could see some impressively accurate bombing of the ground targets. The huge formation wheeled slowly about and set course back for England.

'Bandits! Twenty plus coming in behind the big boys!'

'I see them. Tally-ho!'

This was the Germans' first sight of the 'Forts' and they made the mistake of coming straight in at them without any height or speed advantage.* The

*The Luftwaffe soon learnt. A year later, in the first great massed attack on Schweinfurt, the German fighters exacted a terrible carnage on the US bomber formations when they were 'naked' and exposed. It gave rise to a telling remark by Curt Lemay (General Curtis Lemay), one of the group leaders that day. David Scott-Malden, the Royal Air Force's liaison officer with the US 8th Bomber Command, heard it. At the debriefing afterwards, the group commanders were asked to give their views, but no one spoke. Eventually, Lemay was ordered by the General to give an account. 'Sir,' he said, 'the A2 officer briefed us the Germans had 1127 fighters operational within range of our route. They had no abortive sorties that day,' and sat down.

first two ran into a tremendous concentration of fire from the 'Forts' gunners... I found myself in a whirling dog-fight with a dozen more milling around my section.... I closed with another head-on but at the crucial moment my cannon jammed and I only had the four machine guns. Twisting round to face every attack as it came, the fight drifted over the French coast....

The 'Forts' had crossed out again heading for England and our job was done. Levelling out over the sea I looked ahead and saw three fighters going my way. I closed on them thinking they were three of ours when I realized it was a lone Spitfire being chased by two 190s. So engrossed were they in closing in for the kill that they committed the fighter pilot's cardinal sin of forgetting to look back. I edged close in behind the second 190 ... pressed the button and fired a long burst from my machine guns, the only armament I had left. I saw strikes on his wings. He broke hard left and I had enough time and enough ammunition to get in another burst at the leading one before he too dodged left in a screaming turn.

I came alongside the Spitfire and recognized it as from 403 Squadron. Together we flew back to England, the other pilot giving me a thumbs-up and grinning from his cockpit. Later that evening he telephoned ... to say ... his aircraft had been damaged and he was incapable of taking evasive action. Had I not suddenly appeared literally out of the blue he would have been a goner. He was very grateful....

Dieppe

Colonel Bernard Dupérier, *La Vieille Equipe* (Edition Levrault-Berger, Paris, 1951)

2300 hours, 18 August 1942
This is it!

At 5.30 this afternoon all pilots were called to the briefing room. The atmosphere was quite different from that of other days. The controllers were there from the ops room as well as many other officers who normally never come. Eventually Group Captain Lott (Air Vice-Marshal G.E. Lott) and 'Dutch' (Group Captain P.H. Hugo) arrived followed by Albert, all three laden with maps and papers. At once there was complete silence.

'Well now,' said the CO, 'at dawn tomorrow a Canadian Division will land at Dieppe.'

Wild enthusiasm greeted these words. Lott smiled, waited a moment, held up his hand for silence and then added: 'And at 11 o'clock all will re-embark.'

So much for our hopes! You could see in the eyes of my pilots intense disappointment. What! They were going to land, take a town – and all that for nothing. They didn't understand, did not wish to understand or admit that, once having set foot on our soil, they could deliberately leave again.

The Group Captain then gives details of the operation. Before dawn, Hurricane squadrons will bomb and machine-gun the German coastal positions. At the same time Bostons will lay smoke while commandos land to the east and west of the town to take out the defending gun batteries.

Passing through a channel cleared by minesweepers, a whole fleet of special craft will land the Canadian regiments, and supporting tanks, right on the Dieppe beaches. And the names of these regiments which were going to renew their blood ties with the Old Country give us a thrill. The 'Fusiliers de Mont Royal' and the 'Royal Canada' will be here tomorrow on the soil of Normandy. Montcalm's soldiers are coming home.

'Dutch' takes up the briefing and says he expects a maximum effort from all of us. Take off for the first sortie will be at 4.35 a.m. while it is still dark.

Finally, the Group Captain tells us that the aerodrome is now completely sealed off and the telephone cut off except on his own personal order.

Leaving the briefing, I take Labouchère and Mouchotte to my office where we are joined by Herrera, Tilly and Day. We quickly draw up the order of battle for each sortie. Then we go to the mess for a light dinner and spend the rest of the time on the lawn nattering away in the calm of the evening.

Once again I am struck by this obsession we all feel, pilots, fitters and riggers, who only live for the day when we get back to France after chucking the Huns out of our country or burying them in it.

When, just now at the briefing, Lott had told us that those in difficulties could belly-land on the racecourse where the commandos would pick them up, I had seen such looks and smiles on my pilots' faces that I had to warn them in no uncertain terms that no landing would be permitted that could not be proved to be absolutely essential.

19 August 1942
A disturbed night. Impossible to sleep under such strain. You've got to have lived a night like this to know what 'armed watch' really means.
3.20 a.m.

I get up, I put on my No. 1s. If I'm shot down today, I must be properly dressed.
3.45 a.m.

Down to the mess for early morning tea. All the pilots arrive either one by one or in small groups – most of them having had the same idea as myself and discarded their RAF battle dress in favour of French uniforms.
4.10 a.m.

Leave for the aerodrome. A wonderful starry night, black as pitch. At dispersal, those flying on the first sortie are nervously checking their equipment. The others scarcely hiding their envy, help as best they can. Outside in the cold night, the fitters are running up engines and doing the final inspection, their torches making small circles of light on the dark outlines of the Spits.

As the 'Wingco.' is flying with us, I shall be leading 'B' Flight, blue section. I am a bit worried at the thought of a wing take-off in total darkness.
4.30 a.m.

The pilots get into their machines and one by one the engines start up. The sparkle of the red and green navigation lights shows the aircraft parked all round the field.
4.38 a.m.

I see 'Dutch' taxi out slowly on the south side of the runway and as the thunderous roar of twenty-four engines shatters the calm of the night the red

and green lights skip along to line up at the end of the field.
4.45 a.m.

Both squadrons are airborne and formed up and have been joined by the other two from Fairlop. With all navigation lights on and the night just beginning to fade far away to the east, we set course for France.

Gradually, the lights are switched off as it becomes possible to see the other aircraft against the dawn. A great reddish halo guided us to Dieppe where we arrived at the appointed hour over the beaches.

I think it will be a long time before those who took part in it forget that first sortie. 1500 metres below us the battle was raging. Frightful flashes of explosions tore apart the dark curtain of the night still covering the coast. Tracers from automatics seemed to weave threads of gold or purple through the scene which gradually emerged from the dark covered with a pall of smoke. And day broke with a grey and pink dawn out of which rose a monstrous, blood-coloured sun.

But we were not there to enjoy this fantastic sight as the sheaves of tracer so forcefully reminded us. The FW 190s made their first appearance coming in and out of the dark and at once our chaps turned to meet the challenge. It was a quick pass and the Huns evaded but not before Boudier got one of them. He disappeared in a spin and, though we could not count it as a confirmed, he probably never got back to base. For my part, I got in a short burst at a FW 190 which was attacking Yellow Section. This chased it off and it disappeared but I don't know if I hit it.

At 5.50 a.m. the relieving squadrons arrived and we returned to base via Beachy Head.

Editor's note: In the course of the day, 340 made four sorties over Dieppe, shot down 3 Dornier 217s and damaged 5 more as well as 1 FW 190 – at a cost of 1 pilot and 2 aircraft.

An Eagle Fell

Vern Haugland, *The Eagle Squadrons: Yanks in the RAF 1940–1942* (Ziff-Davis Flying Books, 1979)

The three Eagle squadrons, Nos 71, 121 and 133 which, with their American volunteer pilots, had built such a formidable reputation with the Royal Air Force were transferred to the United States Army Air Force in the autumn of 1942. The experience they took with them was important to the US 8th Fighter Command in which they formed the nucleus of the famed 4th Fighter Group.

The changeover period, for 133, was marred by hideous catastrophe. On 26 September, it was ordered to move from its base at Great Sampford, on the northeastern periphery of London, down to Bolt Head on the south Devonshire coast. The squadron, recently equipped with the latest Spitfire IXs, was to fly in a wing escort mission to Brittany, in southwest France, in support of the 97th Bombardment Group with its B-17s. The primary target, 120 miles to the south across the English Channel, was the Focke-

Wulf maintenance plant at Morlaix on the northern coast of the Brest peninsula.

It looked, on the face of it, to be a straightforward operation. But everything went wrong – the briefing, the timing of the rendezvous with the B-17s which was missed, the weather and the leadership. As a result an entire Eagle squadron perished.

The unit's able commander, 'Red' McColpin (Major-General Carrol Warren McColpin), had been called to London that day to deal with the transfer. His stand-in was Gordon Brettell, an English flight lieutenant, 'a good planner, impetuous to a fault, but a great guy'. He was light on 'leadership experience'. The principal cause of the disaster, however, lay elsewhere as Vern Haugland readily explains.

Bob Smith* agreed.... It all seemed very elementary. Simply fly south, pick up a bunch of B-17s on their bombing mission, escort them back to the friendly shores of England, and buzz off home again. It sounded like a piece of cake....

Perhaps because of last-minute haste, the weather briefing was incomplete and in error. Bob Smith said, long after the war:

'The big hooker was the metro briefing. Some clown masquerading as a weather officer forecast a 35-knot head wind at 18,000 feet, our mission altitude. Instead, we had a 100-knot tail wind – a 135-knot bloody streaking catastrophe.

'We never found out if it was stupidity or carelessness or whatever – or even who was really responsible. Whoever the thick-headed incompetent son of a bitch was he can take credit for twelve Spitfires, brand-new, destroyed; five good fighter pilots down the tubes; and assorted types of grief for the rest of us. He should have gotten the Iron Cross and a pension from the Third Reich....'

...The bombers, far ahead of the Spits, unwittingly had crossed the 75-mile cloud-blanketed Brest peninsula and, far south of their assigned target, were racing across the Bay of Biscay towards Spain ... whisked along by tail winds of 100 to 115 miles an hour. The lead navigator for the B-17s later estimated that the bombers had reached the base of the Pyrenees on the Spanish border before turning back. But turn back they finally did, meeting the Spitfires which also immediately swung north toward home.

The squadron eventually let down through the overcast, breaking cloud at some 3000 feet above the ground. Haugland continues:

...Smith recalled that ... a southerly coast line soon came into view. 'What we didn't know was that this was the south coast of the Brittany peninsula, not the south coast of England,' he said. 'The low altitude made it impossible to see enough of the region to identify it....'

...Someone said: 'There's a city off to starboard'... That city should have been Southampton or Plymouth or Portsmouth or anywhere except France. It was Brest – wall-to-wall anti-aircraft guns and odd fighter bases here and there.

No self-respecting fighter squadron is going to fly over a friendly city in a

*R.E. Smith baled out, evaded capture and returned to England.

loose, unimpressive formation.... No way. Tighten it up! Wing to wing! Nose to tail!

That's what we did – close formation over Brest at about 2500 feet. What a target! Those German gunners must have had a hundred casualties, stepping on each other, trying to get off the first shot. One of them finally did....

...In London, Squadron Leader McColpin, newly commissioned a major in the US Air Force, listened in shocked incredulity to a German radio broadcast about his squadron's destruction and rushed back at once to Great Sampford....

'All my queries were blunted both by the RAF and the USAAF, either because of security in the case of the RAF or ignorance in the case of the US officials,' McColpin said. 'There was no first-hand information at all at the base.

'As far as I could determine, the fighter wing as a whole lost six other aircraft, from which four pilots were rescued off the English coast, in addition to our twelve. Another four crash-landed, out of fuel. All in all, 22 of the 36 fighters taking off were lost or wrecked.

'As I recall, eighteen B-17s started the mission, and none was lost. Those that failed to complete the mission simply aborted.

'No bombers or fighters were intercepted by the enemy at altitude. The only action came when 133 Squadron descended through the clouds. No doubt the Germans couldn't believe anyone would be flying a combat mission in that area in that weather....'

Of the Eagle pilots who survived this dreadful experience a number, including the leader, Gordon Brettell, became prisoners in Stalag Luft III at Sagan in Silesia. Brettell took part in the 'Great Escape' at the end of March 1944. Recaptured, he was one of the fifty to be executed by Himmler's Gestapo and the SS on the direct orders of Hitler.

'Immortal is the Name'

Quoted by Vern Haugland in *The Eagle Squadrons: Yanks in the RAF 1940–1942*

Lord, hold them in thy mighty hand
Above the ocean and the land
Like wings of Eagles mounting high
Along the pathways of the sky.

Immortal is the name they bear
And high the honour that they share.
Until a thousand years have rolled,
Their deeds of valour shall be told.

In dark of night and light of day
God speed and bless them on their way.
And homeward safely guide each one
With glory gained and duty done.

Anonymous

Loss of a Brother

The Galland Papers, Bonn, (Bad Godesberg), 1982

The losses multiplied – on both sides. Grief came to the Galland family, as they had always rather anticipated it might, when, on 31 October 1942, Paul, youngest of the four brothers, all of whom served as pilots in the Luftwaffe, was killed in action returning to France from an attack on southeast England.

The family called him 'Paula', 'Paulinchen' or 'P.G.' and he was a favourite. This was manifest in the letter, dated 18 November, which the next brother, Wilhelm-Ferdinand ('Wutz' to the others), wrote, soon after Paul's loss, to Theodor Lindemann ('Uncle' Theo) who was himself a pilot in Jagdgeschwader 26 during the war, and a close friend.

Dear 'Uncle' Theo,
Your letter confirmed the deep sentiments and enduring relationship on which our friendship rests. Thank you, my friend.

I cannot describe just how unbearably saddened I am by Paul's loss. The days I spent with my shattered parents were the most painful I have ever experienced. Until then I never imagined what it would mean to lose this brother and friend. You know what he meant to me in my life, dear Theo.

The days when I have realized that I would never see the sunny P.G. again, have caused me physical pain. However, the work and the operations with the squadron, and, not least, the kindness and splendid comradeship of my officers, have done me good. I now try to think of the good Paula with pride and loving admiration. Whenever I fly over the Channel – over his grave – I see his face, happy and shining, and with that inner contentment which, to me, he will always have, having made the supreme sacrifice.

I would give anything to be able to sit down with you now by the fireside and pour out my heart.

It now seems quite clear that P.G., having sized the situation up correctly, saved the life of a Jabo, by turning and attacking the three Spitfires which were chasing and shooting at him. He fell making this attack after having shot down a Boston in flames a little while before.

I cannot write any more about it now; we will talk everything over later....

Farewell, old 'Uncle' Theo....
Your
Wutz.

Over!

Air Vice-Marshal J.E. 'Johnnie' Johnson, *Wing Leader* (Chatto and Windus, 1956; Hamlyn Paperbacks)

The permanent staff at Kirton Lindsey welcomed us back with lavish hospitality ... for they regarded 616 as their own squadron....

Each of the Auxiliary squadrons had its own honorary air commodore, and ours, the Marquis of Tichfield,* drove over from his country seat in Nottinghamshire to welcome us back at a guest night in the mess. After dinner the conversation turned to shooting. Were we interested, Tichfield inquired?

...Two days later there was a phone call from Welbeck. Lord Tichfield would like the squadron to provide four guns next Thursday. Bring plenty of cartridges and some lunch. Yes, it would be all driven game, no walking....

Our new CO was a natural shot and would lead the expedition to Welbeck. Nip† could hit a pheasant and was quite safe with a loaded gun. As the fourth member I selected Jeff West,‡ newly commissioned and keen to see some of our country life. He had never shot game before, but he could bring down Messerschmitts, where the same basic principles of deflection shooting held good....

...Personally, I found my own game and wild-fowling experience to be of the greatest value. The fighter pilot who could hit a curling, down-wind pheasant, or a jinking head-on partridge, or who could kill a widgeon cleanly in a darkening sky had little trouble in bringing his guns to bear against the 109s. The outstanding fighter pilots were invariably excellent game shots....

We were confident that Jeff would be all right once we had given him some elementary training, so we began on the clay-pigeon range, where he proved to be an above-average shot. Then we explained how the beaters brought the birds to the guns and told him of the safety factors which applied not only to the beaters but also the guns themselves....

We reported at Welbeck. It was a crisp, firm, winter day which promised well for the sport. The customary draw for positions took place and for the first drive Jeff found himself between our host and a famous amateur golfer, and I was next along the line of guns. The horn sounded and the partridges swung across the sugar beet, fast, jinking and very low. The first few coveys came straight at Jeff, and he went into action. The guns on either side fell silent and with good reason, for Jeff had his sights on the enemy and swung his gun from front to rear at shoulder height! His lordship and his retinue – loader, under keeper and dog handler – took suitable avoiding action, as did the amateur golfer.

After the first drive I had strong words with the New Zealander and suggested to our host that since West was very inexperienced at this sort of thing it might be wise to put him well behind the guns as a stopper. Lord Tichfield never batted an eyelid and said he was sure Mr West would soon pick up the drill and that we had better get along to the next drive. For me the day was not improved when one of our more enterprising airmen, acting in the temporary capacity of loader, decided to take a hand in the proceedings and brought down a wild duck!

We continued to shoot at Welbeck until the end of the season, and for us those days were some of the happiest of the war years. We always shot until it was nearly dark. Some very old men acted as beaters, for the youngsters were away at the war. One cold, wet afternoon when the light began to fade,

*Now the Duke of Portland.
†Philip Heppell.
‡From New Zealand.

we finished a drive and our host addressed the beaters, who were bunched together:

'Would you fellows mind if I asked you for another drive?'

The simple reply came straight back from an aged countryman:

'We'll beat till midnight for 616, m'lord.'

'Aircrew'
31 December 1942

Fighter trail etched white on blue,
Bomber captain, seaplane crew;
Like Gods; these men, though wrought as we
Are brushed with immortality.
Their youth not lost to age as years go by
But grandly, in a moment, soaring high
through conquered sky.

Harold Balfour

PART FIVE

Gathering Onslaught

1943 was the year when, globally, the pendulum of war began to swing back towards the Allies. There were still, surely, areas of real anxiety – the Atlantic; Burma, with the continuing Japanese thrust; the stunning extent of Germany's advance towards new weapons – rockets and jet propulsion; the pressures on the Eastern front. But, overall, the Allied juggernaut was moving forward and the power build-up was unmistakable....

The great bombing offensive against the Third Reich – by day, now, as well as by night – was mounting; the ending of the North African campaign and the push across the Mediterranean into Sicily and Italy, and, with it, the collapse of Mussolini, muted the Soviet Union's demand for 'a Second front now'; the brilliant US Navy and Marine Corps flyers' performances in the Pacific, following the victories at Midway and in the Coral Sea, set new standards in carrier-borne warfare – these were the factors which now brought victory seriously into prospect.

Terrible battles were still to be fought; casualties would mount as the Allied effort increased and the enemy found its back getting closer to the ultimate wall....

1943 was the year, then, when the Lion, the Eagle and the Bear were striking out in strength at resolute foes.

The Great Bombing Offensive

Marshal of the Royal Air Force, Sir Arthur Harris

There are no words with which I can do justice to the aircrews who fought under my command. There is no parallel in warfare to such courage and determination in the face of danger over so prolonged a period, of danger which at times was so great that scarcely one man in three could expect to survive his tour of thirty operations.

Quoted by the Polish Air Force Association in *Destiny Can Wait* (William Heinemann, 1949).

Few would question, in retrospect, that the most effective, the most dominant air force commander of World War II was the then Air Chief Marshal Sir Arthur Harris, Air Officer Commanding-in-Chief, Bomber Command. No air commander in history has shouldered for so long as great a day-to-day, operational load – and such power of destruction – as this granite-hard, decisive and determined character whom the Royal Air Force called 'Bert'.

By the opening of 1943, within twelve months of taking over, Harris had riveted his style and character upon his Command. The offensive policy which he pursued had been laid down by the War Cabinet and the Chiefs of Staff before he arrived. An Air Staff paper of 23 September 1941, left no one in any doubt:

> The ultimate aim of the attack on a town area is to break the morale of the population which occupies it. To ensure this we must achieve two things; first, we must make the town physically uninhabitable and, secondly, we must make the people conscious of constant personal danger. The immediate aim is, therefore, twofold, namely, to produce (i) destruction and (ii) the fear of death....

Five months later, on the eve of Harris's accession to the most powerful operational chair in the land, the Air Ministry introduced an important refinement which concentrated the aim. It confirmed that Bomber Command's principal objective must now be:

> to focus attacks on the morale of the enemy civil population, and, in particular, of the industrial workers.
>
> Air Ministry letter S.46368/DCAS, dated 14 February 1942

The Chiefs of Staff fine-tuned this directive still further almost twelve months later. On 21 January 1943, they underscored the principle that the attacks should embrace the general disorganization of German industry.

What, then, of the figure who was now carrying and, until the end of the war would endure without respite, this hideous burden? What did the operational commanders and the aircrews see in this man? What of the ethos he created?

The Mahaddie Lectures

Group Captain T.G. 'Hamish' Mahaddie, extract from 'The Bombing Years', delivered in Winnipeg, Manitoba on 5 November 1982

'It might be called the "Giant's Awakening.... Only the oldest of "chiefies" and warrant officers could recall the squadron commander from the old "Mespot" days, who had reduced his squadron's bombing average from an accepted 200 yards to within 25 yards of the target. It is of interest that the CO's two flight commanders at the time were Robert Saundby (Air Marshal Sir Robert Saundby)* and Ralph Cochrane (Air Chief Marshal The Hon. Sir Ralph Cochrane)† a formidable team by any standards....

'"They have sown a wind and they would reap a whirlwind." That was the Harris message. The Command took increasingly to this distant figure. Not for him two berets and a pocket full of cigarettes. Bert Harris seldom left his HQ, unless it was to welcome home his Dambusters. But we knew where he was and what he was doing. And there was this warming, compulsive

*First, Senior Air Staff Officer and then, subsequently, Deputy Air Officer Commanding-in-Chief, Bomber Command.
†Air Officer Commanding, 5 Group, Bomber Command.

feeling that something was stirring, something was happening.... It was simply the confidence we all had....

'Early in the Harris days came the 1000-bomber plan. This was to be seen as a turning point in the war. Historians will cast it in the mould of an El Alamein, a Stalingrad. In fact, it was the greatest confidence trick of the war....

'Consider the facts. Harris (at the time) had barely 300 front-line aircraft. There were more aircraft in the training establishments and almost as many in reserve, tucked away in hangars; there, they were to stay, until the Command decided that the 1000-plan be mounted. The whole, and sole, aim of the C-in-C, and his staff, was to put over a thousand aircraft into the air.

'For the first time there was a concerted plan. Aircraft were pushed through the target in 90 minutes; 600 acres of Cologne were devastated. In recent times I have researched, with the Boffins, the losses attributable to collisions in the air. The figure of 1 per cent of losses due to collision was a cock-shy estimate. They feared it would be more. In fact, it was nothing like it. Oddly, crews became very, very conscious of timing and sticking to the right altitudes and headings. The use of the new navigational aid, known as Gee, helped considerably....

'Slowly, delightful little stories (about the C-in-C) began to spread. Bert Harris was known to be very unhappy about being chairborne. In the Air Ministry, during his time as Deputy Chief of the Air Staff, he maintained a running battle with one of the more senior of the civil servants.... If they met in the corridor what would pass for a growl at the Battersea Dogs' Home implied "Good Morning" – and never a pause for conversation.

'On one occasion the Chief encountered this rather uncivil servant head-on. In place of his customary growl he offered a question. "And what aspect of the war effort are you retarding today?" he inquired. True or false, it was heady stuff.

'I've never been quite sure whether the Chief really sensed the warmth and feeling that the whole Command had for him, or, indeed, the degree of understanding there was of his incredible task, and the way he countered the critics, and fought his endless battles with the other Services. I recall the signals that he sent to us. They were Churchillian in content. There was a famous one addressed, as was his custom, to squadron commanders. "Tonight you go to the Big City – to Berlin. You have an opportunity to light a fire in the belly of the enemy and burn his black heart out."

'The troops knew that that message hadn't been sent by My Lord Bishop, Dr Bell, of Chichester. It came directly to them with the stamp of the Commander-in-Chief. As the Squadron Commander read it out at briefing it became something personal for each one of them....'

Talented Australian

Another formidable and well-illuminated figure was now playing on the Bomber Command stage. The formation of the Pathfinder Force on 15 August 1942, with Don Bennett (Air Vice-Marshal D.C.T. Bennett) at its head, made an ever-increasing national impact as 1943 rolled on towards

1944 – the two great bombing years of the war. Harris was resolute in his opposition to the setting up of this special Force. In a trial of some strength he was overruled.

Air Chief Marshal Sir Arthur T. Harris, Despatch on War Operations 23 February 1942 to 8 May 1945. Addressed to Under-Secretary of State for Air on 18 December 1945

'I was entirely opposed to creaming off the best crews of all the groups in order to create a *corps d'élite* in a special group. This could be calculated to have a bad effect on morale in the Command as a whole, and furthermore, human nature being what it is, it would undoubtedly be difficult to extract the majority of the best crews out of the groups, because, naturally enough, not only would the groups want to retain their best personnel to take command of flights and squadrons, but the best personnel themselves would strongly object to leaving squadrons in which they had half-completed a tour of operations and in which they had been looked up to as the best crews, in order to be sent to another squadron and start again at the bottom as new boys. In this view I had the unanimous support of my group commanders who agreed with me that the best system would be to form inside each group special target-finding squadrons which could be used as group markers for smaller operations and, combined with those of one or more of the other groups for bigger operations. That, in fact, was the logical development of the current tendency in all groups to send the best crews first in order to improve the chance of the remainder being led correctly to the target. However, Gee had failed as an accurate bombing aid and the promise of Oboe and H2S was being continually postponed, and it was therefore obvious that urgent steps would have to be taken for improving visual finding and marking methods in the interim. The Air Ministry, however, insisted on the formation of a separate Pathfinding Force as a separate group – yet another occasion when a commander in the field was overruled at the dictation of junior staff officers in the Air Ministry. In the outcome the Pathfinder Force, although it did the most excellent work, nevertheless displayed all the handicaps and shortcomings which I had anticipated and which are referred to above, while in the latter part of the war my contention that each group should find and maintain its own Pathfinding Force was proved infinitely the superior method when tried out in No. 5 Group. . . .'

Bennett himself has confirmed the Commander-in Chief's opposition.

Air Vice-Marshal D.C.T. Bennett, *Pathfinder: Wartime Memoirs* (Frederick Muller, 1958)

'Bert Harris was blunt, honest and to the point, as always. Roughly, the gist of his conversation with me was that he had opposed the idea of a separate Pathfinder Force tooth and nail. . . .

However, he had been given a direct order from the Prime Minister through the Chief of the Air Staff, and since it was forced upon him he insisted that I should command it, in spite of my relatively junior rank. I was to be promoted to acting group captain immediately, and as a group captain could not command such a force, I should do so in his name as a Staff Officer of Headquarters, Bomber Command, and I should therefore have a subordinate headquarters to handle the Pathfinders at a station of my choosing convenient to the aerodromes which I also had to choose for the establishment of the force....'

Against such a background, it was plainly going to require an exceptionally well-qualified, talented and resolute officer to lead this hand-picked force. Bennett's background and ability for the job were unmatched in the Service – as Harris himself generously recognized. None was better placed to judge than Hamish Mahaddie. With an 'undoubted' operational record and a broad Service experience behind him, he served Bennett until the end of the war. No one could pull the blinds down on Hamish's vision.

The Mahaddie Lectures

Group Captain T.G. 'Hamish' Mahaddie, extracts from 'The Bombing Years' delivered in Winnipeg, Manitoba, on 5 November 1982

'... Bennett came to Bomber Command via Point Cook, the Australian Cranwell. A short service commission in the Royal Air Force; senior captain in Imperial Airways; operations executive and architect of the wartime Atlantic Ferry which, with Lord Beaverbrook's spirited support, he successfully established against ill-founded opposition from the Air Ministry.... This was his pedigree background. Known to Harris as "the most efficient airman I have ever met," Bennett had earned his title, "the Boy Wonder", in his Imperial Airways days.

'After leaving the Royal Air Force pre-war, this dedicated and concentrated character amassed an enormous amount of flying experience. He held virtually every known certificate associated with flying – 1st class navigator's licence, a GPO wireless licence, and all the licences required by an aviation engineer – including an "X" licence. He made his mark as the author of accepted text books on air navigation.

'International fame spread with his corporate flights in Mercury which culminated with the record run from Dundee to Walrus Bay, near Cape Town, and from the first commercial flights across the Atlantic and to the Middle East. All this unique flying background he brought with him to Bomber Command – as a pilot officer, acting wing commander.

'He was posted to No. 10 squadron in 4 Group. There, he did a dozen sorties before being shot down by *Tirpitz* in Alten Fjord, in northern Norway. He evaded capture, walked across the mountains into Sweden where, within a month, he had talked the Swedes into returning him to the United Kingdom. Harris then appointed him to be Commandant of the Pathfinder Force, and, later, Air Officer Commanding, No. 8 (PFF) Group.

'Characteristically, on the day that the Pathfinder Force was formed Don Bennett mounted the first Pathfinder attack on Flemsberg. It was an absolute shambles and delighted the critics. Thereafter, there was little improvement in the PFF ability to find and mark a target until the early months of 1943, when the H2S and Oboe techniques were beginning to come into play.

'Small-scale Oboe attacks had, of course, been mounted from December 1942, but it wasn't until March 1943, that the real Harris offensive got under way. With the operational expansion came more experience with

H2S, sharper navigation, better bombing techniques, better crews.... Yes, better crews, not only in the PFF but also in the main force.

'The manning of the Pathfinders had actually failed during the first six months. But, after that, from March 1943, we took care of our own manning. I became Don Bennett's horse thief and found that the groups (other than 4 Group) were not bothering to earmark good crews for the Pathfinders. Some even used PFF ... as a dumping place for crews. All this had to change.

'Selection was now made from a very careful study of the Command bomb plot. After each raid, HQ Bomber Command, issued a plot in the form of night photographs which were taken when crews dropped their bombs. I always claimed that the advent of the night camera cut down the number of heroes in the Command by quite some 50 per cent. So, when I came to examine the bomb plot I found the evidence produced a pattern. As in most things in life, it was the same story. There were the crews (they were *always* the same crews) who were near the centre of the target; there were the crews (they were *always* the same ones) who dropped their load around the perimeter; and then there were the crews who *never* got on the plot at all.

'The recruitment of good crews thus became very important to the Pathfinder priority. I recall having a terrible argument one time with Guy Gibson because he thought I was after his 617 Squadron crews. We were having tea together with the AOC, 5 Group, Ralph Cochrane, in his office. When we came out, Guy turned to me and told me in no uncertain manner to leave his crews alone.

'"You can keep your hands off Searby!" he said.

'I called my office at once and instructed Jimmy Rogers, our P4 (aircrew postings) to transfer John Searby to 83 Squadron. Not long after this, he led the complicated Peenemunde raid and gained one of the PFF's best DSOs.

'I was saddened by the Gibson affair because I believe that our clash at 5 Group – and in the Black Boy pub in Nottingham later that night – rankled with him and that his subsequent refusal to come to my station – Warboys – for a Mosquito conversion contributed to his end. I would like, however, to pay this tribute to his leadership of the Dams raid. I knew every squadron commander in the Command at that time and I can recall very few, if any, who could have carried out that raid in the manner of Guy Gibson. We had the same end in view – he wanted to keep his best crews and I was certain that the best were only just good enough to be Pathfinders.

'This was the only serious conflict I had with crew selection.... But 1943 was the difficult year because of the free hand I had been given and the flood-tide of effort stimulated by Command and boosted by Bennett. Greater lifted weight was being left in the right place with greater accuracy. The Command was thus set fair to mount the offensive in deadly earnest.'

Edwards on Bennett

Peter Firkins, *The Golden Eagles* (St George Books, Perth, Western Australia, 1980)

Of the famous Australian commander of the Pathfinders, Edwards [Air Commodore Sir Hughie Edwards VC] says: 'He was not popular with group commanders. All groups had to cough up above-average crews to form the Pathfinder Force. Rightly so, but that and the implication that the groups were unable to find and bomb the target was hard to take.... He certainly held ... strong views on most things....

He once summoned me to his headquarters and offered me a squadron.... I said I'd be prepared to take a Mosquito outfit. It seemed fairly cushy to be dropping a 4000-pounder on Berlin from 30,000 feet in a fast aircraft. He rang back later and said I could have No. 156 (Lancaster) Squadron, but I turned it down on the grounds that I was already operating on Lancasters....

He finished by saying he could not give me a Mosquito squadron and would I not mention to my AOC [Sir Ralph Cochrane] that he had been lobbying me. I never did!'

'Bomber Pilot'

Chosen by David Scott-Malden from *Rhymes for Everyman* (Peter Davies Ltd)

With the wings of a bird, and the heart of a man
he compassed his flight,
And the cities and seas, as he flew,
were like smoke at his feet.
He lived a great life while we slept,
in the dark of the night,
And went home by the mariner's road
down the stars' empty street.

Ernest Rhys

'Digger'

Sir Douglas Bader: on Friday 20 August 1982 – just a fortnight before he died – talking about the author of the next story.

'"Digger" Kyle (Air Chief Marshal Sir Wallace Kyle) ... my greatest friend and closest contemporary.... Extraordinary how his life has turned the full circle....

'In 1927, as a schoolboy, he went to Government House, in Perth, Western Australia, for an interview for Cranwell. We arrived at Cranwell the next year and were cadets together. Then he went to 17 Squadron and I to 23.

'... "Digger" did all sorts of jobs in the Service.... Bomber Command throughout the war ... C-in-C of it afterwards. Then, in 1975, forty-seven years after that first visit, he went back again to Government House in Perth – as Governor of Western Australia.... I saw him there.

'Dear chap, "Digger".'

First 'Daylight' on Berlin

Air Chief Marshal Sir Wallace Kyle, Tiptoe, Lymington 1982

'On Saturday 30 January 1943, Mosquitoes of 105 and 139 Squadrons, based at Marham, in Norfolk, where I was then Station Commander, made the first daylight bombing raid on Berlin. They attacked twice that day, at 1100 and 1600 hours.

We were operating regularly by day at low level against targets in occupied Europe with occasional, deeper penetrations into Germany itself. We were independent of the main bomber force and although we had, of course, been given a general operational directive, the detailed planning, including the selection of specific targets and the timing of the raids, was left to me and the squadron commanders.

Now and then, however, we would be given a particular task by Bomber Command Headquarters, and these were usually prompted by special intelligence.

We operated largely in heavily defended areas, but, on the whole, the losses were relatively light. The high speed and manoeuvrability of the Mosquito was a very important factor and, in many instances, it was the first attack on the target and, more often than not, we went in at dusk and withdrew at low level in the gathering darkness.

It was about 1500 hours on 29 January when I received a telephone call from Group Captain Elworthy (Marshal of the Royal Air Force Lord Elworthy), then Group Captain Operations at Bomber Command. I had just stood both squadrons down after a very heavy week. He told me that the C-in-C wanted Mosquitoes to attack Berlin the following day at 1100 hours precisely, when Göring was scheduled to speak at an open-air mass rally. What did I think?

This was, to put it mildly, rather a surprise, not so much because of the target (Berlin), but because of the timing which would restrict our tactical freedom of action. It meant that we would be exposed to fighter attack both during penetration and, again, on withdrawal when the advantage of surprise had gone. I said that we would be lucky to get away without severe casualties. Elworthy, who knew the operational form well enough, agreed and offered to represent this to the C-in-C.

Before long, he came back to confirm that we were to go. There followed a pause. 'You haven't heard all of it yet. We want you to attack again at 1600 when Goebbels will be speaking at a similar rally.'

The conversaton which followed was short and pithy.... But there was nothing for it but to get the squadron commanders together as quickly as I could and work out a detailed operational plan.

We decided to send three aircraft from 105 Squadron on the morning raid and three from 139 in the afternoon. The choice was made by tossing a coin. Squadron Leader Reynolds and Flying Officer Sismore led the first attack and Squadron Leader Darling and Flying Officer Wright, the second.

The 'Met.' people were, understandably, fairly non-committal because of the exact timing, but there were no problems from a flying point of view either en route or at base.

We thought we would get away with the penetration phase by staying at low level until the last moment before climbing to 25,000 feet for the bombing run. The route chosen was pretty well direct and we were able to stay under radar cover as far as Hanover. Withdrawal was to be by diving at maximum speed to low level, heading north to Norway, without too much regard for Swedish air space, and then directly back to base. This was about the limit of endurance for the Mk IV Mosquito in the low-high-low profile.

The morning attack went exactly to plan and bombs were dropped at precisely 1100 hours. Even the heavens were on our side because, at the very last moment, after Sismore had told Reynolds it would have to be an ETA release, he spotted a hole in the otherwise complete cloud cover. It was plumb over Berlin and they could see the lakes and Spandau. The bombs hit the northeast area of the city. The tactics had worked and all aircraft returned safely. It was a superb piece of accurate flying and navigation. (*We* had no radar aids then.)

We were apprehensive about the afternoon attack because the defences had obviously been stirred up. We certainly didn't think we should follow the same penetration and withdrawal routes. After a lot of discussion with the crews we decided on the simple solution of reversing the routes.

We calculated the run in would have maximum surprise, especially using the fringe of Swedish air space; and most of the withdrawal phase would be in dusk conditions over Germany and Holland at low levels.

Sure enough, we got away with the penetration, but the flak was more intense and Squadron Leader Darling was picked off after bombing.*

In the bar that evening we had the satisfaction of listening to a tape recording of the air-raid alarm which accompanied the first raid, and the cancellation of Göring's speech by the announcer. As Goebbels' speech and rally were also disrupted by the second attack, we felt we could claim success.

We were well aware that these raids could have little material effect. But, until then, Berlin had not been bombed by day and, as the night raids were building up, we all reckoned that a few bombs dropped in daylight would have a significant effect on morale This, plus the impudence of the whole affair, made an obviously hazardous operation exciting and stimulating.

That, I believe, was the reaction of the crews; and success turned this feeling into one of great jubilation.'

Uninvited Guests

His Honour Judge Douglas Forrester-Paton QC, Great Broughton, Cleveland, 1983

'By 1943, the Luftwaffe was heavily committed on the Russian and North African fronts. Attacks on London were rare and relatively light. But Ju 88s

*Squadron Leader Darling and Flying Officer Wright crashed near Magdeburg and, initially, were buried there. In 1949 they were reinterred in the British Military Cemetery in Berlin.

and He 111s were regularly sent over to drop mines in the Thames Estuary. To repel this threat to our shipping regular patrols over the estuary and its approaches were flown by 11 Group's night fighter squadrons – among them No. 29 Squadron, based at West Malling, and flying Beaufighters. One of 29's pilots was South African-born Flight Lieutenant 'Johann' Strauss, and I was his navigator – really an airborne radar operator.

One fine night in March 1943 Johann and I and several other crews from our flight were airborne – either on patrol or doing practice interceptions. So far as I remember, the ground radar stations, who controlled us, gave no hint of enemy activity that night. When it was time for us to go home we could see a mass of searchlight beams converging over West Malling; evidently someone was lost and had asked for a searchlight homing. (The signal for this was to fly round and round flashing your white identification light.) Flying Control kept us orbiting the airfield so that the lost sheep could return to the fold.

Meanwhile, a single-engined plane landed, taxied towards the control tower and was guided by the duty ground crew to the usual parking place for visiting aircraft. The pilot switched off and began to climb out.

"Refuel, Sir?" shouted one of the erks. The reply was a stream of foreign words. "Another bloody Pole!" said the erk to his mate. "Not bloody likely," said the other pointing to a Maltese cross on the fuselage.

The "lost sheep" was a Focke-Wulf 190, one of the latest, with an empty bomb-rack under each wing. They grabbed the pilot and led him, unresisting, into the control tower.

Just then a second FW 190 landed. By this time the RAF regiment had been alerted and when the second visitor appeared to be about to depart in haste the armoured car crew opened fire with their machine gun and set the plane ablaze.

The Flying Control Officer thought it was about time he got his Beaufighters down, and gave them permission to land. We were the last in and as soon as we touched down all the runway and approach lights were switched off. Within seconds, a third FW 190 crash-landed in an apple orchard on the approach to the runway. The pilot was not seriously hurt; he was taken prisoner and brought in complaining vehemently about a "dirty English trick – switching off the lights when a fellow was trying to land".

He might have spoken differently if he had known that in rescuing his comrade from the burning aircraft one of our firemen was killed when the plane exploded. That was the tragic feature of this otherwise comic episode. West Malling captured three enemy pilots and captured or destroyed three enemy planes, but lost one brave man.

What led to this extraordinary affair? Evidently the Germans thought it was time the Londoners received a few more bombs. Being short of bombers in the West (or afraid of losing them) they chose a squadron of FW 190s based in northern France and armed them with bombs. They gave the pilots a flight plan which would have brought them over London – if the westerly wind had not been much stronger than the "met." men predicted.

Some of them never saw London and finding themselves over open country at the limit of their range they turned for home, jettisoning their bombs over Essex. Most got home, but one pilot mistook the Thames

Estuary for the Channel and went for the Pas de Calais. Failing to find his base he resorted to the standard German procedure for requesting a searchlight homing, which just happened to be the same as ours. The other two pilots, also lost, saw their chance and also followed the lights.

It was all very natural, given the original mistake. But one thing I cannot understand to this day: we had several Beaufighters flying in the vicinity – why did not the ground radar stations pick up these "bandits" and send us after them?'

Family Honour

The Galland Papers, Bonn (Bad Godesberg), 1982

Early in 1943, Oberstleutnant Wilhelm-Ferdinand Galland was given command of II Gruppe of Jagdgeschwader 26, his brother's (General Adolf) old command. The family pride in the appointment, and the concern of the elder brother for the younger, is manifest in this letter of advice:

11 January 1943

Dear Wutz,

... You will be watched critically as a commander, particularly in the early days. It is quite easy to follow a leader who did nothing simply by doing something. But, again, I beg of you, work like a lion for your Gruppe. Everyone, right down to the most junior airman, should be able to feel that here is a real leader who genuinely cares about each man; if they all feel that, they will go through the fire for you.

Dedicate yourself totally to the Gruppe and do not allow yourself to be deflected by other interests. There will be difficult things to do, and hard tasks to face; but you must grasp your responsibilities and discharge them. Nothing runs on on its own. Things can't just be taken for granted. Every detail must be attended to, thought through and, if necessary, discussed. Remember, success has to be earned. Men will take a great deal – and give much – if they know they are well led. Avoidable mistakes must never be made for want of careful thought, clear orders and efficient direction.

You should not become too familiar with the Staffel captains and the officers whom you are commanding. But that does not mean you cannot become their best comrade. Concern and consideration for subordinates did not cease with the Cavaliers. Junior officers will always recognize and respond to this approach.

Make it a principle to ensure that those under you feel that you, their Gruppe Commander, understand their problems and are as concerned about the most junior man as for the most senior. Be approachable to everyone in the Gruppe and be prepared to listen. And let those who serve you feel that you would never demand of them more than you would yourself be ready to undertake. This will be very important, particularly on the Eastern front.

Individual successes in the air should be subordinate to the overall, collective success of the Gruppe. Enough rubs off on the leader, anyway.

It may well be that, in the East, the Gruppe's first missions will be independent of, and quite separate from, the rest of the Geschwader. In such cases, careful and detailed planning and briefing will be essential. It will all be very new, thus everyone must be brought fully into the picture – and feel totally involved – otherwise there will be terrible mistakes.

Lastly, a word about your new Kommodore. There can be but one attitude on your part – complete loyalty and obedience. Unless this is so, the rest is impossible. It is the *only* stance you can adopt. There is no alternative. He carries full responsibility; he can only lead successfully if he is able to rely totally on his subordinate commanders. If there is not this loyalty and dedication to his leadership, the whole Geschwader will fail.

That form of parliamentary consensus, with its relatively free and loose accommodations, is 'out'. It cannot be tolerated. There is but one judgement and that is the Kommodore's; and his decision goes for the whole Geschwader. For my part, I could never have rested easily if my own Geschwader, which won a name for itself when I led it, had not given me complete loyalty.

So now you know, Wutz, what I – and the family – expect of you. I am always here to offer help and advice if you feel you need it. Meanwhile, I wish you every success in one of the finest commands an officer can have. I know you will do it well. Go flat out, Wutz, and good luck.

Your brother,
Keffer*

Wilhelm-Ferdinand Galland was killed leading his Gruppe on 17 August 1943. He was shot down by P-47's near St Trond, close to the German frontier. He had 50 air victories to his credit. The youngest brother, Paul, had been credited with 17 when he fell, ten months before.

Staff Interlude

Air Vice-Marshal F.D.S. Scott-Malden, Colegate, Norwich 1982

At this juncture in the war, many aircrew who had completed one or, perhaps, two tours of operational flying were passing their six months' rest period on the staff at Command or Group Headquarters. A select and privileged few served on Group Captain T.N. McEvoy's (Air Chief Marshal Sir Theodore McEvoy) Operations Staff at Fighter Command.

Apart from the priceless experience of seeing a master of the craft at work, they found their duties enriched by a flow of rare minutes, limericks, verse and the rest which adorned the secret files. David Scott-Malden remembers some of the gems.

'I recall a minute of Mac's ... when someone submitted for clearance a lecture to the Turkish Staff College entitled something like "Air Defence – Concentration of the maximum potential resources for resistance of air attack into a comprehensive interlinked system of fighter and AA defences".

Mac cleared the contents and added "I have taken the liberty of

*Keffer was the nickname that Adolf Galland's father gave to his distinguished son. He was Keffer to the family. (Ed.)

changing the title to 'Air Defence' on the theory that your title, meaning little in English, might mean even less when translated in Turkish."'

And, again, Scott-Malden recalls:

'I presided over a file called "Suggestions from civilians for the improvement of air defence." Mostly they were drivel, but occasionally one was whisked out, stamped Top Secret, and hurried away. A letter arrived in good working man's English from one Robinson, leader of a rescue squad on the South Coast, and saying something like "What the bleeding 'ell are you doing about these bastards wot keep dropping bombs all along the coast? If it's airydromes you're short of, me and my mates have got picks and shovels and will soon knock you up some."

Mac simply wrote on the minute sheet:

If Robinson's Rescue-men try,
Our fighters will darken the sky:
They'll have strips to alight on,
At Bournemouth and Brighton,
And Ramsgate and Romsey and Rye.'

Then there was the sad occasion at the end of the war when Mac was at the Air Ministry. A member of his staff, one Arnold Wall, who had lost an eye to a sniper in Waziristan and had spent much of Hitler's war at the Air Ministry, went to him and dolefully announced that he didn't even qualify for a Victory *medal! Mac wrote a little verse to console him.*

Now that we all have medals, stars and such
In glittering rows from clavicle to crutch,
We'll gladly let our aching optics rest
On Arnold's sombre, unreflecting breast.

Consumer Protection?

FOR SALE: *Unique bargain*: Second-hand, first-rate, almost incredible AUSTIN 7, family saloon. Excellent condition; guaranteed fully aerobatic, untaxed, unserviceable. It's the tops! PRICE £1 or NEAR OFFER. Positively last day. P/O F.D.S. Scott-Malden. Late 611 Squadron. Sic transit....

(Notice on mess notice board. Royal Air Force Station, Digby, Lincolnshire, 1941.)

The Atcherley Brothers

No two brothers in the Royal Air Force achieved greater notoriety for their deeds and misdeeds than the Atcherley twins, David (Air Vice-Marshal David F. Atcherley) and Dick (Air Marshal Sir Richard L.R. Atcherley), widely called 'Batchy'. The stories about them, true or false, are still being told today; but behind this improbable, joint façade lay two accomplished and original officers.

'Look-alikes', they played the 'mistaken identity' card to the full.

Air Chief Marshal Sir Christopher Foxley-Norris, *A Lighter Shade of Blue* (Ian Allan, 1978)

On one occasion in 1943, David Atcherley was flying home from North Africa for an important operational conference. The weather over southern England was very bad, and he had to be diverted to the only airfield that was still open, Portreath, in Cornwall. As was, and remains mandatory during flying, an ambulance and a fire tender were standing by the Air Traffic Control Tower. David, now going to be very late for his meeting, taxied in at break-neck speed, leapt out of the aircraft and into the ambulance whose engine was running, and disappeared in a cloud of dust toward the nearest main line railway station, Redruth.

The Station Commander was naturally furious. He addressed an indignant letter to Group HQ. It passed up the line to Command and to Air Ministry. The letter despatched a stern rebuke in writing to Air Commodore Atcherley; but unfortunately it was addressed in error to the other twin. Batchy hastened to reply:

Sir,

I have the honour to acknowledge your letter of 23 November, whose sentiments I entirely endorse. To remove, without authority or permission, the emergency ambulance from an operational station, thus hampering the flying programme and imperilling the lives of crews already airborne is quite unforgivable; and fully merits the tone and content of your communication.

I have to inform you, however, that I personally was in no way involved in this incident. I can only presume the officer concerned may have been my brother, Air Commodore D. Atcherley.

I have the honour to be, Sir,
Your obedient servant
etc, etc.

PS. Personally, I always take the fire engine on these occasions.

Somewhat earlier, David had had a spell in the north of Scotland, commanding the Royal Air Force station at Castletown, not far from John o' Groats.

One day a storm-beaten naval vessel had gone ashore . . . and after a heavy pounding on the rock below, had become a total loss. . . .

Aerial reconnaissance, however, indicated the survival of a gun of considerable calibre installed on the vessel's stern. An intrepid climbing party with appropriate cutting gear was despatched; and before long, the otherwise almost non-existent ground-to-air defences of RAF Castletown, were complemented by the same gun, supplied with what ammunition could be salvaged. Unfortunately, although it was designed to move both in the vertical and the horizontal planes, the latter capability was eliminated by it now being mounted in concrete outside the station HQ.

Daylight was almost continuous in those latitudes and of a late evening David would lead parties of apprentice artillery men from the mess to practise discharging this formidable piece. Its new and over-rigid emplacement had regrettably left it pointing directly at the local lighthouse; the occupants of which displayed considerable alarm and despondency if one failed to apply maximum elevation to the gun during such exercises. Unhappily, David subsequently overplayed his hand by indenting to the Admiralty for a fresh supply of ammunition when the first batch was exhausted. In a matter of days a band of indignant matelots arrived to retrieve their weapon; and RAF Castletown, once again lay wide open to enemy air attack.'

Group Captain Duncan Smith has recorded another sensitive incident involving the two brothers.

The Duncan Smith Papers, Carpow, By Newburgh, Fife, 1982

'David and Dick not only shared many idiosyncrasies and mannerisms but also an uncanny telepathy that often worked in unusual ways....

'When Batchy was Station Commander at Kenley (in Surrey) in the middle of the war, he caught a packet from some Focke-Wulf 190s over the Channel and was seriously wounded in the arm as well as losing a finger of his left hand....

'As he floated down in his parachute, he noticed his torn and bleeding arm and that his wrist watch had been shot off. He got into his dinghy wondering about his chances of survival.... Fortunately two minesweepers had heard the cannon fire, seen the Spitfire crash and then the parachute coming down.... They picked him up.

'Batchy told me, years afterwards, the sequel to the story.

'"A strange thing happened that day," he said. "At the precise moment when I was hit in the arm, my brother, David, who was sitting in his office at Fairwood Common, where he was then the Station Commander, felt a piercing pain in his left arm. He shouted to his Adjutant to get through to Fighter Command and ask the Duty Controller in the Operations Room whether I was OK. He was told that as far as was known, I was 'in the pink' at Kenley minding my own business.

'"Find out for sure," my brother insisted, "I think Group Captain Atcherley has had a serious accident and hurt his arm rather badly."

'"Twenty minutes later the Sector Controller at Kenley telephoned David and told him that I was overdue, missing from a reconnaissance flight over the Channel."

'It was not until Batchy had been rescued from the sea and delivered to the casualty clearing station at Dover Castle that David learned the details of his brother's wounds. The centre of pain he had felt in his arm was in the exact place that Batchy had been hit by the enemy's cannon shells.'

Of the two, it was Dick Atcherley who may have had the greater propensity for practical – and often irreverent – fun. Joe Cox (Air Vice-Marshal J. Cox), a contemporary, is a connoisseur of 'vintage Batchy'.

Air Vice-Marshal J. Cox, Caversham, Reading, 1982

'When Dick, then an Air Commodore, took over as Commandant of the RAF College at Cranwell, he at once drew his own parachute. Typically, he went along to the Parachute Section personally to draw it. As was often his custom, he was in "mufti".

'An AC* was packing a parachute as he entered. "Good morning," said Dick, "I want to draw a parachute for myself."

'The airman pushed a loan card across the counter. "Fill that in," he said.

'Dick complied. Against rank he wrote "AC".

'The airman ran his eye over the completed form, muttering the details as he went. "Atcherley R.L.R. ... number...." He came to rank. "AC?" he queried, "AC*1* or 2?"

'"Air Commodore," said Dick.

'Unconcerned, the airman looked up,

'"Now that'll be the day, mate, that'll be the day."'

Editorial licence – and Cox's *first-hand* witness – may just allow the inclusion of Batchy's rejoinder to his Personal Staff Officer, Alan Hollingsworth (Group Captain Alan Hollingsworth), soon after taking over as Air Officer Comanding-in-Chief, Flying Training Command. Cox was his Senior Air Staff Officer:

'Alan had taken the opportunity to raise with Dick the subject of his filing system.

'The C-in-C looked up from his desk. "There's nothing wrong with it, Alan, is there?"

'"Not so far as I know, sir," retorted the PSO, "but it's your *personal* filing system. At present it's being run as it was for your predecessor. I felt you might care to consider some changes."

'Dick paused. "Alan," he said, "I'm a simple chap and I like things done in a simple way. As far as my filing is concerned it's as simple as this – If I write a letter to the Archbishop of Canterbury on the subject of fucking, it goes on the fucking file."

'That was the end of it!'

The Masters

By now, the cream had risen to the top. In all the roles, in each of the combatant countries, the true exponents were there to see. Some had been lost after scaling the heights; others would still emerge. But the majority of marks had been made.

Comparisons can be invidious, but few would deny that in Bob Stanford-Tuck (Wing Commander Robert R. Stanford-Tuck) of Britain, and Colin Gray (Group Captain C.F. Gray), of New Zealand, the Allies possessed two of the most enlightened examples of the fighting art.

When Bob was shot down by flak in northern France and taken prisoner, after no more than two years of actual combat, he had 27 enemy aircraft credited to his guns. God knows what he would have collected had he been able to engage in the last three full seasons of the war. But by the time he left the stage he was a principal figure in the cast.

As for Colin, he came out of much the same mould of marksmanship. Moreover, he lasted the war with one of the longest sustained runs the

*Aircraftman (1 or 2 – 1st or 2nd Class).

northwestern and southern theatres of operations had seen. With 27½ victories chalked up against his name, he was top of the New Zealand league and high up in the international table.

Their views are thus important. Here they are.

Go for the Belly

Wing Commander Robert R. Stanford-Tuck, Sandwich Bay, Kent, 1982

'Much has been written about the tactics employed in the great air battles of the 39/45 war, but little has been said about the techniques used by some of the more successsful combatants in the fighter squadrons of the Allied and Axis forces.

Having been fortunate to survive many of these air battles, and, indeed, discuss them with others equally fortunate, I am happy to have the opportunity briefly to record my views.

Fighter Command of the Royal Air Force was a small force in comparison with the Luftwaffe, but it was composed of a very tightly-knit breed of young men, highly trained, very fit, dedicated, and of competitive spirit. Beyond this, they were equipped with those two fine aircraft, the Spitfire and the Hurricane. Such was our position at the real opening of hostilities.

In my case, I was thoroughly experienced on my Spitfire. I had flown it for several hundred hours, done every aerobatic in the book, a lot of high altitude flying and, most important of all, I had completed a considerable amount of air-to-air and air-to-ground firing of the eight Browning machine guns. Lastly, I had flown many hours at night, all of which left me with a deep-rooted confidence in the aircraft.

The first enemy aircraft, an Me 109E, fell to my guns during the Dunkirk campaign, and in three days' fighting over the beaches, the tally rose to 7 destroyed. This short experience gave me a wonderful opportunity to iron out any small faults in my techniques of delivering attacks. It was of tremendous value to me in the great air battles which were to follow.

I did my flying training at Grantham in 1935 in company with my great friend, "Sailor" Malan, who turned out to be probably one of the greatest fighter leaders of the war. Like me, he had been a cadet in the Merchant Service and it became obvious to both of us that if one was not flying accurately, it was no good pushing the firing button and hoping for the best. If you opened fire with your "bank and turn" indicator showing a bad skid either way, your bullets would go nowhere near where you thought you were aiming as you looked through the graticule of your gunsight, just in front of your face. Hence, accurate flying was essential at the moment of opening fire.

When attacking large formations of bombers, the ideal situation was to have the sun behind and be in a shallow dive with good overtaking speed. The tactic then was to pass slightly below the bombers, climb fast, and, when well within range, shoot up into the belly of the aircraft where there was little, if any, protective armour plate. If you were not shot down in the process of delivering an attack of this sort, it always produced instant results.

However, if you hit him really hard, the ever-present danger was from debris breaking off and flashing at you. I well remember a JU 88 which I had plastered shedding a complete oleo leg and wheel, which missed my Spitfire by a few feet.... Certainly a very frightening experience. During two years of continuous air combat I had many near misses from debris and these became more frequent after we were fitted with two 20 mm cannons.

Air fighting, often at great heights, is a "cat and mouse", stark and brutal business, but all I can say is that I would not have missed it for all the rice in China. However, I would certainly not wish to repeat it!'

Get in Close

Group Captain C.F. Gray, Waikanae, New Zealand, 1982

'What did it take to make a first-rate shot? Perhaps one should consider, first, what it took to be an effective fighter pilot because there were many factors involved in the destruction of enemy aircraft other than just being able to shoot straight.

Of course, it helped to have a good idea of what was required in deflection shooting but, even so, the scope for error was considerable – getting the range wrong (an aircraft was always much further away than one thought); setting the wing span of the enemy aircraft incorrectly on the reflector sight; skidding, slipping, judging the "angle off" wrongly – you name it.

Also, in a Spitfire (less so in the Hurricane but still a problem), the reflector sight ring cut across the top of the engine cowling; the target, therefore, disappeared the moment the pilot began to pull through to allow the necessary deflection. He was thus left shooting at a point in fresh air which could be 50 yards or more in front of the target. (Not surprising when you consider that a bullet from a Browning gun was travelling at 2240 feet per second and an enemy aircraft, crossing at right angles at 300 miles per hour, would travel 147 yards in one second!)

I learned my lesson in deflection shooting early on during the Battle of Britain – on 24 July 1940 – when attacking an Me 109 over the Thames Estuary. The pilot started to turn and, being somewhat excited, I overreacted and pulled through to more than twice the angle off which I'd originally intended (and which I thought was overgenerous in the first place). To my surprise the aircraft burst into flames and the pilot jumped out!

The lesson was simple – double the number you first thought of and add a bit for good measure. I think it was an art some pilots instinctively acquired; there wasn't much chance to practise.

But the overriding factor in success was the range of opening fire and the determination to press home the attack as close as possible. After all, we only had a total of 14 seconds fire-power with machine guns and even less with cannons. If you were right up the back-end of an enemy aircraft even quite large errors didn't matter.

Then again, the most effective fighter pilots all seemed to have excellent eyesight and quick reaction to enable them to get in, get close and get out. Also they had the ability to fly accurately. This was important in an aircraft like the Spitfire where the torque* built up alarmingly in a dive and the skid and slip needles disappeared into opposite corners of the cockpit unless trimmed out.†

Perhaps the final element in this matter of effectiveness was the opportunity to prove it. Chance is a fine thing and although there were times during the war when enemy aircraft were relatively easy to find (for me, this meant, for example, Dunkirk, the Battle of Britain, the North African campaign etc.) on other occasions they appeared to be as scarce as hens' teeth.

Although my log book records between 700–800 operational sorties during the whole period of the war I saw enemy aircraft only on about 10 per cent of those occasions, and had the opportunity to engage them on even less.

So, to sum up, the essentials for success were excellent eyesight, speed of reaction, accurate flying, determination to press home the attack really close, but, above all, to have the good fortune to be in the right place at the right time!'

Wing and Group Leaders

A new breed of wing and group leaders was emerging in Fighter Command and in the US 8th Army Air Force in Britain. In the offensive operations over Europe in 1943 and '44, two men, in particular, were to come into their own – 'Johnnie' Johnson, with his Kenley wing of Canadians, and Don Blakeslee (Colonel Donald J.M. Blakeslee), at the head of the 8th's 'Fighting Fourth'.

Johnson, whose personal score of 38, was to exceed that of any other leader among the British and Commonwealth air forces in the second half of the war, brought with him to the job one among several advantages. From personal observation and experience, he had learnt to recognize leadership when he saw it. The examples were still fresh in his mind.

JOHNSON OF KENLEY

Air Vice-Marshal J.E. 'Johnnie' Johnson, Hargate, Buxton, Derbyshire, 1982

'One fine summer's day in the middle of August 1940, I joined 19 Squadron at a grass airfield, Fowlmere, near Cambridge. A few weeks previously they had received Spitfire Is fitted with the new 20 mm cannon to replace the near-obsolete ·303-inch machine guns. When the cannon fired properly they were very effective against enemy bombers, but they seldom worked correctly, and the pilots were angry, vociferous, and frustrated because often the cannon failed when they had been well positioned to destroy enemy aeroplanes. Also, the squadron had lost pilots because they were unable to defend themselves.

*Tendency of an aircraft to pull one way or the other.
†I.e. adjusted with the pilot's trimming controls.

'The faulty cannon had been reported by their Station Commander, Woodhall, to their Group Commander, Leigh-Mallory, and to the Commander-in-Chief himself, "Stuffy" Dowding; and so it was that on the afternoon of my arrival a small passenger aeroplane landed at Fowlmere and the sparse, unsmiling, Dowding got out of the aeroplane accompanied by a fairly senior staff officer. They were greeted by our Commanding Officer and walked across the grass to a Spitfire where the armourers laboured on a faulty cannon. Dowding said courteously:

"Squadron Leader, will you please tell me what is wrong with your cannon?"

'However, before the Squadron Leader could reply the eager staff officer interrupted with his explanation. Dowding listened patiently and when the staff officer paused for breath, once more asked the Squadron Commander for his version. Again, the chairborne officer, anxious to impress, broke into his spiel, whereupon Dowding turned to him and very quietly said:

"I want to hear what the Squadron Commander has to say."

'With the staff officer silenced, the Squadron Commander explained the frustrations of trying to fight with faulty weapons, and later that day our original Spitfires, with their old machine guns, were returned to the squadron, which was then able to play its part in the great air battles....

Then one Sunday afternoon, some months later, I was flying with Douglas Bader over Lille when we were bounced and heavily attacked by a gaggle of Messerschmitt 109s. Our cohesion as a fighting unit was soon lost and a series of individual dogfights took place. During the melee two or three of our pilots were shot down, and I found myself with a couple of yellow-nosed 109s astern who took occasional shots at me. Since the Spitfire could turn inside the angular 109 I could, for the time being, save my skin; but I could not go on turning over much longer because my fuel was getting low and sometime, somehow, I had to make the break for home. Suddenly, out of the corner of my eye, I saw a solitary Spitfire heading towards the Channel and about 1000 yards away. Still turning, sweating and frightened, I found the breath to say:

"It's Johnnie here. There's a Spitfire ahead and I have 109's on my tail. I want help."

Douglas Bader said:

"Douglas here, Johnnie. I'm turning starboard. Join up. Pretend you haven't seen them! And we'll get one each."

'My two stories illustrate the suberb leadership we had at all levels in Fighter Command. The strands of leadership flowed from the top, from Dowding, through our simple chain of command to the Group Commanders, Park and Leigh-Mallory, and to the young fighter leaders like Douglas Bader, Bob Tuck, Mike Crossley, Don Kingaby, Victor Beamish, Frank Carey, Archie McKellar, Jim Hallowes, H.M. Stephen, Finlay Boyd, "Sailor" Malan and "Dutch" Hugo from South Africa, "Jamie" Jameson, "Hawk Eye" Wells, Colin Gray and Alan Deere all from New Zealand, Gordie McGregor, Willie McNight from Canada, and Witold Urbanowicz from Poland. This shining leadership produced that "priceless pearl", high morale, which made men bigger than their normal selves. High morale – the greatest single factor in battle.'

BLAKESLEE OF THE 'FOURTH'

Vern Haugland, *The Eagles' War* (Jason Aronson, New York, 1982)

... Blakeslee became the most skilful of all air combat directors.

Of ... his methods, Peterson* has said: 'Don Blakeslee could take a hundred fighters and organize them in the air, tell them what to do and how the battle was being fought, like no one else. He could have had ten times as many as the dozen or so kills credited to him if he had wanted personal glory. Instead, he just sat there and handled the whole thing and let the other pilots get the credit.

'He was marvellous ...'

'Backroom' Role

Marshal of the Royal Air Force Sir Dermot Boyle, Correspondence, 31 March 1980, Sway, Hampshire

From the flying training schools of the United States, Canada, Australia, Southern Africa, and Britain, a snowballing flow of pilots, navigators, wireless operators and air gunners was now feeding into the squadrons.

Instructing was a lacklustre, 'backroom' role, and many Allied aircrew, after operational tours, tried to avoid it. But it was an essential ingredient in the build up of air superiority. A first-rate chief flying instructor in a training school was as important to the cause of victory as the able leader of an operational squadron.

'The fact that Douglas (Bader) thought as much of Pearson† bears out one of my best-established theories. A pupil will normally worship his first instructor, and this carries an enormous moral responsibility for the instructor because everything he does and says in the air and on the ground will automatically be what the pupil will afterwards emulate.'

The Power to Influence

Derek Bielby, Stokesley, North Yorkshire, 1983

'My first instructor, George Murray, was a Canadian citizen ... and ... a flight lieutenant. He was smoothly dressed in well-pressed clothes. He wore tinted spectacles when he flew as a protection against the fierce prairie sun... His flying gloves were a well-worn pair of Red Indian gauntlets with leather fringes and dripping with decorative bead patterns. On his left breast, just below the pilot's wings, nestled the purple and white slanting stripes of the ribbon of the Distinguished Flying Cross. He was a God-like paragon of effortless flying ability....

*'Pete' Peterson (Major General Chesley G. Peterson) was Don Blakeslee's predecessor as commander of the US 8th Army Air Force's 4th Fighter Group and a former Eagle Squadron Commander. (Ed.)

†Flight Lieutenant Wilfred J. Pearson, who taught Douglas Bader to fly at Cranwell. Bader never forgot what he owed to hm in obtaining the most comprehensive and precise flying training in the world. (Ed.)

'Towards the end of the course we came together again when he gave me the Flight Commander's check. This time we were flying a brand new, Canadian-built Anson Mark II with Jacobs engines.... It was during a low-flying exercise on this test that Flight Lieutenant Murray "cut" an engine by closing one throttle while I was flying at about 50 feet. "Now," he said, "your reaction please."

'Mindful of the old Mark I's inability to maintain height after "losing" an engine, I trotted out the standard procedure of closing the "good" throttle, keeping the aircraft straight and attempting a mock, wheels-up landing as best I could among the trees of the Assiniboine River whose topmost branches we were by now bending with our slipstream.

'Such an undignified prospect didn't appeal to my instructor. "I've got her," he announced and opened up the "good" engine to full throttle and trimmed the aircraft to fly straight and level. With complete control he curved away from the river and brought the aircraft in over the vast stubble fields of the prairie. Picking his spot, he lowered the undercarriage and announced that he could then land softly ahead. He did not, of course, set the aircraft down, but opened up the "dead" engine and climbed away.... A telling lesson, beautifully demonstrated by a flying maestro, whose skill seemed supernatural to an admiring pupil.

'Later, while serving with a Coastal Command squadron in southern England, I was idling through the *Aeroplane* one day and on the page headed "Service Aviation" I found this entry under "Bar to the DFC":

Acting Squadron Leader G.B. Murray DFC ... now on his second tour of operations ... and ... in command of his flight ... has inspired all ... by his own high standard of ability and courage....

'I wasn't surprised.'

Brotherly Inspiration

John Spicer, Chester-le-Street, County Durham, 1983

'... Among us in Canada, where we were training under the Empire Air Training Scheme, was an air gunner – Paddy Finucane – a flying officer at the time, and the younger brother of the lately-deceased "ace". He dedicated himself to getting his (pilot's) wings and avenging his brother. He was a fine man and we went through everything together – Elementary Flying, Service Flying Training, but, after that, he beat us back to England where, I understand, he was soon posted to a squadron....

'The thing about Paddy was that wherever he went his late brother's uniform jacket, complete with "gong" ribbons and insignia, went with him.... He seemed to keep it with him as a talisman and an inspiration....'

Canadian at Large

Murray Peden QC, *A Thousand Shall Fall* (Canada's Wings Inc., Ontario, 1982)

... I reached London about two that afternoon, and, recalling that it was

Tuesday and still within banking hours, I squandered three shillings on a taxi and dismounted at 9 Waterloo Place, Pall Mall, the London office of the Bank of Montreal. From past experience I was not optimistic about the possibility of drawing money out of the bank before I had put any in, but deemed it worth a try.

In two minutes I was sitting in the manager's office... but before I had even finished the truthful portion of my story, the manager rang for the accountant and, without asking to see my identification card, instructed him to open an account in my name and bring in a £5 advance. As I thanked him sincerely for his kindness he smiled and said:

'My boy is in the RAF. He has been training these last months in the States and Canada, and his letters are full of nice things people have done for him over there. I'm pleased to get the opportunity to treat a Canadian or an American the way they've been treating him. Are you sure now that five pounds will be enough?'

... I then decided to go and see some sights, starting with Buckingham Palace. I elected to travel on the underground and soon found myself sitting next to a window.... My eye was drawn to two signs, one authorized, the other somewhat anti-establishment, but definitely relevant. The transit authority sign had reference to the woven fibre material which covered the train windows completely, except for a small elliptical hole in the centre. Through this hole, if one craned one's neck sufficiently, the platform signs could be seen. The material had been installed as a safety measure to prevent shards of glass flying about in a blast.... The authorized sign pictured, with strong reprobative overtones, a passenger scraping the fibre off the glass with his penknife. Below the picture appeared this prim chastisement:

Pardon us for our correction
But this is here for your protection.

A heavily inked arrow led my gaze 8 inches to the right, below an irregularly shaped porthole, recently hacked through the fibre by penknife or nail file, to the scrawled rebuttal of an unrepentant rebel:

That's OK Jack, but for your information
I want to see my f—ing station.

There was something about the idiom that stamped it as Canadian composition beyond the shadow of a doubt....

'MacRobert's Reply'

In 1922, Alexander MacRobert, a Scot, son of a machineman in a paper works in Aberdeenshire, and grandson of an innkeeper, was created a baronet in the style of Sir Alexander MacRobert, Bt, of Cawnpore and Cromar. He had made a fortune in India after starting with a humble job in the Cawnpore Woollen Mills having left his employment in Aberdeen when his salary was £100 per annum.

MacRobert's achievements in the East were many and these were matched by his benefactions both in India and at home. After his first wife died, he later married Rachel Workman, of Worcester, Massachusetts, a lady of proven courage and spirit. There were three sons of the marriage. All three were killed flying, the last two in action with the Royal Air Force. The baronetcy thus became extinct within twenty years of its creation.

When her third, and only surviving, son, Iain, a 24-year-old Pilot Officer of Coastal Command, and the 4th Baronet, was killed, Lady MacRobert made a gift of £25,000 to the Royal Air Force. She sent it direct to the Secretary of State for Air:

> Let it be used where it is most needed.... I have no more sons to wear the MacRobert Badge* or carry it in the fight.... If I had ten sons, I know they would all have followed that line of duty.
>
> It is with a mother's pride that I enclose my cheque.... With it goes my sympathy to those mothers who have also lost sons, and gratitude to all other mothers whose sons so gallantly carry on the fight....

The gift was used to buy a Stirling bomber. It was named 'MacRobert's Reply.'

Lady MacRobert didn't stop there. Shortly afterwards, she gave another £20,000 to buy four Hurricanes. Then, in April, 1943, she set up the MacRobert Trust in which she made a notable and enduring direction to the Trustees:

> ... that they shall always maintain, in memory of my three sons, Alastrean House, formerly known as the House of Cromar, with its gardens and grounds for such charitable or benevolent use of the Royal Air Force (including the use of the said House as a Rest Centre for flying personnel) or for such use of the Royal Air Force Benevolent Fund as may be approved by me in my lifetime and, on my death, by the Trustees....

Here, overlooking the lovely hills of Deeside, serving and retired officers of the Royal Air Force, and their families, still continue to enjoy the comforts of an agreeable Scottish estate.

Rachel MacRobert left a message for those who visit Alastrean House:

> In other days this house had another name, and we had some very happy gatherings of friends. You know the story of my three sons. I am sure they would like to think that flying men – their comrades of the RAF – are the guests, and greeted in the name of MacRobert. You would have liked my boys. They were Hospitality and Goodwill personified. They had a way of making you feel at home....
>
> July, 1943

... And Then There Was None

In daylight, on 3 May 1943, a small force of twelve escorted Ventura medium bombers of No. 487 (New Zealand) Squadron, led by the New Zealander, Squadron Leader Leonard Trent, attacked a power station on the northwestern periphery of Amsterdam. The raid had a special purpose. The underground network had revealed that tough and patriotic Dutch workmen, employed in powerhouses all over Holland, were awaiting a lead

*The MacRobert Badge (not to be confused with Crest) was a frond of bracken and an Indian Rose crossed.

from Amsterdam to go on strike as a 'protest' against Allied bombing by day. This assault was to light the fuse.

To coincide with this raid another force of Boston bombers was planned to make a diversionary low-level attack on a steel works at Ijmuiden.

Of the formation of Venturas, one turned back with engine failure before reaching the target area. The rest of the squadron was mutilated. The operation was a disaster for two prime reasons – one unpredictable, the other the result of downright incompetence by 11 Group and the Hornchurch wing leader.

Trent, although hard hit at 12,000 feet as he began his run up to the target, pressed his attack with customary Kiwi tenacity in the face of an unrelenting ground and air defence. He was the last of the eleven to go. For his conduct he earned the right to add the Victoria Cross to the Distinguished Flying Cross he already wore.

His own story lays bare the cause of the catastrophe, and it underlines yet again the inescapable responsibility which rested upon the fighter leaders who covered these daylight bombing attacks on precision targets. One mistake in timing or position could provoke calamity.

Group Captain Leonard Trent VC, *Royal Air Force Flying Review*, 1955

'We all roared away low level across Norfolk to pick up our close escort of fighters from Coltishall.*

We continued flying at 100 feet to avoid detection by the enemy radar. We hoped to get within 10 miles of the Dutch coast before being plotted; at this point we did a full-power climb to bombing height. In this way, we were usually able to complete the bombing run, and be homeward bound before the German fighters were in a position to attack....

The German governor of Holland had decided to pay a state visit to Haarlem that particular afternoon, and fighter reinforcements had been brought from as far as Norway and France. Unfortunately, Haarlem was only 5 miles left of my track to Amsterdam, and midway from the coast. *Even more unfortunate, a fighter diversionary wing made some mistakes and got twenty minutes ahead of schedule. Instead of drawing any enemy fighters down towards Flushing this wing succeeded in bringing up over 90 Focke-Wulfs and Messerschmitts. This formidable array was therefore high in the sky when our puny force appeared on radar* (editor's italics).

Suddenly things began to happen. Thomas, from the astrodome, shouted: 'Here's a whole shower of fighters coming down out of the sun; 20, 30, 40. Hell's teeth, they're 109s and 190s!'

... As I turned, I could see them coming in one after the other – about a dozen had attached themselves to my box and the same number were attending to Flt Lt Duffield and his box.

Duffield was hit in the first attack and went down with smoke pouring from his tail. Two of his crew were badly injured but he regained control near the ground and limped off home unnoticed by the enemy. My Number Five was set on fire and fell away....

*Led by a Canadian from Alberta, Wing Commander H.P. Blatchford, who was killed in action that day. (Ed.)

The next attack came from the starboard beam ... giving our gunners a clear field of fire ... as the fighters ducked under the formation.... But my Number Four had gone the way of Number Five....

This had all happened in something like two minutes, which seemed like two years.... I now realized something had gone seriously wrong, and the top cover could not help.... We were on our own in a hopeless situation.... From then on it was just a succession of attacks, and I was amazed that we had lasted so long.... My Number Three had pulled out with the port engine in flames.

And so my Number Two, who was flying as if it was a practice over Norfolk, was the only one left as the first of the built-up areas appeared in the bombing sights.

My mouth was dry as a bone.... "Left, left, five seconds to go". That bombing run lasted a lifetime ... then a jubilant "Bombs gone". I looked up from the instrument panel and discovered my Number Two had gone. We were on our own.

As I reached for the lever to close the bomb doors ... there was a fearful bang and, horror of horrors, all the flying controls had been shot away.

"We've had it, chaps; no controls, bale out, abandon aircraft, quick...."

... We were taken to an officers' mess in the middle of Amsterdam and to a military hospital for a further dressing and a stitch or two, then on to a huge military barracks where the cell doors of solitary confinement clanged behind us.

I sat on a narrow, hard bed and stared into space for hours, wondering what had gone wrong. Had I made the right decisions? What a shock it was going to be for my poor wife. These disturbing thoughts accompanied me to Dulag Luft.... Another fortnight of solitary confinement, with no books, then the interrogation. A few days later, about fifty new prisoners who had been gathered from all points of the compass, were entrained for Stalag Luft III, the main air force prison camp....'

'The Airman'

He laughed at death,
Pursued him with a kiss
Climbed to the skies
Pursued him to a star
But death, who never had
Been wooed like this
Remained aloof, afar

With spurt and gleam and
Brightness like the sun's
He circled death as with a
Wheel of flame
But death, capricious,
Sought those other ones
Who had not called his name

He mocked at death
Pursued him into hell
Mocked him afresh, then
Crashed to burning space
But death, grown gentle,
Caught him as he fell
Nor let him see his face.

Quoted by Peter Tory in his column in the London *Daily Mirror* of 14 September 1982. The author is unknown. The poem was sent by a reader to Tory after he had earlier quoted High Flight by the Canadian, Pilot Officer J.G. Magee of the RCAF (see page 131) which is on a similar theme. Publication did not prompt information regarding authorship.

The Guts of a Gunner

Peter Firkins* City Beach, Western Australia, 1982

'The first raid on Düsseldorf, during the historic "Battle of the Ruhr", went in on 25/26 May 1943, with 686 bombers attacking, but heavy cloud, which extended up to 20,000 feet, threw the bomber stream into confusion and the raid was a failure. The next attack, a fortnight later, on 11/12 June was, by contrast, overwhelmingly successful. Conditions were perfect with a bright moon, and the Pathfinders laid their flares with great precision, so that the main force was able to devastate the central city area.... However, the night fighters were particularly active, and 38 bombers were shot down.

It was on this raid that the role of the gunner was exemplified by the courage of Flight Sergeant N.F. Williams, rear gunner in a Halifax of No. 35 Pathfinder Squadron, who had already won the DFM and bar on previous operations.

Williams' crew had just started their bombing run when, out of the haze came blazing trails of cannon and machine-gun tracers as a night fighter attacked from slightly above the port quarter. Williams swung his turret round to reply, at the same time instructing his pilot to "corkscrew port" and begin the violent evasive action....

As the Halifax was thrown into a steep dive to port, another night fighter came up from underneath and raked the bomber with cannon and machine-gun fire, wounding Williams in the stomach, legs and thighs, paralysing him from the waist down which he described as though "the world had fallen on top of him". Apart from his wounds, Williams' turret was riddled and one of his four machine guns put out of action.

The mid-upper gunner was also wounded, temporarily blinded by a bullet that furrowed the back of his head. A shell pierced the starboard wing petrol tank, setting it on fire; blazing petrol streamed from the wing, making the Halifax visible for miles.... Both fighters attacked again out of the haze, but despite his wounds and pain, Williams returned their fire as well as

*Firkins, author of *The Golden Eagles*, *Strike and Return*, *From Hell to Eternity*, etc. In this specially written piece, he is able to draw on his own extensive experience of the air gunner's often testing lot. (Ed.)

giving his pilot instructions to continue violent evasive action. "Dive port, steady, dive port, steady, climb starboard, steady, climb starboard, dive port." The manoeuvre was repeated again and again.

As the two fighters pressed their tactical advantage with great resolution, Williams asked that the wounded mid-upper gunner be relieved in his turret by the bomb aimer. As the fighters came in again, the Halifax banked away bringing the second aircraft into Williams' line of fire. After a long and accurate burst, the Australian saw the fighter explode into two flaming sections which broke up and cascaded earthwards.

The crew shook off the other fighter long enough for the bomb aimer to return to his own compartment and drop the bombs. He, in turn, was relieved by the wireless operator, who throughout the attacks had been giving first aid to the wounded mid-upper gunner.

The second aircraft now came in again, attacking finally from directly astern. Williams, with extraordinary courage and judgement, held his fire until the enemy was closing right in for the kill. Then he gave a long, accurate burst. Pieces flew off its fuselage and wings and other members of the crew saw it roll over and disintegrate into a ball of fire, eventually plunging away out of sight....

The Halifax crew now assessed the shambles that surrounded them after the attacks.... The fire which had broken out in the wing, had died down, and as the pilot dived it was extinguished completely. Course was set for base.

With Williams in dreadful distress, members of the crew tried to extricate him from his turret, but the doors had jammed and he was forced to remain there until his pilot had made a perfect crash landing at base. He was then chopped from the turret, and, with the mid-upper gunner, taken away to hospital, where he remained for two months.

When told he had been awarded the Conspicuous Gallantry Medal, he replied modestly: "Any of the other chaps should have received it."

With his CGM and Distinguished Flying Medal and bar, Williams became the most highly decorated NCO in the Royal Australian Air Force. With 4 enemy aircraft destroyed, 2 probably destroyed and 2 more damaged from upwards of fifty operations, many of them as a Pathfinder, Williams was Bomber Command's outstanding gunner of the war.

He had attracted extraordinary punishment. He was attacked by fighters no less than twelve times in his first seven operations and on more than half his missions his aircraft had returned damaged either by flak or night fighters....'

Retribution

Wherefore by their fruits ye shall know them
St Matthew 7:20

A most fearful retribution was now being wreaked upon the German people for the earlier misjudgements and decisions of Hitler and the other Nazi leaders. Take Hamburg.

Air Vice-Marshal D.C.T. Bennett, *Pathfinder, Wartime Memoirs* (Frederick Muller, 1958)

Towards the end of July, 1943, we achieved what I regard as the greatest victory of the war, land, sea or air. This victory was in the Battle of Hamburg. The first raid was on 24 July, when a fairly big force of about 700 and 800 aircraft attacked, using Parramatta ground marking, and everything went according to plan. All of these aircraft were four-engined heavies, and with the relatively reasonable range to Hamburg, they were all carrying a big load of bombs. The result was absolutely staggering. The fire-fighting services failed completely, and enormous infernoes raged on the ground. So great were these fires that the smoke reached 20,000 feet, and it was subsequently discovered from German reports that the wind speed, due to this violent convection, reached about 100 miles per hour on the ground, blowing bicycles along the streets and, it is said, even blowing over pedestrians who were attempting to run for better cover....

... On the following night we reverted to Essen, in order to fool the German defences, but I sent a few Mosquitoes along to Hamburg just to ring the alarms and make the frightened people of Hamburg frightened once again. On the 26th I did the same with another handful of Mosquitoes, just to keep their nerves on edge. Then on the 27th the C-in-C laid on another big one, and once again it was wonderfully successful. A few Mosquitoes kept the pot boiling on the 28th, and then on the 29th we went there in force again. The final raid was on 2 August, four days later, when we finished off the job....

The Air Officer Commanding-in-Chief, Bomber Command, in his Despatch on War Operations, quoted from a German secret document entitled 'Fire Typhoon – Hamburg, Night 27/29 July 1943'

'The cause of the terrific damage lies in the fire-storms. The alternative dropping of blockbusters, HEs and incendiaries made fire-fighting impossible, small fires united into conflagrations in the shortest time and these in their turn led to the fire-storms. To comprehend these fire-storms, which go beyond all human imagination, one can only analyse them from a physical meteorological angle. Through the unison of a number of fires, the air gets so hot, so on account of its decreasing specific weight, receives a terrific momentum, which in its turn causes other surrounding air to be sucked towards the centre. By that suction combined with the enormous difference in temperature (600–1000 degrees centigrade) tempests are caused which go beyond their meteorological counterparts (20–30 degrees centigrade). In a built-up area the suction could not follow its shortest course, but the overheated air stormed through the streets with immense force, taking along not only sparks but burning timber and roof beams, so spreading the fire further and further, developing in a short time into a fire typhoon such as was never before witnessed, against which every human resistance was quite useless.'

Reichminister Albert Speer, Minister for Armaments, during interrogation in 1945

'The first heavy attack on Hamburg in August, 1943, made an extraordinary impression. We were of the opinion that a rapid repetition of this type of attack upon another six German towns would inevitably cripple the will to sustain armament manufacture and war production. It was I who first verbally reported to the Führer at that time that a continuation of these attacks might bring about a rapid end to the war.'

Stygian Darkness

General Jean Calmel, *Pilotes de Nuit* (Editions de la Table Ronde, 1952); *Night Pilot* (William Kimber, 1955)

The captains of aircraft saw in nightly excursions a strange amalgam of impressions:

... The night revealed to us the most terrifying and the most monstrously beautiful spectacles.

I have seen a whole city ablaze in brilliant scarlet surrounded by the innumerable beams of searchlights like immaculate swords trying to stab the sky. We were in that sky with our bombs in their bays, bringing more and more thunder and fuel to rekindle the brazier below, while flares of violent colours – blue, red and green – were dropped by our markers to show precisely what still dark quarters we had to illuminate with our death-dealing cargo....

... The night, however, is not entirely hostile. It is, on the contrary, full of friends who are making the same marvellous efforts for you as you are making for them. Their eyes go unceasingly from their dials, which are faintly lit so as to avoid dazzle, to their windscreens, through which they scrutinize the Stygian darkness in desperation. With sustained attention, with all their muscles and their intelligence, and thanks to their intensive training, these pilots enabled a mission, that *tour de force*, to be accomplished: to fly throughout the night side by side, invisible to each other except over the target, where the Titanic gleams of fires on the ground brought them to life again for a few seconds.

The secret of this success for us night pilots lay in following with absolute precision the course, the speed, the altitude and the time given to us before we left....

The Big City: the Question Marks

Air Vice-Marshal D.C.T. Bennett, *Pathfinder, Wartime Memoirs* (Frederick Muller, 1958)

When we ultimately did get photos, the results on Berlin were said by the C-in-C to be disappointing. What he did not seem to realize was that what I had been saying throughout the whole series of attacks was perfectly true, however unpalatable it might be. ... The crews had been worried by the defences of Berlin, and had been doing everything they could to be as defensive as possible. In particular, they had been encouraged to believe that height was their salvation...

Unhappily, someone had decided (I am not quite sure who was

responsible) to add 3000 lb to the all-up weight of the Lancaster to increase the bomb load on Berlin, which was of course a considerable range from England. In fact, the range effect alone knocked about 1500 lb off the pay-load of a Lancaster and about 2000 lb off the pay-load of a Halifax compared to the short target such as Hamburg or the Ruhr....

The idea of offsetting this by putting up the all-up weight of the Lancaster by a further 3000 lb was in my view quite wrong, as the rate of climb with a normally-loaded Lancaster was already sufficiently slow to be extremely worrying to a captain approaching hostile defences. The result was inevitable, and I reported it raid after raid, only to be told that we were imagining things!

What I reported was that senior Pathfinder crews at interrogation were constantly saying that they had seen scores of bombs being jettisoned in the North Sea on their outward-bound flights. Experienced Pathfinder crews do not misidentify bomb bursts. More particularly they do not misidentify 'cookies' (4000 lb) with their light cases. These cookies could not be dropped 'safe' because they always exploded in any case. Thus the aircraft which were jettisoning these bombs gave away the fact that they were doing so and, incidentally, letting go of the most valuable bomb in the whole of their load. The net result was that of the bombs which left England, a very large proportion did not reach Berlin.

The captains concerned felt justified in acting in this way as they were quite unable to climb, and indeed, in many cases, were apprehensive of icing in the clouds ahead. In their view it was thus a question of necessity.... By being too ambitious, we in fact dropped less.

Another thing which seriously reduced the effect on Berlin, and which also was reported by Pathfinders, was that the bombing was more scattered than on other targets. Whether this was due to the weight of the defences, or the psychological effect of believing that a vast area of targets lay beneath, it is hard to say. In any case, the results were not as good as they should have been.

Nevertheless, Bomber Command achieved a very great victory, and achieved it by sheer hard fighting in bad weather, and against the strongest defences which the enemy could muster.... It is surprising that our loss rate was not higher. Bomber Command itself averaged something over 6 per cent, and Pathfinders, which started badly and at times reached as much as 13½ per cent lost on one raid (1 January 1944), averaged just over 5 per cent. The total area destroyed in the main battle of Berlin was about 5500 acres. Hardly disappointing!

Berlin Exponent – and the Enduring Spirit of his Crew, and Squadron

Alec Wales, Montville, Queensland, 1982

No pilot in Bomber Command surpassed the Berlin record of Alec Wales, the Australian captain of a Lancaster of 460 Squadron, Royal Australian Air Force. First as an NCO and, later, as a commisssioned officer, he made a mark on the capital that few, if any, could match. It's not hard to see why.

. . . I went to Berlin fifteen times in all – twice on my first tour, on 16 and 17 January 1943 and thirteen, plus one 'abort' on my second. The 'abort' was on 26 November 1943 – my fourth trip on that tour of which the first three had also been to Berlin.

Just after take off, the aircraft (not my usual one) became practically uncontrollable, directionally. We returned to base using aileron and engine power alone for directional control. After landing, the ground crew could find nothing to cause the problem. Naturally we were all apprehensive, fearing that it was thought we had no wish to continue.

Group Captain Hughie Edwards, who, in my opinion, was a great leader and who had the ability to command respect, admiration and loyalty, met us and took us to the mess, where he bought beers and allayed our fears.

Not long after, it was found that the starboard rudder was half-filled with water. Sabotage? . . .

Berlin, as I remember it, was not heavily defended in January 1943, but in November and December 1943 and January, February, March 1944 the defences were greatly improved.

The sky above 20,000 feet was always lit up with flares with a wall of flak up to 18,000 feet and night fighters above that.

I do remember one trip when we saw one Me 110 shoot down four Lancasters in about as many minutes, and we ourselves nearly crashed into an Me 109 as we left the target. He passed directly in front of me at a distance of no more than 10 yards. The gunners were unable to get him in their sights as he was gone in seconds. I don't know who was most surprised – he or us! He was so close, I could see him looking straight at me!

I was fortunate enough to come through my two tours safely, due mainly to having such a fine crew. I had the same bomb aimer, rear gunner and wireless operator on both tours and have always felt that a well-trained and disciplined crew was the key to survival, plus of course, Lady Luck. . . . I still keep contact with the members of my crew who are still alive. Roy Canvin, my first navigator, passed away a few years ago and I keep in touch with his widow, as I do with the widow of my rear gunner, Alec Sanders. Just recently I was invited over to Perth for the 70th Surprise Birthday Party for my bomb aimer, Ron Lawton. I met quite a few other 460 types. . . .

Peenemunde
17 August 1943

Constance Babington Smith, *Evidence in Camera* (David & Charles, 1957)

Peenemunde, the centre of Germany's top secret rocket development programme, was the place-name of Bomber Command's most important single attack of the war. Well ahead of the Allies in rocket and jet propulsion, Germany was staking much on the early introduction of these new and potentially match-winning weapons.

The raid was the outcome of a finely-concerted, combined operation. It embraced courage and special talents – astute and brave intelligence work;

highly perceptive and tenacious photo-reconnaissance flying; clever and alert photographic interpretation and, finally, a minutely-planned and resolutely discharged night attack by Bomber Command on this critical Baltic target.

It was Constance Babington Smith, then an officer in the Women's Auxiliary Air Force, who started things moving with an acutely observant assessment of PRU's pictures. Here, in this brief passage, she exposes the facts.

On 15 May 1942, Flight Lieutenant D.W. Steventon flew his Spitfire high above the western shores of the Baltic, on the way to cover Swinemunde after photographing Kiel. Far below and ahead lay the island of Usedom, with its long belt of woodland facing the Baltic, and separated from the mainland by the River Peene. He happened to notice ... there was an airfield at the northern tip of the island, with quite a lot of new developments nearby, and he switched on his cameras for a short run....

.... I remember flipping through the ... photographs.... Then something unusual caught my eye, and I stopped to take a good look at some extraordinary circular embankments. I glanced quickly at the plot to see where it was, and noticed the name Peenemunde....

Meantime ... General Dornberger and Werner von Braun were working day and night at their rockets, and the first fully successful launching of ... the V-2 took place at the experimental station in the woods on 3 October 1942; while in December an early version of the flying bomb, the V-1, was launched from below a large aircraft over Peenemunde....

Soon the decision was made to attack.... On the night of 17 August 1943, when Bomber Command made their famous raid, 40 aircraft were lost, but considerable destruction was caused. We know now that it seriously delayed the whole programme....

Sir Malcolm McAlpine, the eminent engineer, who was asked to comment on the damage assessment photographs, said: 'It would be easier to start over again.' According to General Dornberger, the German engineer in charge [also] shared his view.

'Unidentified'

This is the tale of the Gremlins
Told by the PRU
At Benson and Wick and St Eval –
And believe me, you slobs, it's true.

When you're seven miles up in the heavens,
(That's a hell of a lonely spot)
And it's fifty degrees below zero
Which isn't exactly hot.

When you're frozen blue like your Spitfire
And you're scared a Mosquito pink,
When you're thousands of miles from nowhere
And there's nothing below but the drink –

It's then you will see the Gremlins,
Green and gamboge and gold,
Male and female and neuter
Gremlins both young and old.

It's no good trying to dodge them,
The lessons you learnt on the Link
Won't help you evade a Gremlin
Though you boost and you dive and you jink.

White ones will wiggle your wingtips,
Male ones will muddle your maps,
Green ones will guzzle your Glycol,
Females will flutter your flaps.

Pink ones will perch on your perspex,
And dance pirouettes on your prop;
There's a spherical middle-aged Gremlin
Who'll spin on your stick like a top.

They'll freeze up your camera shutters,
They'll bite through your aileron wires,
They'll bend and they'll break and they'll batter,
They'll insert toasting forks in your tyres.

That is the tale of the Gremlins
Told by the PRU,
(P)retty (R)uddy (U)nlikely to many,
But fact, none the less, to the few.

An unnamed PRU pilot
Believed stationed at St Eval, Cornwall
circa 1941

Spirit of Poland

Polish Air Force Archives, Polish Institute and Sikorski Museum, London

On 22 August 1943 a Polish Spitfire pilot, Flying Officer Kurytowicz, was hit by a Focke-Wulf 190 during an offensive operation to northern France. He baled out over the Channel and spent four days and three nights in his dinghy being buffeted about in the heavy seas. As his hopes of rescue began to fade, he decided to leave a last message. He took off his shirt collar and wrote on it:

'Did not want to be taken prisoner. Baled out in Channel on Sunday. It is Tuesday today. Have been afloat two days already. Have seen rescuers but haven't been seen. Can see many aircraft. The torch does not work. Nothing to attract their attention with. . . .

'Third day. Can see land on horizon. Am afraid will be blown away during night.

'Fourth day. Can see the land but still far away. Many aircraft overhead. But I am not seen. Can see four ships ... heading for me? Maybe it's help. It's so difficult to die slowly....

'I have been spotted. I thank God. Tears of my mother and wife have saved me.'

'Tears of my mother and wife' is a Polish expression meaning that their prayers had saved him.

Flying Officer Kurytowicz was picked up by a Royal Navy minesweeper, HMS Aggressive, *commanded by Lieutenant J.D. Hastie.*

Impact of the 'Eighth'

Laddie Lucas, *Five Up, A Chronicle of Five Lives* (Sidgwick & Jackson, 1978)

Throughout the last months of 1943 we were seeing almost daily at Coltishall evidence of the gathering strength of the daylight attacks of the US Eighth Army Air Force as they began to penetrate deeper and deeper into the heart of the German fatherland. We could bear witness, too, to the murderous treatment this onslaught was receiving at the hands of Göring's steadily increasing fighter force.

Apart from seeing it all at first hand from the missions which the wing was able to fly in support, there was the additional proof of shot-up P-47s and P-38s and, occasionally, of damaged B-17s and B-24s which, at their last gasp, had found sanctuary in ... [this] ... haven after struggling back across the sea.

The fights started soon after the bombers crossed into enemy territory and continued all the way in to the target. With the bombs gone, the Fortresses and Liberators, some already heavily scarred, had to brace themselves again for renewed and repeated attacks as these great sky fleets fought their way back to the coast and the North Sea.

For sustained courage, and as an example of how the impossible can be achieved by improvisation and determination, the United States aerial offensive against Germany of 1943 and 1944 was never surpassed in any theatre of the Second World War....

Schweinfurt

Elmer Bendiner, *The Fall of Fortresses* (G.P. Putnam's Sons, New York, 1980; Souvenir Press, London, 1981 and by Pan, London, 1982)

1 PRELUDE

... I woke to a flashlight beam in my face ... and a voice that said, 'Briefing at four'. I recollect the flow of obscenities that followed with the wistful sentimentality usually reserved for such homey things as the smell of a fire in a hearth or the warmth of a wool blanket on a cold night.

'What son of a bitch was that?' 'It's the middle of the motherfuckin' Goddamned night.' 'Shut the fuckin' door – it's colder'n a witch's tit.' It was the routine indignation, the customary riot of Americans responding to military orders. I liked it then. I like it more in retrospect. I lay a little while, as was, and is, my habit, seeking to come to terms with the day ahead, particularly this day of our first battle. I was never one for leaping out of bed. I must persuade my legs to reach for the floor...

Bob was up and bouncing. He bubbled and yiped like a boy before a picnic. 'Benny, get up, you bastard. Gonna sleep all day? Oh, Mama, it's cold.' He slung his towel over his shoulder and dashed out toward the latrine, shouting at the moaning forms he left behind him, 'Goddamned son of a bitch, get up.' It was as innocent, as joyful, as the wag of a tail.

I chose to wash and shave at the mess hall, where there was a modicum of warmth. I dressed and put on a tie. I remember that, and yet it seems a very formal dress for battle. Over it went the thick-ribbed yellowish olive drab sweater and the coveralls with capacious pockets to hold all sorts of things from rations in case of disaster to spare pencils. (I was always afraid of being caught without a pencil in the heat of battle. It was far more important than my pistol, which I hung from my belt as an ornament like the lieutenant's bars I pinned to my collar tabs.)

The sun was not yet up, but the sky was fading. On the way to the mess hall, amid the hushed sounds of men walking and the clink of mess gear, I rode my bike into a ditch and thought it ignominious, like falling off a horse en route to Armageddon...

2 ATTACK AND WITHDRAWAL 17 AUGUST 1943

It was 1314 when we left the coast of England. We could see no vapour trails of friendly fighters as called for by the script. They may be up ahead, someone said, or far behind. There was nothing we could do about it and that was a comfort....

I remember that while we were still over the Channel we spotted a B-17 far below us, heading for home. Actually eleven planes aborted before we reached the enemy coast. Any reason for an abort seems good enough at the time, whether it is prompted by engineering symptoms or more personal considerations. We felt no resentment, a bit of envy perhaps, countered by a feeling of tolerant superiority.

We were somewhere between Antwerp and Aachen when I was aware of the first rocket attack. It seemed to come in from seven o'clock over *Tondelayo*'s left wing. I remember seeing a brownish object tumbling and then bursting into an orange-yellow flash and an enormous black cloud. *Tondelayo* reared like a frightened horse.

There was a running chronicle from the tail and the waist describing the fall of planes from the high group where six of the 379th had been assigned. 'Plane in flames, chutes opening.' 'Plane falling like a stone – no chutes.' So it went. We droned on. When I was not at the gun I was scribbling in the log the time, place and altitude of flak, of rocket bursts, of kills and fallen comrades, of headings and checkpoints.

Can I have been so detached? Not to feel the heat of battle or the clutch of panic or the numbing chill of altitude? ... I noted the events of battle like a metronome timing the music without hearing it. I do remember looking down somewhere after Eupen and counting the fitful yellow-orange flares I saw on the ground. At first – so dense am I – I did not understand them. Here were no cities burning. No haystack could make a fire visible in broad daylight 23,000 feet up. Then it came to me as it came to others – for I remember my headset crackling with the news – that these were B-17s blazing on the ground.

The afternoon was brilliant, but, as I remember it, the earth was sombre, smudged, dark green and purple. In the gloom those orange-yellow fires curling black smoke upward were grotesque. I was as incredulous as I had been when first I saw a fuselage red with the blood of a gunner's head blasted along with his turret. As we followed that trail of torches it seemed unreal. I see it now as a funeral cortege with black-plumed horses and torches in the night.

In England monitors heard the German pilots gathering from all over France and Germany to ambush our homeward flight.... All across Germany, Holland and Belgium the terrible landscape of burning planes unrolled beneath us. It seemed that we were littering Europe with our dead. We endured this awesome spectacle while we suffered a desperate chill. The cartridge cases were filling our nose compartment up to our ankles....

At last we came to the blessed sight of soaring Thunderbolts above the Channel coast. It was 1669 by my watch. The afternoon sun warmed the portside window.... On the field at Kimbolton ... I felt exhaustion creeping in beneath the excitement like death beneath a fever.

We had been in the air for 8 hours 40 minutes. We had been in incessant combat for close to 6 hours. It had been 14 hours since we had risen in the predawn. In that time 60 B-17s had been shot down, 600 men were missing. The first major strategic air battle of the war had been fought.

3 POSTSCRIPTS

... The professorial Captain of Intelligence confirmed the story. Eleven unexploded 20-mm shells were in fact found in *Tondelayo*'s tanks. No, he ... could not say why.

Eventually [he] broke down. Perhaps it was difficult to refuse ... the evidence of a highly personal miracle.... Or perhaps ... the truth ... was too delicious to keep to himself. He swore [the crew] to secrecy.

The armourers who opened each of those shells had found no explosive charge. They were as clean as a whistle and as harmless. Empty? Not quite, said the Captain....

One was not empty. It contained a carefully rolled piece of paper. On it was a scrawl in Czech... Translated, the note read: 'This is all we can do for you now.'

'Prison Camp'

Chosen by J.D. Rae – ex-POW, Stalag Luft III, Keri Keri, Bay of Islands, New Zealand, 1982

Day follows day in dull monotony,
  The sun hangs heavy in the changeless sky;
Dust devils eddy down the sandy road,
  The long, drab rows of huts lie mute within
The shadows of the all-encircling wire –
  And this is life.

The hours silent to eternity,
  The days stretch into weeks, the weeks to years;
Time ages, yet the features do not change;
  Time sweeps along on feet that never move –
Feet fettered by the wire's weightless bond,
  With night comes sleep.

And sleep brings dreams to flaunt these timeless days,
  And life runs sweetly as it did before –
Bright eyes, sweet lips, cool drinks, good food, soft beds –
  The thousand fantasies of vanished peace –
Till morning light returns with hopeless hope.

POW Stalag Luft III, 1943
Author unknown

Stalag Luft III

Pat Ward-Thomas, *Not Only Golf* (Hodder & Stoughton, 1981)

Much has been written and filmed about life in prisoner of war camps. . . .

It was a life often noisy, dirty and afflicted with extremes of heat and cold. Food frequently was inadequate and accommodation grossly overcrowded. It was a life of nostalgia and dreams which at times demanded all the reserves of patience and tolerance that a man could muster. In all the dragging days the only solitude was in sleep, unless one was fortunate enough to be sentenced to a few days in the cooler.

Twice I was sentenced and recall the relief of being alone in a single room with no one's snoring or stirring to disturb me. One could settle peacefully to reading or writing and I managed the greater part of *The Forsyte Saga*. Five days in the cooler was pure holiday, ten a shade too long. . . .

Many of us did not speak to a woman, even to the point of saying 'Good morning' for over four years. Fortunately women were rarely visible; no beauties swayed enticingly beyond the wire. . . . Gradually one was able to think of women as creatures of dreams. All the same it was interesting to observe how, after some while, one became aware of the attractions of other young men. Weary eyes would scan new intakes of prisoners, much as they might a group of girls suddenly appearing in a bar, but this was light-hearted enough....

The camp had its little theatre where plays were produced. The Germans allowed clothes and costumes to be hired, on parole of course, from agencies. Some of the female parts were most convincingly played, stirring thoughts of other times. One young man took a large camp by storm with his seductive rendering of a popular song. People talked about it for days but in ordinary garb he was quite inconspicuous. Erotic illusions swiftly melted once make-up and dresses had been discarded.

When men are totally deprived for years of even the minutest contact with women, homosexual inclinations are inevitable. At best there were never less than eight, and later ten, people sharing a room about 25 by 15 feet, sleeping on two-tiered bunks. Nowhere in Stalag Luft III were there private places for assignations and amatory tendencies had no opportunity for physical expression.

Close, sometimes intense, friendships would develop but a severely limited diet, while providing enough energy for games, study and other activities, helped to sublimate thoughts of secret pastimes, even on Christmas Day in 1943....

The Germans were remarkably understanding. The Stalag Luft camps were run by the Luftwaffe who seemed to pursue a velvet glove policy in their treatment of British, Commonwealth and American prisoners. Their determination to prevent escape was no less evident but rarely were there instances of the petty nastiness, arrogant stupidity and hysterical screaming sometimes encountered at camps controlled by the army. Possibly someone in the Luftwaffe realized that such behaviour usually produced an adverse reaction from the prisoners, and that if they were not harassed they would give less trouble....

Courage of the Underground

Wing Commander J.M. Checketts, Christchurch, New Zealand, 1982

By late 1943, the secret underground movement in occupied Holland, Belgium and France was operating a wonderfully effective service for Allied aircrews who had been shot down, evaded capture and were bent on returning to their home units. The sustained courage of the organizers of this 'system', and their assistants, working under terrible risks, was scarcely credible.

Two brief experiences of the striking results of the network, one of Johnny Checketts, a New Zealander, the other of Sigmund Sandvik, a Norwegian, will suffice.

First, the New Zealander:

'I was at 10,000 feet, south of the Somme Estuary, and some 15 miles inland, when the FW 190 hit me.... As I floated down in my parachute I could see vehicles heading towards the place where they expected me to land.

My legs hurt, the left knee was numb. My trousers were burned almost up to my waist and were still smouldering. The right sleeve of my jacket was gone and my mae west was charred and smoking. I managed to beat out the smouldering fabric.

I landed in a stubble field. The pain in my legs was severe. I tried to hide my parachute and mae west in the wheat sheaves. An elderly woman came up to me and asked my nationality. She saw my RAF Wings, said "Angleterre" and gave a piercing whistle.

I felt awful now. A youth came up on a cycle and the woman signalled to me to get on the carrier. I did, and was taken towards a spinney about ¼ to ½ a mile away. The rear tyre of the cycle was half flat and the ride was very bumpy. I was dumped near the spinney and the lad made off as hard as he could.

I heard a motorcycle approaching and hid in the undergrowth. The machine passed and returned in about 15 minutes and I saw that the rider was a German. He appeared to be carrying my parachute on his bike. Shortly after, a man came up to me from the spinney and signalled me to follow him. He led me into the centre of the wood and made me lie down. It was now about 2100 hours.

My burns and wounds were very painful, but my helper covered me with leaves and brush and left me. Daylight started to fade and I think I must have lost consciousness because the next I knew a man was lying beside me whispering "intelligence" and pressing my face down into the ground. He motioned to me to keep quiet and pointed to my left. I could just discern another man in the darkness and my guide whispered "Boche". I kept dead still.

After a few minutes we slowly and quietly crawled away from the spinney and across fields to a farmhouse. There, the lady bathed my burns, dug some splinters of metal from both knees and painted the lot with mercurochrome.

We continued travelling and we now had a cycle. I could not pedal so my helper had to push it; we went on like this for about two hours. Eventually, we arrived at a small house around 0400 hours to be greeted by my helper's wife and another man who turned out to be a RAF pilot, Flying Officer Baddock. He had been shot down about six weeks earlier. I was now stripped, bathed, spoon fed because of my burns, given a strong drink of fierce Calvados and put to bed.

So ended the high escort for the Marauder Bombers to Seaquex on 6 September 1943, and so, also, did my Command of No. 485 (NZ) Spitfire Squadron at Biggin Hill.

My French hosts nursed and cared for me, putting me into an escape line via which, after many close calls, I arrived back in England seven weeks after being shot down.'

Now, the Norwegian:

'. . . I came down in a small forest in northern France to which I had tried to direct my descent by parachute.

After about ten days of hiding by day and walking southwards by night I was exceptionally lucky to make contact with a member of the French

Underground. From then on, the organization took over and efficiently and pleasantly helped me on my way.

Early in October 1943, I moved to Paris, staying in various apartments for about seven weeks. I was never more than about five days in each place. However, I began to feel that not enough effort was being made to get me out.

Being young and impatient, I broke a golden rule of this type of existence. Late one afternoon I left the apartment where I was staying to visit a man whom I had been told was one of the top organizers. I had lived in his apartment several weeks earlier.

When I rang the doorbell I was bodily lifted into the hall by a huge German. In the apartment were two or three more Germans. I had French identity papers, but my French was very poor, which could certainly not be said of at least one of the Germans.

My efforts at speaking German fared no better. Eventually, I was told that the owner of the apartment had been arrested and that they had found two American air crew in the apartment. I then admitted to flying with the RAF.

I was taken down to the street outside where a fairly large German car was parked. I did not at once get into it as I was told, so one of the Germans grabbed me by the neck, and literally, chucked me into the back. The door on the far side happened to be open and I virtually shot through it coming to rest on my feet, on the opposite side. Automatically, I started running across the street and down various sidestreets. It was getting dark, they started shooting, but I wasn't hit.

As there was a curfew in Paris at night, and not knowing my way around the city, I spent the night locked up in a public lavatory. Next day, I managed to find my way back to the apartment where I had been staying. The family had been told of my arrest and they were just about to go into hiding when I turned up.

They were, of course, angry at my lack of discipline, but pleased that they did not now have to leave. I was moved the next day to the south of France. After a rather strenuous walk across the Pyrenees with other American, Canadian and British aircrew, some of us reached the British Consulate in Barcelona.

From there, there was no difficulty about getting back to the UK.'

Torch and Dagger Work

Hugh Verity, *We Landed by Moonlight* (Ian Allan, 1978)

Another clandestine service was being worked in great secrecy to the occupied countries of Europe and Scandinavia. The landing and picking up of agents in small, out-of-the-way fields by night, was one of the most hazardous duties of the war, as dangerous for the pilot as for the 'passenger'. If ever there was a case of two – or more – people being 'in it together' this was it. The qualities required were not those possessed by ordinary mortals.

Hugh Verity, (Group Captain H.B. Verity) had had a sound record as an operational night-fighter pilot when he elected, during his rest period from flying, to switch to this hideous role. The few friends who knew what he was up to questioned his sanity. His answer was to become one of the most successful – and intrepid – of these undercover operators.

It was late summer and approaching midnight. I was duty intruder controller in the intruder operations room at Headquarters, Fighter Command.... I was sitting with my assistant on a little platform in a bare room deep down below ground level. We had a sloping table below us, painted with an outline map and a reference grid.... This included the main bomber bases of the Luftwaffe in France and the Lowlands, as well as their probable targets in the UK. Two lively WAAF girls plotted any movements of either air force across the Channel and the North Sea.

Beside me sat my assistant, a Y-service officer (Y was short for Wireless Intelligence). He was told on the telephone any news we could pick up about the movements of the German bombers. My object was to 'scramble' (order off) long-range night fighters from bases in southeastern England to prowl round the German bomber bases at about the time they were taking off or – and this was easier to estimate – at about the time they returned to land after bombing England.

This was one of the nights when bad weather grounded both German and British bomber forces. On some of these nights I had noticed single RAF aircraft plotted as they crossed the Channel outwards and then, a few hours later, homeward-bound. These lonely singletons were called 'specials' and I had not been told what they were.... I gathered they were secret and not part of my private intruder battle.

I was contemplating just such a lonely single plot crawling south across the table when Dusty Miller came in for a chat. He had been the Senior Flying Controller at RAF Tangmere, near Chichester. We had a cup of coffee and I asked him about the 'special'. He told me that it was a cloak-and-dagger Lysander. There were a few of them in a detached flight operating from Tangmere during moon-periods to fetch agents from occupied France. They were commanded by a rather eccentric character called Squadron Leader Guy Lockhart. A previous flight commander called 'Sticky' Murphy had been ambushed in Belgium one night during the previous winter. I asked Dusty to tell me about it.

The agents, he said, were trained by the pilots to find suitable fields and describe them by wireless. They laid out miniature 'flare-paths' consisting of three pocket torches, tied to sticks, in the shape of an inverted 'L'. When the Lysander arrived they signalled a pre-arranged morse letter with a fourth torch. If the pilot did not see the correct letter flashed his orders were to return to base.

On this trip to Belgium Sticky had found the field without difficulty. Not only was there a moon but there was also a little snow on the ground. He was put out to find that the signals flashed towards him did not give the right letter. He circled round weighing up the situation. Should he just go home? Wasn't there a possibility that the agreed letter had been garbled in the messages by a primitive portable wireless? Perhaps the agent was in danger and needed to be rescued. He decided to land, acknowledging the flashing light by throttling back, momentarily, twice.

Because he was not quite sure what the reception committee might consist of – and as it was a clear night – he did not land on the 'flare-path' but well away from it. He taxied round and flashed his bright landing lamp towards the first torch. The beam lit up what seemed to be a company of German soldiers in their helmets. He turned through 90 degrees and took off without delay.

At this moment they started firing at him and as he got airborne he felt a bullet go through his neck. The Lysander was still in working order as far as he could see and he set course for Tangmere. He always carried with him as a lucky charm one of his wife's silk stockings which he managed to wind round his neck to reduce the bleeding. But when eventually he could call the tower on the radio-telephone he sounded rather drowsy for he had lost a lot of blood.

Dusty said they kept him awake by telling him rather blue limericks on the R/T and he made it eventually though he had wandered off course on the way home.

When he landed at Tangmere they found that the Lysander had been hit by 30 bullets.

A Squadron's Best Friend. . . .

Jaroslav Hlado, Praha, Czechoslovkia, 1982

Dogs were synonymous with squadron life and some became a feature of it. They were loved and fussed over. In manner, they had something in common with the aircrews. Their habits were often endearing, but sometimes downright naughty: maybe that's why they fitted in so well.

Sadness came when master did not return from an operation. 'Old Faithful' could never bring himself to adapt to loss as the squadron did. It took time for him to recover and rally behind a new leader. Coopie, or Krupy as they wrote it in Czech, was just such a dog. A chow, he was devoted to his first owner, a pilot in 132, the Natal Squadron. When he was shot down '. . . the dog was full of grief, refusing food and spending his days lying hidden away, alone under the raised wooden floor of the dispersal hut.'

When he came out of mourning, Coopie eschewed seniority. He chose for his No. 1, one Jaroslav Hlado, a lowly pilot officer in 312, the Czech Squadron. He was probably more discerning than people thought, for Hlado was to become the successful and much-decorated leader of 134, the last Czech fighter wing, which he took to the Continent after D-Day. By then, as No. 1 recalls, Coopie had become the Czechs' mascot:

'As Hlado was unpronounceable for most Anglo-Saxons, Coopie became P/O Hlado's name, his call sign, and the squadron's. Later it was the call sign for all Czech aircraft in 134 Wing. He was a special character and became very fond of his new master. Whenever we moved to a new station, his first act was to sort out his bedroom. After dark he would go there and snore like hell; but no one could pass the entrance. He was quite dangerous.

His snoring? There was one way to overcome it. Just start breathing to the same tempo as Coopie's snores. Pilots soon found this was the best way to fall asleep. His master had learnt this when he was escaping to the UK from Czechoslovakia via Russia.

With his huge body and his dark tongue, he would enter the mess dining room, satisfy himself where "his squadron" was sitting, and then proceed with dignity into the kitchen to get his lunch. There was nothing in King's Regulations about dogs in the officers' mess.

Coopie lived the squadron spirit. Nobody had to teach him anything. If the klaxon sounded for a "scramble", he would go immediately to the door of the dispersal hut and sit there with his snout up. As the pilots rushed out they would touch his nose for luck – and a safe return.

No one knew whether he ever had a love affair. But one morning, quite early, the squadron was returning to Hornchurch from a patrol and there was Coopie in the middle of the airfield, looking very tired and moving slowly. Pilots were told to choose their landing path carefully and watch out for "Coopie dog".

When he got back to the dispersal hut, it was plain he had been seriously wounded; his neck was badly torn and he was stained with blood. Nobody knew what had happened, but it looked as if he had had a terrible fight with his arch-enemy, the "B" Flight dog, one of whose paws had been nearly bitten off. Both had to stay with the vet. When Coopie came back, his neck was still painful; he would only let his master treat it.

When the second front opened, the chow had to be left behind. He went to live with the WAAFs on a balloon site; he wore a red ribbon round his neck and was kept spotlessly clean.

Then, one day, when he was "on duty", he was rushing to "help" the girls bring down the barrage quickly, and was killed.

But his master kept his name till the end of the war when he returned to Czechoslovakia; it was the way Coopie was remembered....'

35 Wing Chronicler

Much of the Tactical Reconnaissance role – Tac-R, for short – was entrusted to No. 35 Wing of the Royal Air Force. As thoughts began to turn towards the long-awaited invasion of Europe, the Wing's news-gathering missions ranged deeper into enemy-occupied territory. The pilots became good reporters.

The Wing published a news sheet which soon built a reputation for its light sallies of wit and humour as well as for the squadrons' more serious needs. Its editor was the Chief Intelligence Officer, Laurence Irving (Squadron Leader L.H.F. Irving) who, as a World War I pilot, had served with the old Royal Flying Corps in France and Flanders. Irving had gone back to France again in 1939 as a staff officer with the British Air Forces.

Son of Henry Irving, he had inherited much of his father's talent for the arts – as the Wing's *News Sheet* often bore witness....

One of the squadrons had been detached briefly to York for an exercise with the army during a period of particularly disagreeable winter weather:

Now is the winter of our discontent
Made glorious summer by this sun of York
Richard III (who can't have known the place well).

168 Squadron was standing by, one time, to go to the Armament Practice Camp at Llanbedr, near Harlech, in Wales:

"This must be the place,' mused the editor, 'of which the poet sings':

A P/O who came from Llanbedr
Straight into the ground took a Hedr
'How delightful,' thought he,
'This is Negative "G",'
But, in fact, he could hardly be Dedr.

Intelligence reported an operation on 8 September 1943:

168 Squadron carried out one Tac-R sortie to Cambrai/Epinoy....

F/O Brennard reported that on crossing the beach on the way home he obtained a visual of a belle baigneuse (bathing popsy), large areas of whom appeared to be well tanned. Unfortunately no photos were taken. F/O Brennard returned safely.

An editorial note was included in *News Sheet* No. 11 of 11 September 1943:

Henceforward No. 35 Wing *News Sheet* will appear weekly unless operational exercise or intense social activity justify more frequent publication.

Those who find it entertaining or boring are earnestly requested to contribute to its columns, the entertained that they give their share of entertainment and the bored that they can only blame themselves if it is dull.

News Sheet No. 47 of 26 August 1944 referred nostalgically to the Lyons Tea Shops of other days, staffed by the familiar 'Nippies', with their neat white caps, aprons and short black skirts:

Per Ardua ad Suburbia

It is reported that in the mess of a certain high formation, waitresses crowned with caps like 'Nippies' serve the illustrious members –

Time was the brave deserved the fair,
But now protagonists of valour
Are living in the open air
In gentlemanly squalor.

Exalted staffs now claim these rights
And, with ingenious patience,
Transplant suburbia's delight
To camps in swell locations.

Where midinettes in linen caps
Serve luscious food and viands,
And so persuade these martial chaps
They're back at home in Lyons.

Before this, however, a Movement Order had been received on 30 March

1944 for one of the units to leave the Wing's base at Sawbridgeworth, in Hertfordshire, not far from the towns of Bishop's Stortford and Much Hadham. It moved, it was recorded, 'without any regrets because, at that time, the whole aerodrome and precincts were comparable to a quagmire'.

In *News Sheet* No. 29, two verses saluted 4 Squadron's departure:

Move on, you jolly campers!
Farewell to Sawbridgeworth!
Where everything is clampers,
The dullest place on earth.

Move out from winter quarters!
If they exist as such
The joys of Bishops Storters –
We haven't Hadham Much.

Source: Air Chief Marshal Sir Theodore McEvoy, ex-Group Captain, Operations, Fighter Command, 1942–43

North African Finale

ROMMEL – VERDICT FROM THE AIR

Peter Atkins, *Buffoon in Flight* (Ernst Stanton, Johannesburg, 1978)

... Colonel 'Blackie' strode into our tent and announced in his brusque way, 'Peter, you're grounded, tour expired.'

I protested forcefully but the CO was adamant ... I had hoped to see out the final battle for North Africa and, equally, would have liked to see out my century of operations.... I felt I could go on for quite a time yet and, hopefully, went to see the squadron medical officer, Denis Fuller, to enlist his support. But he merely said, 'You are grounded and that's that.'

The long advance had been a series of frustrations. It had taken five months from victory at Alamein to victory at Mareth.... We seldom, if ever, got to grips with Rommel. Until he chose to stand and fight at Mareth, Rommel's retreat had been like a wily old fox slipping away from a pack of leaden-footed hounds in a chase that should never have happened....

Out in the Desert we were often accused of having an exaggerated respect for Rommel. I don't think this was so at all, but we did have a grudging regard for him as a brilliant tactical general who was finally defeated by overwhelming numbers. When the odds were more or less even, he towered over the Desert scene and I believe we had no general on our side with either his flair or ability.

Lieutenant-General H.J. Martin and Colonel Neil D. Orpen, *Eagles Victorious* (Purnell, Cape Town, 1977)

For the Commonwealth, the capitulation of the Axis forces in Tunisia on 13 May 1943 marked the end of the most anxious phase of the war since the Battle of Britain. Now an ultimate Allied victory seemed assured. With the SAAF among them, the Allied air forces in the Mediterranean area in mid-

1942 were about to reap the reward of nearly four years of unremitting effort, often against seemingly insuperable odds....

Adapted extracts from *Eagles Strike* by J. Ambrose Brown (Purnell, Cape Town). Reproduced by courtesy of The War Histories Advisory Committee

VICTORY OVER THE GULF OF TUNIS

Before the curtain finally fell on the air fighting in North Africa, and activity moved across the Mediterranean to Sicily and Italy, a remarkable last scene was played out over the Gulf of Tunis. It provided the South African Air Force, in the form of 1 Squadron (Major D.D. Moodie), with its Spitfires, and 7 Wing (Lieut.-Colonel D.H. Loftus), composed of Nos 2, 4 and 5 Squadrons, with their Kittyhawks (P-40s), with one of the great aerial coups of the Mediterranean theatre.

Intelligence* had given exceptional advance warning of the enemy's intention to fly in troop reinforcements early in the morning of 22 April 1943. It was a last ditch attempt to aid the crumbling Axis forces in Tunisia.

Rendezvous for Loftus' and Moodie's aircraft was to be over Hergla at 0755 hours.

... The two South African formations joined forces in what was to be ... a mass downing of transports which caused Kesselring (finally) to cease daylight supply running....

... [The South Africans] ... left the coast at Nabaul and ... a course alteration brought them over the sea to the island of Embra.... Visibility was poor. When still 8 miles from the island, the great (enemy) transport(s) ... were seen flying southwest, low over the water. They were six-engined.... Me 323s, each capable of carrying a load of 14 tons or 100 troops on their two decks.

These high-wing [aircraft, each] with a wingspan of 181 feet, were flying in a great vic formation ... with a smaller vic ... inside (and behind) it....

Major Moodie ... detached two of his sections to attack, while he stayed up with the remainder of his squadron, and the (six) Poles [who were flying with them], to give cover against the escorting Me 109s, Me 202s and Re 2001s – about a dozen in number. The Spitfires peeled off, with cannons hammering, to draw first blood. Almost simultaneously the Kittyhawks turned ... and a great slaughter ensued.

It was a rare battle in the annals of air fighting: 36 Kittyhawks and 8 Spitfires in combat with the gunners of [the] ... huge transports and their ... escorts. The opaque mirror of the water below was broken by the splash of plunging, blazing machines; the air streaked with cannon and ... machine gun ... fire ... as the fighters wrought their destruction. The Italians, at the rear of the transport formation, abandoned their charges. As Lt. D.T. Gilson dashed down towards the air fleet, four Macchis dropped their long-range

*The British, like the Americans with the Japanese in the Pacific, had, since earlier days, been breaking the Enigma code by which the German High Command issued orders to their forces. The intelligence gained, known as 'Ultra', enabled the Allies to get wind of some of the most secret operations and plan accordingly. This was one such example. (Ed.)

tanks and dived away, leaving him to sweep over the [great troop-carriers] and down first one, then another in a ... mass of flame.

Lt. G.T. van den Veen (accounted for) three Me 323s and Lt. M. Robinson destroyed two escort fighters....

Meanwhile No. 7 Wing's Kittyhawks ... [now] ... fell on the transports. Major J.E. Parsonson, who was ... [awarded] ... an immediate DSO for his leadership in this raid, led No. 5 Squadron into a head-on attack against the 323s. He set two alight and they broke in two on impact with the sea...

For No. 4 Squadron, Lt. H. Marshall and Lt. F.M.F. Green each destroyed two and a half ... [transports] ... which went down in flames and also disintegrated on impact with the water. Four other pilots of this squadron claimed one ... each....

... [The bulk of all] ... the killing was done by the SAAF. The final bag, as compiled by the RAF, was 31 Me 323s, 7 Me 109s, 1 Me 202, 1 Re 2001 and 1 Me 109 damaged; the South Africans claimed to have destroyed 30 323s. These figures differed markedly from the Germans', but, whatever the exact totals ... [the outcome of the engagement put an end] ... to all further daylight reinforcement flights....

Air Power

THE SICILIAN EQUATION

James Doolittle (Lieut.-General James Doolittle), Commanding General, Twelfth Army Air Force, North Africa, 1942 and Fifteenth Army Air Force, Italy, 1943, in a memorandum (1943) on the Sicilian operation: 10 July-17 August 1943:

'A part of Strategic's* contribution to this operation† was the careful herding of a large part of the Axis air force into a small area of eastern Sicily and then destroying it there. The method was simplicity itself. There was a limited number of airfields in Sicily. We started on those in the west, destroyed as many planes as possible on the ground (the concentration was high and the dispersal poor) with fragmentation bombs, then "post-holed" the fields with demolition bombs until they were no longer usable. The remaining enemy planes were forced to move eastward, securing operational fields in the Gerbini area. An all-out air offensive then destroyed many of these planes and drove most of the rest out of Sicily. The almost complete freedom from air opposition experienced by our ground troops spoke highly for this operation and for the effective cover furnished by the Tactical Air Force.'

THE ITALIAN FACTOR

The invasion of the Italian mainland by British forces on 3 September 1943, opened the decisive phase of the Allied attack on what Churchill had, earlier,

*Northwest African Strategic Air Forces (NASAF) under Doolittle's command.
†Capture of Sicily, code name 'Husky'.

euphemistically called 'the soft underbelly of the Axis'. 'Soft' the Italians may have been, but there was nothing particularly pliable about the Germans' resistance at Salerno, Anzio, Cassino and other fierce-sounding points of battle. But Doolittle – and the Luftwaffe – saw the Italian factor in terms of a different strategy.

James Doolittle in a paper, dated 3 September 1943, to Brigadier General Hoyt Vandenberg, Deputy Chief of the Air Staff in Washington:

'The air battle is becoming interesting.... If we are successful in invading Italy, we can establish air bases and attack vital military manufacturing facilities in Southern Germany.... Apparently fearing this ... [the enemy] is fighting desperately. His fighters are far more aggressive than we have previously found them. He is using aerial bombing and rockets extensively. Yesterday, a half-dozen rocket planes got through ... and attacked the bombers. Sixty to seventy enemy fighters attacked the escort – they *always* attack now – and shot down six. Three more are missing....

Every day brings its air battle and while we are more than breaking even, the Hun has the advantage ... [over our escort fighters], fighting over his own territory.... Our losses will increase until the Hun is knocked out....'

Göring's 'Contempt'

Group Captain W.G.G. Duncan Smith, *Spitfire into Battle* (John Murray, 1981; Hamlyn Paperbacks, 1983)

... I decided to go along and interview the rescued German.... Through an interpreter I found out the Luftwaffe was in disgrace with Göring.... A few days later a captured enemy Order of the Day was passed from Desert Air Force to Wing Intelligence. Its ... text read as follows:

Together with the fighter pilots in France, Norway and Russia, I can only regard you with contempt. If an immediate improvement is not forthcoming, flying personnel from the Kommodore downwards must expect to be reduced to the ranks and transferred to the Eastern front to serve on the ground.

[signed] Reichsmarschall Hermann Göring

I do not think these sentiments ... (lack of will to fight) ... were general.... At Salerno, Cassino and Anzio German ... pilots put up a stiff resistance.

The Same Old Story?

Air Vice-Marshal Brian A. Eaton, Red Hill, Australia, 1982

In Italy I commanded No. 239 Wing, consisting of (the following squadrons): Nos 3 (RAAF), 112 and 260 (RAF), 5 (SAAF) – all with Mustangs – and Nos 250 (RAF) and 450 (RAAF) both with Kittyhawks.

My Wing Commander Flying was George Westlake and at one period, somewhere south of the (River) Po, the Spitfires of Duncan Smith, Brian Kingcome and 'Cocky' Dundas were getting all the publicity, so we concocted the following song, sung to the tune of 'It's Still the Same Old Story'.

It's still the same old story,
The Spits get all the glory,
And all we do is die;
And detrimental things apply
As Flak goes by

Here comes the Eighty-Eight,
My throttle's through the gate,
I hear my Wingman cry:
'This bloody Kittyhawk won't fly'
As Flak goes by

Twenty and forty are all the same to me,
Eighty-Eight's plenty – that's plain to see,
My Wingman's just bought it –
Thank Christ it wasn't me,
That's a thing I can't deny

It's still the same old story,
The Spits get all the glory,
And all we do is die;
And detrimental things apply
As Flak goes by.

Salerno

Group Captain P.H. Hugo, Victoria West, South Africa, 1982

'I took 322 Wing to Milazzo on the north coast of Sicily on 6 September in readiness to cover the landing at Salerno (Operation Avalanche) on the 9th. It was there that we had the unique experience of witnessing the surrender of an enemy aircraft in mid-air.

It was a Savoia Marchetti 79 of the Regia Aeronautica. When intercepted it streamed a white parachute to indicate surrender. The antics of this big aircraft, trying to cope with the oscillations of the dragging parachute, plus the shaky landing on our small airstrip, caused great merriment. I had some difficulty, as the Commanding Officer, in accepting the Italians' surrender with customary decorum.

But then the question was how to dispose of the aircraft. We hadn't room to keep it where it was. My Wing Commander Flying, Lieut.-Colonel Laurie Wilmot, of the South African Air Force, one of the ablest wing commanders flying I ever came across, assured me he had flown JU 52s in the SAAF and could handle this; what he didn't tell me was that his experience of JU 52s consisted of taking the controls for ten minutes one time while the pilot had a cup of coffee. I had no hesitation in agreeing to let him take the aircraft over to Catania.

The take off was spectacular – to say the least – and so, I believe, was the landing at Catania; but the aeroplane remained intact. The Air Officer Commanding, the Desert Air Force,* who saw the landing, was suitably

*Air Vice-Marshal Harry Broadhurst.

impressed – by the ruggedness of the SM 79....

With our 90-gallon, long-range drop tanks on our Spitfires, we patrolled Salerno continuously* for a week and on 15 September we landed on the small strip which had been bulldozed out of the olive groves opposite the beachhead. It was called Asa strip.

Asa was an airfield in name only. It was very short and ran directly from the beach eastwards straight inland, more or less in no-man's land between the sea and the hills where the Germans were dug in. My ground personnel, who were already ashore, had been digging vigorously, an occupation we all entered into wholeheartedly after the first German 88-mm shells started to come over. The speed with which all ranks disppeared into the ground would have done credit to an antbear....

With the Navy on our western doorstep, adopting a sensible shoot-first-say-sorry-afterwards approach, and with the Germans, to the east, always reminding us of their presence, we had some very lively flying....

All good stories have a dragon. In this case it was a German railway gun which had its lair in a tunnel in the hills. It was run out by its crew, usually at most inconvenient times, and after four or five shots was then run back again under cover, before there was time for retaliation.... It was called Big Bertha and flourished until an irate American general, who had had his pneumatic mattress punctured by shell splinters, decided to put an end to it.

He laid on an air strike of medium bombers which changed the topography of the hills round the tunnel and permanently sealed off the entrance.

But the Germans had the last laugh; they evacuated the gun out of the far end of the tunnel on the other side of the hills.

It was a long time before I came across Big Bertha again. It had been caught in the end by fighter bombers, just south of Naples. When I went to see it, some wag (he must have been a Salerno veteran) had chalked on it 'GRAND CHARLES *NOT* BIG BERTHA'.

It turned out to be a French gun made by the Le Creuseot works.'

Bang away, Bang away, Lulu?

The Honourable Mr Justice Drake, Shrewsbury, 1983

'Farmer' Giles and 'Quack' Drake flew together as pilot and navigator of one of the Beaufighters of 255 night-fighter Squadron. On 'ops' they were known as a happy and reasonably efficient team;but it was in the mess, on nights off, that they led the activities. 'Farmer' was a natural pianist 'by ear'; hum or sing a tune and he could repeat it on the piano at once and without fault. 'Quack' had the memory and sturdy voice for songs – air force songs, naval songs, barrack room songs, old rugger songs.... So, night after night, in the mess, Farmer tickled the ivories and Quack led the choruses: 'Bang away, bang away, Lulu....!', 'An airman told me before he died....' – all the bawdy, unprintable songs sung lustily until the late hours.

*Milazzo to Salerno was a sea crossing of some 170 miles – quite an operation to maintain continuous cover in the face of the enemy. (Ed.)

October, 1943; the Adriatic, off Bari; the last fighter patrol of the night from 0300 till dawn. The Germans hadn't been active lately; a successful sneak raid by Ju 88s on Bari harbour – then nothing – nothing at all for nearly a month. It was boring and tiring, stooging up and down over the dark sea for half the night with no 'bandits' to chase. Farmer and Quack had to fight off sleep: another hour and a half to go and nothing but silence broken by an occasional vector from Base Control. 'Rutban 20 ... steer 150 degrees ... no customers tonight, I'm afraid.'

Farmer decided a song or two would keep off sleep. 'Come on, Quack,' he said, on the intercom. 'Let's wake ourselves up by a rousing rendering or two!' They did. Half an hour later, when they rested their voices, they were fully awake and ready to watch the dawn coming up.

About a year later, an RAF Intelligence Officer, reading through the German intelligence reports captured at a recently overrun German listening post in Yugoslavia, was astonished to find several printed pages of 'unprintable', bawdy songs.

A week or so afterwards a group captain had a chat with Squadron Leader 'Farmer' Giles, DFC. He showed Farmer the captured German papers:

Bang away, bang away, Lulu,
Bang away good and strong,
What shall we do for a good —
When Lulu's dead and gone!

'I suppose,' he said 'that if you'd kept awake by talking to your navigator about the working of the secret radar, or discussing squadron movements, you'd have faced a courtmartial for negligence in leaving your intercom on to "Transmit". As it is, I'm tickled to death to think of the hours those chaps must have spent trying to discover what new code was involved in all of those four-letter words. And if it was some German WAAF who could understand such basic English to take down all of those songs, her face *must* have been red!'

Effect on the Soviets

Oberst Hans-Ulrich Rudel, *Stuka Pilot* (Euphorion Books and Transworld Publishers)

In the second half of July 1943, the resistance in front of the German divisions on the Eastern front stiffens; hedgehog after hedgehog has to be overcome and progress is only very slow. We take off daily from morning till night, and support the spearheads of the attack which have advanced northwards across the Pskoll river far along the railway from Bjelgorod.

One morning on dispersal we are surprised by a strong formation of IL II bombers which had approached our aerodrome unobserved at a low level. We take off in all directions in order to get away from the airfield; but many of our aircraft are caught on the ground. Miraculously, nothing happens; our AA guns on the airfield open up for all they are worth and this evidently impresses the Ivans. We can see normal 2-cm flak ricochetting off the armour of the Russian bombers....

Our advance has been halted all along the front. We have seen too clearly how this has happened: first the landing in Sicily and afterwards the putsch against Mussolini, each time our best divisions have had to be withdrawn and speedily transferred to other points in Europe. *How often we tell one another during these weeks: the Soviets have only their Western Allies to thank that they continue to exist as a militarily effective force!* [editor's italics].

Niet!

Air Commodore H.A. Probert, RAF (Retd), Head of the Air Historical Branch (RAF) of the Ministry of Defence, 'The Royal Air Force and the Soviet Union, 1941–45', *Air Clues*, July 1981

... There was no change in the general atmosphere and when Air Marshal Babington joined the [Moscow] Mission in July [1943] his first reaction was: 'I do not like the look of this job – just about everything is either faked or phoney.' Two weeks later he saw 'no prospect of establishing the contacts and the frank dealings with the Soviet military and administrative staffs which are now so conspicuously absent and without which no progress can be achieved.' As an example he quoted the lack of response to repeated requests for information about the Ploesti oil refineries and their air defences with a view to air attacks that would be of obvious benefit to the whole Allied cause. Finally, at the end of his short sojourn, he concluded: 'We do not know enough about the Soviet war machine, and its masters are not disposed to enlighten us.'

Air Commodore Roberts took a similar line in October 1943 with the cynical comment: '*The Soviet Army Air Force is a securely fastened door to us – I suggest your best source of such interesting gen is our friend the enemy*' [editor's italics]. In December he reported that apart from a Lieutenant Colonel who met his Intelligence Officer there was no Soviet Air Force officer who would meet him to discuss air matters; in six months since his arrival he had seen nothing of the Air Forces of the Soviet Union. The reasons, he surmised, were the traditional suspicion of foreigners, the desire to conceal their inferiority to the RAF, the Army General Staff's reluctance to let the air force learn the advantages of an independent air force, and the fact that the Hurricanes being supplied were obsolete....

Eastwards, in Southeast Asia, however, the tide of battle had yet to turn.

The War Lord of the Arakan

Marshal of the Royal Air Force Lord Cameron, London, 1983

'In 1943, the war situation in Burma was desperate. The Japanese were threatening Kohima in the north. They were attacking Calcutta with carrier-borne aircraft. The chance of a Japanese advance into India was very real. Morale in the 14th Army, and its associated air forces, was not good. "The Forgotten Army" was the phrase which was used and certainly Burma got very little coverage in the British press. But by spring, 1944, things began to change dramatically. Lord Louis Mountbatten had been appointed Supreme Allied Commander and a new team of personalities had entered both the staff side and the operational squadrons. The Japanese took a heavy defeat at Kohima, though it was a close run thing.

The Arakan front had always been one of the vital areas in the defence of Calcutta. The Japs had penetrated beyond the port of Akyab and had operational airfields in and around the Akyab area and a carrier force in the Indian Ocean. On the Allied side our air forces were based on two hard strips at Chittagong and Cox's Bazaar – (I never found out who Cox was but the title had an air of Kipling about it). There were also a variety of strips made out of hardened paddy fields....

With the injection of new blood, one, Group Captain George F. Chater came to Cox's Bazaar to take charge of the Wing there. The Wing consisted of eight Hurricane squadrons – later to be equipped with P-47 Thunderbolts, two squadrons of Vultee Vengeance dive-bombers and various communication aircraft. The role of the Wing was air defence and among other tasks, to give close support to the army. The Thunderbolts were later to provide long-range bomber escorts.

George Chater was a South African who had come to the UK to join the Royal Air Force just before the war. He had won a DFC during the Battle of Britain and was an outstanding pilot.

I was a squadron commander in his Wing so I had the opportunity to see him work at first hand. He was an ebullient and enthusiastic character with great leadership capability and he turned his Wing, within a few weeks, from a tails-down organization into a highly operational formation. He did this by his personality and by making a point of getting to know everybody – most of them by their Christian names, and by showing them he could fly an aircraft. He was very hard on staff officers who tried to tell his Wing to do something that was beyond its capability. He was not popular with the staff, but he built up a really operational force of high morale which would do anything for him.

He loved India and Burma and all the opportunities these countries gave him for his favourite sports. He had a variety of guns, and nights and days off were spent in the jungle after big game or feathered eatables. He had a marvellous bearer, a Ghurka, who had been with him all his time in India and Burma and who looked after him with great care. (He could also produce an excellent curry, even for breakfast, if required.)

The Arakan coast was not exactly overpopulated with pretty girls, but those that were about usually congregated from time to time around the Chater bungalow on the beach at Cox's Bazaar. They came to pay tribute to the War Lord of the Arakan.

At that time the Wing was operating in the Arakan coast area and into the Kaladan Valley, in the heart of Burma. It was nasty and inhospitable jungle and the Japs did not treat captured aircrew too well.... But there were relaxations.

There were jeep races which Chater organized. The squadrons tuned up their favourite jeeps and raced them round a marked course. (What *would* the Air Ministry have said!) The betting was usually heavy as there was not much to spend money on unless it be the infrequent trips to Calcutta. There were shooting expeditions and swimming parties; there were all sorts of rivalries between the squadrons; there were concert parties raised by the squadrons and always a great feeling of competition which kept them on their toes.... This was the basic leadership, but George Chater encouraged his air and ground crew in such a way that they barely knew that it was happening.

He was one of the great leaders of the last war in a theatre which did not hit the headlines. When the war ended there were many aircraft and staff cars which seemed to be under his control but which did not appear on any inventory. They just "appeared" and were all used to good effect.

George Chater was a wartime character for whom, sadly, there did not seem to be a place in the peace-time air force. It was sad, in a way, to see the staff officers, who had managed to remain in that role for much of the war, getting the jobs when it was all over. But George will always be remembered by all who served with him. A great character who earned well his title, the War Lord of the Arakan.

We salute him.'

Busting the Japanese Naval Code

George T. Chandler, Pratt, Kansas, 1982

By now, a sudden and unpredictable event had intervened in the Pacific to bring the US Navy an unexpected bonus and, in its train, an interesting reaction.

'In the spring of 1943, Admiral Yamamoto, the Commander-in-Chief of the entire Japanese Navy and the architect of the attack on Pearl Harbor, was making an inspection trip of Pacific combat areas.

Through its deciphering of Japanese coded naval messages, our naval intelligence had advance knowledge of the time the Admiral would arrive at Kahili airfield on Bougainville Island (in the Solomons). On this information, a mission was scheduled for our 339th Fighter Squadron to intercept and destroy his aircraft. It worked perfectly, our P-38s were there waiting for him.

However, when our pilots returned to their base on Guadalcanal, the naval Intelligence Officers made the mistake of telling them they had destroyed Admiral Yamamoto. Immediately the gossip was all over the island telling how he had been intercepted and killed by the navy's success in breaking the Japanese code.

This drove the Intelligence Officers in Hawaii to the brink of insanity, fearing that the Japanese would have their suspicions confirmed by publication in the US press of letters and stories coming out of Guadalcanal. The result was the most stringent censorship I personally saw throughout the Pacific war.'

Colonel Gregory 'Pappy' Boyington, USMC (Retd), *Baa Baa Black Sheep* (Wilson Press, 1958)

Man, oh man, what excitement in our tent area the afternoon these P-38s came back from Kahili and told us about the rendezvous with Yamamoto's transport as it was circling for a landing. They had gotten him all right. . . . This must have been a horrible blow to Japan, not only to lose a great admiral, but also to know we could crack their code. . . .

We found out later we weren't supposed to know about this top-secret mission. . . .

Mine Executioner

Peter Firkins, 'The Firkins Papers', City Beach, Western Australia, 1982

William Ellis Newton was born in St Kilda and, after leaving Melbourne Grammar, was one of the first to apply for enlistment in the Royal Australian Air Force on the outbreak of war in 1939. After completing his pilot's training, he had a period of instructing before being posted to No. 22 (Boston) Squadron on its formation in the middle of 1942.

He said goodbye to his mother at the end of his embarkation leave: 'Mother, if I don't return, keep your chin up. No tears! Toast my memory in a glass of sherry.'

When he arrived in New Guinea, Bill Newton quickly established himself as a daring operational pilot, earning the name of 'The Firebug' for the way he pressed home his attacks against Japanese targets on the other side of the forbidding Owen Stanley Ranges. . . .

On 4 March 1943, the Japanese suffered a dramatic defeat in the three-day Battle of the Bismarck Sea when 7 transports and 4 destroyers were sunk by the Australian and US air forces. The enemy were desperately trying to reinforce their bases at Salamaua and Lae, but only 900 of the 4000 troops who had sailed disembarked. This engagement was the most significant air victory of the South-west Pacific war and Salamaua now became the principal Japanese base and the next major objective of the advancing Australian troops.

There was no respite for the Bostons and on 16 March they attacked Salamaua's stores and dump areas at low level, leaving the target in flames with smoke rising to 10,000 feet. The squadron had launched its onslaught down the barrels of the enemy's guns and as Bill Newton pulled away, leaving a trail of destruction after bombing with pinpoint accuracy, his aircraft was hit directly four times, damaging both wings, puncturing fuel tanks and disabling one engine.

Despite the damage, the pilot flew the damaged Boston back 180 miles to base where he made a successful landing as a tyre blew out on impact with the strip.

On 18 March, the squadron repeated its attack on Salamaua, Newton's target being a big, Japanese-used building which the CO, Wing Commander Hampshire, had described at the briefing as 'an extremely difficult target'.

Newton, setting the pattern, went in at 50 feet. Hard hit by the ground batteries, he was seen by others in the squadron to clear the target with his aircraft streaming smoke. He flew along the shore line, trying to put as much distance as he could between himself and the target, before ditching his crippled aircraft. Two of the crew, Newton himself and his wireless operator, Flight Sergeant J. Lyon, were seen to free themselves from the aircraft and begin swimming.

They were soon picked up by natives who, it was thought, intended to lead them to a coastwatcher's position high in the jungle above Salamaua. After following rough and narrow jungle trails for what seemed like an eternity, the two became suspicious thinking they were being led to the Japanese. They, therefore, abandoned the native party and decided to make their own way to safety. As they turned back on their tracks they walked straight into a Japanese patrol.

Tragically, the natives who had befriended them were loyal. Had they persevered with them, the two Australians would, most probably, have been led to safety. As it was, the Japanese took them to Lae where, on the orders of Rear-Admiral Fujita, Lyon was executed. However, when Newton's identity was established, he was returned to Salamaua so that those whom he had so recently attacked might have the honour of executing him.

Newton was on his fifty-second operation when he was shot down. The posthumous award of the Victoria Cross was made on 20 October 1943.

The natives who witnessed his execution spoke of his fine bearing, courage and noble character as he faced the final moments of life. A Japanese version of the dreadful action confirmed the impression: 'The precaution is taken of surrounding the prisoner with fixed bayonets, but he remains unshaken to the last. When I put myself in his place, the hate aroused by his daily bombing, yields to ordinary human feelings.'

Several months later, photographs smuggled out behind the enemy lines showed an Australian being executed. At first, he was thought to be Newton; later he was identified as an Australian army sergeant who had been landed from a submarine to pinpoint Japanese strongholds.

The Australian troops in New Guinea swore to avenge the brutal beheading of a hero which was simply one more example to add to the long list of Japanese crimes in the bitter jungle war. However, they were cheated, for Rear-Admiral Fujita committed suicide at the end of the war to avoid being tried as a war criminal. Naval Sub-Lieutenant Komai, Newton's executioner, was killed in action in the Philippines.

When Bill Newton's mother was given the news of her son's loss, she took a long walk along the sea front at St Kilda. 'My wealth', she said, 'is my great pride in Bill. That, I can never lose. My hope is that we shall meet again.'

'Air Raid'
31 December 1943

Last night I heard the sound
Of bombers, outward bound.
I rest in peace while they
Speed bravely on their way.
And as I lay – I knew
That I could pay no due
Except to pray with love
God guard those gallant few above.

Harold Balfour

PART VI

Maximum Effort

The Allied war effort in 1944 was now rising to a crescendo. Memories of the retreats and reverses of earlier years, which had struck so cruelly at the time, had given way to a spirit of confidence and reconquest. In those places where the Allies had elected to stand and fight – often at dreadful cost – the certainty was abroad that the sacrifice and loss had not been in vain. Retribution and revenge were releasing an impulse which the enemy could not match.

In the West, the Channel was bridged and the land forces, supported by superior air power, were driving northwest through France, the Low Countries and towards the Rhine and the Fatherland. The bombing offensive, sustained by the United States' 8th Army Air Force by day, and the Royal Air Force's Bomber Command by night, was progressively weakening Hitler's Third Reich. New airborne devices, products of a brilliant science, had already turned the U-Boat war in the Atlantic.

On the Eastern front, the massive Russian armies, recovering from past disasters and profiting from the development of Soviet air power, were rolling the Germans back in a way that, only two years before, would have seemed barely credible. To the south, if the going up through Italy was stubbornly slow, the cooperation between the air and ground forces reached new standards of efficiency.

In the Pacific – in the Marianas, Saipan and the Leyte Gulf – United States carrier-borne air power was exacting a fearful price from the Japanese, and leaving the world marvelling at its might.

In Burma and the Far East, the Japanese Sun had passed its zenith and now headed on its downward path.

Throughout this period, the Germans and the Japanese were feeling the grip of a steel ring which was steadily being drawn tighter. But each was resolved, within such areas as were left for manoeuvre, to fight it out to the death. Resolution, in the face of defeat, did not waver.

1944 was the year, then, which proved beyond doubt that supremacy of independent air power, and victory in modern war, were indivisible.

COURAGE, PHOBIAS, SUPERSTITION AND FEAR

The most pernicious phrase in the Royal Air Force's wartime jargon was made up of three letters – LMF, alias Lack of Moral Fibre.

Its spectre haunted operational commanders. For a 'certified' victim, it meant demotion and ignominy. Yet, astonishingly, there was little of it. In the worst part of Bomber Command's hideous offensive in 1943 and '44, less than half of one per cent of all aircrew actively engaged were affected by it. For the whole of the Royal Air Force's operational commands, the total was even less than that. And of the LMF cases which were considered, no more than 67 per cent were classified.

Self-discipline, leadership and the well-charged spirit of the Service were, of course, the key.

Ministerial View

Harold Balfour, *Wings Over Westminster* (Hutchinson, 1973)

Captain Harold Balfour (Lord Balfour of Inchrye), Under-Secretary of State for Air in Churchill's wartime Coalition Government, brought a rare advantage to the ministerial task. He had fought in the skies over France and Flanders in the First World War and knew, at first hand, the demands which air combat, in all its forms, placed upon the human frame.

... Archie Sinclair* handed over to me another job that yielded no pleasure, much anxiety and searching of conscience. A lot of misgivings were being voiced by MPs inside and outside the House on the handling by the Air Ministry of LMF.... In plain language LMF meant cowardice in the face of the enemy unless some medical reason could be held the cause.... There was a feeling that within the Air Ministry the senior officers, tied to their desks, who decided these 'waverer' cases were old-world 'blimps'. Archie promised Parliament that in future every case which could involve some penalty would be looked at ministerially before any action was taken, and this I was to do.

The files came to me from the Air Member for Personnel with his views and recommendations. The majority of cases came from Bomber Command. LMF was dangerously contagious. One LMF crew member could start a rot which might spread not only through his own crew but through the whole squadron, particularly when there happened to be a lot of inexperiencd crews replacing casualties. Directly an LMF case showed, the first and vitally important step was to remove him from all contact with other aircrew personnel. He would be posted to a depot while his case was being considered....

Always I started out with the hope that the various reports would allow me some grounds on which I could go back to AMP. In reviewing a case my sympathies always started off with the poor fellow as I knew only too well myself what it was like to be scared stiff in air warfare. We did have a few cases where a man just stormed up to his flight or squadron commander and announced that he was through with flying and through with fighting the war, whatever might be his fate. In such cases the usual course, if the man was an officer, was to take his wings away, order him to resign his commission and arrange for the army to pick him up for enlistment, probably in a labour battalion. In the case of an aircrew sergeant his flying badge was forfeited, he was reduced to the lowest rank of AC2 and put on the

*Rt Hon. Sir Archibald Sinclair (Lord Thurso), Secretary of State for Air in the Coalition.

worst fatigues.

There was one case which came to me which I felt I had to refer to Archie. This was not an air crew but a ground staff, self-admitted homosexual caught up under National Service. This man had not wished to be in the RAF and was determined to get out. He told the authorities that he was a confirmed homosexual since he had been seduced by a master at his school. When war broke out he had been living with an artist in Paris. He maintained that he had tried to break away from homosexuality but had failed. Now he had resigned himself to living his life as a pederast. Doctors examined him and reported that there was no reason to doubt and every reason to believe that the story he told was true. The man warned us that if he was posted to some unit then sooner or later there was bound to be a scandal for the temptation of so much youth around him would be irresistible.

We talked over this case wondering what we should do. On the one hand we felt we were trustees for the parents of all our young recruits; on the other we did not want to start a precedent that anyone who laid claim to being a homosexual could get out of the service. We might have to face a whole queue of reluctant recruits forming up to demand similar treatment. So far as the RAF was concerned this man had us beat. We granted him his discharge but made sure that as he walked out of the gate he was picked up by National Service Officers for immediate direction to the coal mines. Whatever faults of judgement I may have made during my time at the Air Ministry and, looking back, there are quite a few, my conscience is clear about dealing with LMF cases. Always I had at the back of my mind the knowledge that right through the war every man who flew had started off as a volunteer....

The Only Answer

Arthur B. Wahlroth, 405 Squadron RCAF, Mount Albert, Ontario.
Extract from *Wellington Pilot*, Canadian Aviation Historical Society Journal Vol. 19, No. 2, 1981

... The third trip, however, was a horse of a different colour. We went out over the North Sea, headed about due east. The Northern Lights were quite bright and silhouetted us from the southern side of the sky, Germany's side. The Captain was in a bit of a flap, as usual, and had me stationed in the astro-dome keeping my eyes on the dark side of the sky in case of a fighter attack; he had me so scared that when he wanted the cabin heat turned on I wouldn't take my eyes from the sky long enough to bend down and operate the controls; somebody else had to come and do it. I figured after a short time, "This is ridiculous, why let this guy make me scared like this?"

I think we were going to Heligoland before turning south to Wilhelmshaven, our target city, flying at about 12,000 or 13,000 feet on oxygen. Suddenly the Captain disconnected the automatic pilot, whereupon the right wing went down and we went into a terrific spiral dive that took the two of us to correct. We put the machine back on 'George',* took a look at the fuel gauges and discovered that the Captain had left the cross-feed valve

*The automatic pilot.

on, pulling all the fuel from the port wing for both engines! I think we jetisoned the bombs, and, by the time we got home, the tanks had levelled out. We came in quite normally.

On the next trip we started out to Kiel, but the Captain figured that we had an excess fuel consumption, so after an hour and a half we were back on the deck. I didn't think too much about this for I was still in the learning stage and didn't know too much about things. But on the next trip I learned plenty.

This was our second trip to Cologne. I cannot remember very much about the German part of the mission, but on the way back – flying up England, everything going well, weather not bad, position known – the Captain suddenly put the IFF* onto emergency frequency (this was the forerunner of the modern transponder). In about five seconds searchlights came on all over the country, going straight up. Then most of them went out except the ones nearest us, which proceeded to dip from the vertical position to show us the way to the nearest aerodrome. We followed them around in a great turn until, directly in front of us, was a brilliantly-lit aerodrome. The navigator and I wondered what was going on, the navigation was good, but the captain was the captain and he landed. Once on the ground I asked him what the idea was. He looked at me. He was in a complete flap and said, 'Just be happy you're on the ground.'

He was taken off flying right then....

Luftwaffe Attitude

Peter Henn, *The Last Battle* (George Mann, 1973)

Leutnant Peter Henn was a successful Luftwaffe fighter pilot who fought in the Sicilian and Italian Campaign.

Herbert was our most skilled pilot.... A native of Breslau, he formed the link between the old and the new generation of aviators. Cool and calculating, never nervous, always amiable and agreeable, a true comrade. With him you could speak openly and say what you thought. I remember once I went to see him one night in Sicily after our first missions.

'Herbert, I've got the breeze up.'

'The breeze up, eh? Don't talk rubbish. You don't know what you're talking about. I've sweated more often with fright in my cockpit than you've ever done in your life, but no one ever had an idea that I was scared. The main thing, you see, is to overcome the obstacle ... to master your own instinct.... It's a tough job and it cost me a lot to learn. Believe me, Peter, I'm just as scared as you are. I tremble like the others, like all of them without any exception... There are some who pretend they are not afraid. It's a lie. There are others who flout death and spit in the face of their fear. Everyone is not like them; and besides, it serves no purpose. In spite of appearances, they too feel the sweat trickling down their backs; and I tell you, some of them are far more frightened than we are. Try and remember this principle, Henn. Realize you're frightened, but never show it. No one will take it amiss if you admit it. On the other hand, if you take a powder at the critical moment,

*Identification of Friend or Foe.

pretending that your engine is losing revs, they'll never forgive you. Never show any cowardice ... I know it's not easy.... Above all, don't worry yourself. The next time we fly stick to me and say to yourself that we're both on the point of soiling our pants. You'll see. It will make you feel better....'

That is a portrait of Herbert. He was irreplaceable. He was the patron of his young comrades in aerial combat, our friend and spiritual father rolled into one.

When we lost him the group lost its moral insulation....

A Leading Navigator's Fear

Peter Atkins, Parkwood, Johannesburg, 1982

Peter Atkins was the leading navigator of 24 Squadron, South African Air Force, in the Western Desert campaign. He was one of the stars of that theatre.

'I cannot accept that courage in the air could or should be confined to the fearless few or to those dumb heroes who simply did not have the imagination to be scared. All who came through a tour of operations exposed the strain of their experience, some more than others; but to the perceptive eye (the squadron medical officer, for example) everyone showed some change of character from the trial.

'Only the highly qualified could recognize individual fears. Some – or most – were frightened of being killed; others had a greater fear of being maimed, while some had a dread of being shot down and kept in a POW cage for years.

'So how does one distinguish between heroes and cowards? In this, I think the Lack of Moral Fibre inquisition failed. After all, those who flew were all volunteers and knew – or should have known – what they were letting themselves in for. If, having done their best, they failed, was it fair or just to demean them and treat them like pariahs?

'... The bravest man I know was my first squadron CO and I only found out about it forty years on. In the Desert we viewed him with awe as a man without fear. He led every "sticky" raid and gave the others to his flight commanders. He would light a cigarette as the bomb doors opened and smoked nonchalantly over the target.

'Recently, while he was recovering from cancer, we were talking of the Desert. "You know," he remarked, "I lost two stone in weight in the Desert."

'Due to dehydration, lousy food and a lack of booze we all lost weight in the Desert – but 28 pounds? His reply explained it. "I was shit scared all the time."

'That, to me, was the real courage!

'I was just as scared as most, but when I achieved the position of leading navigator in the squadron there was a new and tormenting fear. In the daylight raids the leading observer was in charge from the time the squadron formed up until the landing ground was in sight on our return. He was thus responsible for some seventy lives – the number was the same whether it was a formation of eighteen Bostons or twelve Marauders... The fear was of making a mistake that could lead to a squadron catastrophe....

'... Courage, on the other hand, was, I always thought, a question of bracing oneself to do a terrifying job as best one could....'

New Zealand Experience

Group Captain A.A.N. Breckon, Glendowie, Auckland, 1982

Aubrey Breckon was one of New Zealand's best qualified, all-round, operational pilots. He completed a tour with No. 75 (NZ) Squadron in Bomber Command when the unit was formed in the early days of the war; he was in command of 1 (Bomber Reconnaissance) Squadron of the RNZAF, and still flying operationally, in the closing months of the Pacific campaign. Few understood the ordinary aircrew member's problems and feelings more acutely. With his younger brother, Ivan, who flew with similar distinction with 75 Squadron, and then with the Mosquitoes of the Pathfinder Force, the two made an exceptional pair.

'We were lucky,' wrote Aubrey, 'that two of us in the family should survive that kind of war.'

'After briefings, particularly for the exacting missions, I would notice how some aircrew reacted, psychologically, to what might happen to them on the operation or to the possibility that they might not return.... They would try to hide their worried feelings with a laugh or a joke....

Most had girlfriends, and some could not bring themselves to delay until the end of the war before marrying at a young age. It was very sad when the husbands or boyfriends did not return from "ops"....

A number became superstitious and carried good luck charms, photos of wives or girl friends, badges and so forth. I remember an air gunner who lent me a New Zealand greenstone Tiki (a Maori symbol) to make a sketch for a Christmas card. In Maori, the Tiki is called Heitiki which, according to custom, is believed to protect the wearer from harm. The gunner had carried the Tiki on all his "ops". He asked to have it back for the raid on Germany that night. He said it kept him alive....

The selection of crews was all-important; the greater the operational experience, the better, generally, was a captain's leadership. Aircrew were similar to submariners in their attitude to team effort, discipline and leadership. They responded to good leadership.

'One couldn't afford a "dud" in a crew as morale could quickly be affected. On the other hand, close crew friendship, with good leadership, brought out the best in a man, especially during the frightening ordeals. Unfortunately, LMF did occur, but only infrequently. Perhaps there were sound psychological reasons why it should happen.... I used to think that some pilots, who were quite outstanding in their operational training, given time, would make brave and efficient captains when, eventually, they got on to bombing missions, but the opposite sometimes proved to be the case....

'Men reacted differently to real danger....'

Explicable Aversion

R.M. Lucas, Ascot, Berkshire, 1982

'After flying Defiants* for the first years of the war, and to avoid converting on to twin-engined aircraft when my squadron re-equipped with Beaufighters, I transferred to a Canadian squadron to form a Defiant flight. Soon this, too, converted to "twins" and I opted for photographic reconnaissance, flying Spitfires.

'After doing the course, and before being posted to a squadron, I was transferred at the last moment to PR Mosquitoes and so to a conversion to "twins". I went on a training course flying Airspeed Oxfords.†

'It was at this time that I experienced a strong aversion to flying below about 3000 feet, particularly at night although I was an experienced night flyer, preferring night to day flying whatever the weather.

'A few months earlier, I had crash-landed in the hills above Dyce (near Aberdeen) while on a blind-flying exercise, but I was not conscious of this having anything to do with the later aversion.

'In the end, I didn't make the conversion and wound up in A16 on intelligence duties both here and in the Far East.'

Inhibiting phobias of one kind or another were not uncommon in operational pilots. They became manifest in various ways. Robin Lucas's story (he became a squadron leader) was not untypical. It is difficult to believe, however, that this aversion did not stem from his earlier crash-landing – a most unsettling experience.

Welsh Prejudice?

Carel Birkby, *Dancing the Skies*, (Howard Timmins, Cape Town, 1982)

Wing Commander Ira (Taffy) Jones,‡ DSO, DFC, with the military awards of MC and MM as well, served in 74 Squadron in both world wars....

... The ebullient little Taffy, never afraid of sticking his neck out, says: 'I have noticed that big men rarely prove as brave as their build and manner would suggest, whilst small men seldom show cowardice.... Most of the great air fighters in the two wars have been men of small or medium height. Typical in World War I were Ball, McCudden, Beauchamp-Proctor, Willy Coppins, Richthofen, Udent and Fonk. Similarly, I have noticed, no professional footballers or boxers become famous pilots or air fighters, while there is a wealth of Rugby Union footballers among the aces.... However, the big men seem to get on well in staff jobs, and I suppose you can't have it both ways.' (Wow!)

*The Boulton-Paul Defiant was a single-engined fighter with a separately-operated gun turret behind the pilot. The aircraft quickly became obsolete.

†The Airspeed Oxford was a twin-engined training and communications aircraft.

‡Taffy Jones, with the bubbling Welsh spirit and a most unusual record spanning two world wars, was one of the Service's endearing 'characters'. An engaging stutter embellished the telling of his (unlikely) experiences.... He knew a thing or two about courage – and the frailties of man....

A Gunner Shivers

Wing Commander H.R. 'Dizzy' Allen, *Fighter Squadron* (William Kimber, 1979)

'Dizzy' Allen fought through the Battle of Britain with 66 Squadron and later commanded the unit, becoming an experienced fighter leader. He had more deranging incidents to contend with than most. With a critical and keenly discerning intellect, he could size up a pilot quicker than the next man. The editor knows; he once served under him – and that's the test.

.... The most difficult task for a squadron commander was to come to the conclusion that a pilot was suffering from 'lack of moral fibre' – i.e. fear of flying, fear of being engaged with the enemy, or both. The brain is the most complex organ of them all, and slight derangements can have fearful results.... A number of men simply did not know what they were taking on when they enthusiastically applied for flying training....

Bomber operations differed widely in psychological stress from fighter operations.... They had the advantage that there were, generally, five or more aircrew members in the same aircraft. The Germans recognized the importance of such camaraderie, and designed their bombers, in the main, so that each crew member was in touching distance of another. But the shivering rear gunner in a Lancaster would be 30 feet away from his colleagues, desperately lonely, and he knew that more likely than not, he would be the first to die. Bomber operations needed an ice-cold courage. For 8 hours the aircraft would be over hostile territory.... And, of course, the flash of every flak gun impressed on the mind that the shell was aimed at you personally, not at any of the other eight hundred bombers surrounding you....

... I had a doctor established with me in No. 66 Squadron, but he was of little use. In theory the diagnosis of LMF was the joint responsibility of the squadron commander and the squadron medical officer. I didn't even bother to consult him when I met a particularly newly-joined pilot and immediately diagnosed that he was suffering from LMF, nor did I consult my flight commander to whose flight I appointed the suspect pilot. Sure as eggs are eggs, my flight commander saw me within a fortnight and told me that, in his view, this pilot was suffering from LMF. I wrote out the symptoms on a form marked 'Medical Confidential', showed it to the doctor who countersigned it, although he really hadn't the first idea of what it was all about. Then I gave it to the pilot to read and countersign, which he did. He was grounded, and in due course appeared before the psychiatrists who confirmed my opinion and removed him from flying duties.

... I have researched the subject in depth, and I can tell you the good'un from the bad'un at the drop of a hat....

(Temporary) Staff Officer

Desmond Scott, *Typhoon Pilot* (Leo Cooper/Secker & Warburg, 1982)

I don't think any of us temporary staff officers left our mark in the annals of

Daunting reverse. On 10 December 1941, *Prince of Wales* and *Repulse,* capital ships of the Royal Navy, were sunk off the east coast of Malaya by waves of Japanese torpedo bombers. Here is *Prince of Wales* before the attack and after (p. 185 *et seq.*)

Contrast in Burma. Ground crew on a forward airstrip prepare a bomb-carrying Hurricane *(top)* for an attack on Japanese front-line positions. And a USAAF B-24 Liberator *(bottom)* after cutting the Bilin railroad bridge on 13 November 1944

Flashpoint in the Pacific, 7 December 1941. A Japanese Nakajima B5N2 'Kate' torpedo bomber *(top)* takes off from Pearl Harbor. And the US Navy's *West Virginia (bottom),* one of the five battleships sunk in the attack

Target Tokyo. James Doolittle *(top)* (left foreground) talks with Marc Mitscher while the crews gather on board US carrier *Hornet* for the attack on the Japanese capital. And *(bottom)* around 0840 hours on 18 April 1942, a B-25 Mitchell medium bomber revs up for the historic take off

Midway, 4-6 June 1942, the battle that turned the Pacific war. Of the 11 USN Devastator (TBD) torpedo bombers *(top)* launched from US carrier *Enterprise* against the Japanese fleet only 4 returned. And the enemy's heavy cruiser *Mikuma (bottom)*, burning strongly before the final attack which sank her

Gallery. Two girls of Air Transport Auxiliary *(left)* who delivered aircraft to stations and squadrons, Audrey Sale-Barker (Countess of Selkirk) and left, Joan Hughes (p. 299 *et seq.*); Kenley trinity *(below)*, early 1942, left to right: Reg Grant, CO 485 (New Zealand) Squadron, Dick ('Batchy') Atcherley, station commander, and E. P. ('Hawkeye') Wells (New Zealand), wing leader (67 *et seq.*, 235 *et seq.*): *Facing page:* Aristocratic 'ace', Heinrich Prinz zu Sayn Wittgenstein Sayn *(top left),* night fighter (p. 330 *et seq.*) and 'Coopie dog' *(top right)*, the Chinese chow of 134 (Czech) Wing and its leader, Jaroslav Hlado (p. 265 *et seq.*). Meanwhile, the Atlantic battle continued *(bottom)* US carrier-based aircraft attacked this U-boat off Cape Verde Islands, 1943

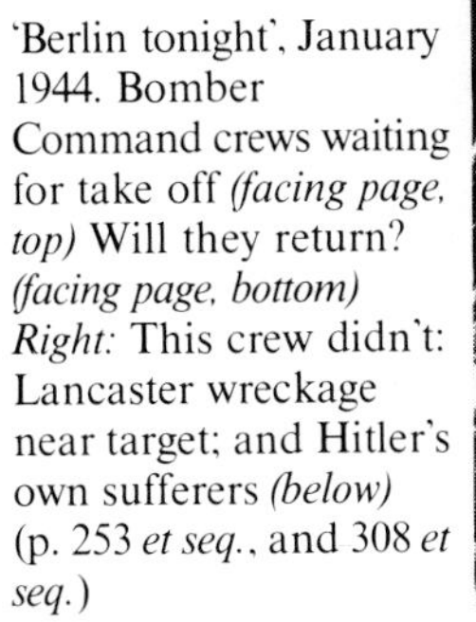

'Berlin tonight', January 1944. Bomber Command crews waiting for take off *(facing page, top)* Will they return? *(facing page, bottom)* *Right:* This crew didn't: Lancaster wreckage near target; and Hitler's own sufferers *(below)* (p. 253 *et seq.*, and 308 *et seq.*)

Precision in daylight. US 8th Army Air Force B-17s hitting the Focke-Wulf factory at Marienburg, October 1944 *(top and bottom)*; and US ground-attack aircraft striking a German supply column near Vicenza, northern Italy, April 1945 *(facing page)*

August 1944. The net tightens in the West. Churchill at Crépon, in Normandy shaking hands with 'Johnnie' Johnson, great wing leader of World War II *(above)*; Typhoon – conqueror of Rommel's armour *(facing page)* – and Harry Broadhurst records it *(facing page, bottom)*. Away to the south the Mediterranean sun shines on a lively Fleet Air Arm quartet of 821 Squadron *(right)* – left to right: John 'Doc' Kevin, David Hone, Lloyd Gibney and David Foster

Facing page: Japanese honour. 1729 hours, 6 January 1945: Ligayen Gulf, Philippines. Nippon kamikaze attack on USS *Columbia* (CL-5). The Ykosuka D4Y4, specially modified for suicide missions, crashed into the main deck, near the aft turret, causing extensive damage (p. 366 *et seq.*)

Above: Japan's 'ace' of all. Zero fighter pilot supreme and undefeated at the close with a bag of 64 enemy aircraft destroyed or probably destroyed from some 200 combats. Salute Saburo Sakai (p. 369 *et seq.*)

No quarter given. On 14 August 1945, US B-29 Superfortresses of the 21st Bombardment Group 'close-carpeted' the Marifu marshalling yards *(top)*, 2 miles east of Iwakuni and the same distance south of Osaka, in a remarkable example of exact bombing. And, away on the northwestern tip of New Guinea, US medium bombers pulverized the Japanese-held base of Sorong and left Jofman airfield *(bottom)* a shambles. Note the airborne bomb (foreground) 'homing' onto the tail of a damaged bomber. The fighter (right) taking off was shot out of the sky by escorting P-38s

Command staff history.... However, I did have the dubious distinction of being the only officer to be ordered from the Command conference room. This happened during a high-powered conference on operational policy ... at which only two of us present had actually flown operationally....

As the meeting was about to close, there came a familiar voice from the far end of the conference table. 'What do you think, Squadron Leader Scott?' It belonged to one of our great war leaders, Air Marshal Sir Trafford Leigh-Mallory. Two rows of balding heads ... turned my way.... It took me a moment or two to ... blurt out a reply:

'I don't, sir! This plan you are adopting is a lot of cock!'

There was dead silence. Then the Air Marshal pointed a finger at the door, which thankfully was at my end of the room, and thundered 'Out!'

Premonition

Desmond Scott, *Typhoon Pilot* (Leo Cooper/Secker & Warburg, 1982)

I had been flying myself almost to a standstill ... and Wally, the station doctor, a good friend to Bruce Hay, an Auckland boy, and myself, ordered me on a week's leave. I went up to Edinburgh ... but after only three days a strong premonition that something dreadful was about to happen decided me ... to ... return to my station....

The train left Edinburgh at midnight.... At precisely 1.30 in the morning I was suddenly woken by a vivid nightmare. I heard Bruce call out 'Boss, I'm on fire!' I could see him trying to leave his Hurricane as it plummeted towards the sea. I stumbled out into the passage and was promptly sick....

Immediately the train pulled into London next morning I phoned Manston and asked Wally if anything had happened to Bruce the night before. There was a long pause, then he answered: 'I thought you would know. He was shot down off Dover at half past one this morning.'

Lost Poet

When John Magee died at the end of 1941 (see page 131 'High Flight'), he left behind an unfinished poem. He was sufficiently encouraged with it to send a rough, first draft to his parents. With the title, 'Per Ardua', it offered tantalizing promise of what might have been....

They that have climbed the white mists of the morning,
 They that have soared, before the world's awake,
To herald up their foemen to them, scorning
 The thin dawn's rest their weary folk might take.

Some that have left other mouths to tell the story
 Of high blue battle – quite young limbs that bled,
How they had thundered up the clouds to glory
 Or fallen to an English field, stained red.

Because my faltering feet would fail I find them
Laughing beside me, steadying the hand
That seeks their deadly courage – yet behind them
The cold light dies in that once brilliant land....

Do these, who help the quickened pulse run slowly
Whose stern, remembered image cools the brow –
Till the far dawn of Victory know only
Night's darkness, and Valhalla's silence now?

Understated Combat

Group Captain Lord George Douglas-Hamilton (The Earl of Selkirk),
'The Douglas-Hamilton Papers', July 1979

Group Captain Lord George Douglas-Hamilton was on his way to take up an appointment as Senior Officer in charge of Administration, Royal Air Force, East Africa, when this incident occurred. The second of the four Douglas-Hamilton brothers, each of whom was at one time a squadron commander, Geordie Selkirk commanded 603, the City of Edinburgh Squadron of the Auxiliary Air Force. This was the squadron which his youngest brother, Lord David, led in the Malta battle in 1942.

... I had rung up Transport Command who had gladly accepted my offer to fly an aeroplane out to Cairo. They apologized for not having a Beaufighter available but offered me a Wellington X with the powerful new Bristol engine replacing the older Pegasus which Douglo* had used to fly over Everest....

I crewed up with three sergeants, a navigator, W/T operator, midship gunner and rear gunner. On 8 September we flew down to Portreath, in Cornwall, from Lyneham and that evening we heard the news that Italy had surrendered. The following day we were due to set off for Morocco.

At the briefing I was told I would only need to fill one tank and then fly sraight for Finisterre; but, being fairly senior, I was allowed to make my own arrangements. I knew German fighters tended to patrol right out into the Bay of Biscay, so I filled up both tanks and set course to a point 15 degrees west, about 1000 miles out into the Atlantic.

It was a beautiful day as we passed over the Scilly Isles and I was standing beside the pilot's seat with 'George' in action, when I saw five aircraft in formation over to starboard. I tried to convince myself they were Beaufighters, but I really knew at once they were JU 88s on patrol. I suspected they were looking for maritime aircraft searching for their submarines. I knew I had more petrol than they did so I turned west for America, putting the boost and revs up to the maximum and dropping down as near as possible to the sea.

The German pilots must have been fairly experienced because they made a

*The Duke of Hamilton (the 14th Duke), eldest of the four Douglas-Hamilton brothers; he was the first man to fly over Everest.

good formation attack, two aircraft coming down in front and two behind. I took standard evasive tactics, pulling the throttle back and turning into the attacking aircraft.

The engagement could not have lasted more than five to ten minutes but it seemed longer; the rear gunner thought he had got one of the 88s which was descending and smoking badly.

It was a blessed relief to get into a cloud; I kept on west until I was sure I was clear.

A cannon shell had come through the windscreen during the attack and torn the sleeve of my jacket; a few bullets had passed through the fuselage. With the windscreen gone the force of the air blew away any coffee I tried to drink – most irritating.

I gradually veered south until I hit the Portuguese coast and then we found Gibraltar covered with what I believe is called the 'Levanter'. Morocco was very hazy and map reading was difficult, but we were able to home on Raz-el-Mar without much difficulty.

After reporting to the intelligence officer who commented 'kite shot up', I spent the night in the local hotel. It was full of French, all of whom seemed so beautifully dressed, sipping expensive cocktails. I greatly enjoyed the atmosphere of peace which we had not known for some years. The next day we flew on to Tripoli with a different aircraft and thereafter to Cairo West.

It might be interesting some day to see the combat report made by the German pilots.

End of a Blockade Runner

J.P. Rennison, *Air Pictorial*, July 1982

'SS *Alsterufer*, an armed supply ship of the German Navy, was returning from a trip to Kobe, Japan, with a cargo of much-needed specialist war materials. Her skipper, Captain Piatek, had hoped to reach France before Christmas, but had been delayed by bad weather. He was now approaching the Bay of Biscay, an area fraught with danger....

His uneasiness was well founded. British Intelligence had been keeping tabs on the ship for some time and her passage around Cape Horn had been noted. Squadrons of No. 19 Group, Coastal Command, had been on stand-by for several days and were impatiently awaiting *Alsterufer*'s appearance as she headed for Bordeaux.

... A Liberator of No. 311 (Czech) Squadron with P/O Oldrich Dolezal at the controls had left Beaulieu (in Hampshire) to join the search for the enemy vessel. Apart from Dolezal the crew consisted of F/Sgt Jindra Hahn, F/O Zdenek Hanus, W/O Josef Kosek, F/Sgt Marcel Ludikar, Sgt Bedrich Prochazka, F/Sgt Ivan Schwarz and Sgt Frantisek Vertl.

Jindra Hahn had picked up a contact on the ASV radar at an unprecedented 60 miles. It had to be *Alsterufer*; there were no other ships in that immediate area. Dolezal decided to approach the target above the overcast for maximum concealment. Suddenly, through a break in the cloud, one of the crew spotted the vessel.

As the Liberator went into the attack, Marcel Ludikar began to transmit an 'Operational Immediate' message to base.... The Captain later described the incident:

'... The blockade runner was travelling at about 15 knots.... She opened fire at us with all guns – it was heavy all right. My navigator, who was also the bomb-aimer, did some fine work. We saw one bomb hit the ship aft of the funnel. Immediately vivid red flames shot out and rose about 200 feet. We flew around for five minutes and saw that the flames had spread the whole length of the ship.'

The Liberator had ploughed through a spectacular array of pyrotechnics, including wire-trailing rockets to deliver its attack. At 800 feet, the first of several rockets were unleashed, the final pair being fired at 600 feet. Two bombs were also released from this height, one of 250 lb and the other of 500 lb. Like fiery darts five of the rockets struck home against the vessel and one of the bombs hit the stern.... It seemed to the crew that the Lib finished its dive by pulling out right over the ship and passing between the funnel and the mast...

It had not come through unscathed; flak fragments had damaged the starboard outer engine, which had begun to run raggedly. After a lingering look at the stricken ship, during which Josef Kosek took some dramatic photographs, Dolezal turned for home. Some time later Marcel Ludikar intercepted a wireless message from a Halifax, saying that *Alsterufer*'s crew had taken to the boats. In fact the doomed ship remained afloat for several hours, the final blow being administered by Liberators of No. 86 Squadron.

A rescue force of six Elbing and five Narvik class destroyers, sent out by the Germans, was put to flight by the cruisers HMS *Glasgow* and *Enterprise* with the aid of Coastal Command aircraft. The German force lost three of its number (Z27, T25 and T26) to the cruisers' guns and the remainder abandoned *Alsterufer* to her fate. Shortly after 8.00 in the evening, she slipped beneath the waves, her last position being 46.32 N 18.35 W.

The survivors gave grudging praise for the tenacious manner in which the attack had been pressed home despite the heavy anti-aircraft fire.

Dolezal and his crew arrived safely back at base after 12 hours in the air.'

Atlantic Battle

The Polish Air Force Association, *Destiny Can Wait* (William Heinemann, 1949)

Operations based on Predannack in Cornwall gave no direct results of U-boat sinkings – the bad weather during the winter months greatly reduced the chances of sighting... But the luck was soon to change, for the aircrews of No. 304 (Polish) Squadon, Coastal Command were now quite at home with the new equipment, using it expertly and with absolute confidence. They could now concentrate more on the U-boats and trouble less about enemy fighters....

And the luck indeed turned for on 28 January 1944, G-304 claimed another U-boat. Here is the story of the sortie told by one of the crew.

'The captain of Wellington "G for George" was struggling to keep awake as he sat at the controls. He had been flying for 7 hours in terrible weather over the Atlantic. The aircraft was icing up. Tired eyes, sunken in a haggard face, dully watched the altimeter....

'The rest of the crew were in little better shape.... Then suddenly; "Contact – five miles eight degrees to port!"

'The whole crew at once became perfectly fresh, as fresh as if they had just left base. Everyone moved like lightning ... a Morse message was being quickly tapped out; the engines were whining; short, sharp orders rang out: "Action stations ... navigator ... gunners ... co-pilot...." "G for George" was telegraphing: "Running down for attack: position 47°00′N-05°38′W: course...."

'The cathode ray tube undoubtedly showed the green glow of a U-boat "blip". The Wellington dived sharply towards the water.

'"Contact; four miles, twenty degrees to port!"

'The pilot corrected the course. "G for George" was slipping through the darkness just above the waves. The radar operator was still directing: "Three miles ... a bit to port, hold it.... Two miles...."

'The red light of the radio altimeter flashed out its warning that the sea was very near.

'"Light down! Attacking...."

'The waves were high and threw their spray just under the wings. The bomb doors were opened and the pilot began to feel for the depth charges release-button and master switch.

'"Contact; one mile ... look out ... keep it...."

'"On Light!"

'The beam of the Leigh Light pierced the darkness in the mad dash over the waves.

'"There it is! U-boat ahead, sir!"

'The U-boat was athwart the Wellington's course. A grey boat on a wreath of foam. It seemed to rush closer to "G for George", and reflected the Leigh Light's powerful beam so that the glare seared the eyes.

'"They're shooting, the bastards! Look out!"

'Evasive action! The front machine guns began to rattle away, darting the golden curves of their shots into the grey outline of the U-boat.... A few feet over the waves, in the hard lines of incendiary and tracer shells, the Wellington pressed home its attack. The pilot gently pushed the release-button....

'The conning-tower of the U-boat flashed by under the aircraft. A Hun was standing on it. The dull explosions of the depth-charges shook the Wellington as the rear gunner let go a long burst. The plumes of water thrown up by the explosions filled the air with spray, and for a space of time the enemy was lost to sight. A sharp turn and "G for George" passed over the U-boat which then disappeared for good. In its place appeared a large and spreading patch of oil, in the middle of which the sea seethed and bubbled as if shaken by underwater explosions. Two solitary flame-floats on the water marked the spot with a feeble flickering light, mournful, bringing out the immensity of the surrounding waste and the darkness of that sombre night.'

... As the U-boat force retreated from the Bay of Biscay to Norway, Coastal Command prepared to concentrate its main effort over more northerly waters.... No. 304 Squadron was, as a result, transferred to Benbecula, in the Outer Hebrides.

... There could probably be no greater contrast to Cornwall than Benbecula. Even those of the squadron who still remembered Tiree felt their blood run cold when they saw their living quarters, the runways plagued by eternal crosswinds, the bare and storm-girt island....

The squadron had been inclined to disbelieve an earlier warning.... 'When you're on Benbecula don't worry if you find yourself talking to the sheep, but as soon as the sheep start answering back – look out.' This was, however, confirmed by the bearded personnel of a Fleet Air Arm squadron who were the only ones to greet the newcomers and who, in 1944, were still devotedly flying their Swordfish. The Poles then learned that their predecessors, some American B-17 Fortress crews, had been transferred on compassionate grounds! With this rather faint ray of hope, No. 304 Squadron installed itself and at once resumed operations....

... The patrol on 4/5 May 1944, had already lasted 6 hours and the crew of 304's 'N for Nun' could hardly keep their eyes open.... Then suddenly ... the radar operator contacted a U-boat at 0315 hours. The crew livened up in a flash, and the Wellington made straight for the enemy. From behind the clouds the moon appeared, and its silvery light threw a long, brilliant reflection on the water, so bright as to dazzle the eyes. Foam dashing against a darkish object could be seen, and, just behind this, another dark object. A double sighting!

The wireless ticked out its message to base as the ponderous Leigh Light was slowly lowered into position.... The U-boats were moving quickly, low in the water and ready to crash-dive at a moment's notice.

The bomb doors opened with a clatter when something flashed inside the fuselage. Flames appeared amidst dense, black, choking smoke....

'I'm attacking,' said the pilot quietly, 'put out the fire, if you can. Light on! Co-pilot, give them a burst....'

The sharp, violet-tinged cone of the searchlight slid across the waves and rested on a U-boat. A tremendous cannonade of flak – cannon, machine guns and pom-poms – was immediately directed against 'N for Nun'.... The pilot's fingers pressed the depth charge release-button. There was a violent bump, a flash, explosions and darkness as the Wellington turned to port out of the barrage and began to circle.

The wireless operator reported briefly: 'The radio short-circuited, but the fire's out. Everything OK now.' And then the rear gunner's voice: 'Charges straddled the Jerry all right.'

A red rocket shot up high in the air from the spot where the attack had been made. By its light the crew saw the U-boat disappear under the waves. Its lines looked strangely distorted and bent. One of the charges must have made a direct hit. The other U-boat cruised around for ten minutes or so and then submerged. There were no survivors to pick up, and it returned to port alone....

... The Germans were now working like beavers to bring more efficient U-boats into action. The radar fitted in Coastal Command aircraft had proved extremely effective. It forced the U-boats to surface only at night for the minimum time necessary to charge their batteries, and even so they were still vulnerable to attack.

It was to overcome this radical disadvantage that the 'Schnorkel' device was developed. It enabled the U-boats to recharge their batteries without losing way and when submerged at periscope depth. It consisted of two tubes within a casing hinged to the deck forward of the conning tower: one tube was for air intake, while the other discharged the exhaust gases from the Diesel engines. The result was that it normally proved all but impossible to detect a U-boat from the air, except when the sea was quite calm. Another novel device installed in the U-boats was a 'search-receiver', which gave warning when airborne radar was scanning the sea in the vicinity. The speed and range of U-boats were also improved.

It can safely be said that these improvements revolutionized submarine warfare. There is no knowing what the outcome would have been if the Germans had been able to use their new U-boats in quantity before, or even during the period of the invasion of Normandy in June, 1944.... [editor's italics]

The Battle of the Atlantic – and the Battle of the Bay of Biscay in particular – enabled the Allies to launch their invasion of the Continent. It has been aptly stated that the way for the invasion 'was paved with the carcases of U-boats'; and, again, it was air power which struck a vital blow at Germany's might. *The U-boat crews were the pick of the German Navy, but, though they fought with great determination and skill, they were thoroughly beaten. Admiral Doenitz declared with pride at Nuremberg that, out of a total U-boat force of 40,000 men, 30,000 did not return, and of these only 5,000 were captured* [editor's italics].

Come Back Safe, Skipper

James Sanders, *The Time of My Life* (Minerva Ltd, New Zealand, 1967)

... She was a quiet, self-possessed and competent driver of the Women's Auxiliary Air Force, and as such rejoiced in or accepted the fact that she was known as 'a Waaf' – that great levelling name that cares naught for home addresses in Mayfair, Limehouse or Glasgow.

She was not beautiful. Nor yet was she plain. And, of course, the uniform did not help. Indeed, as one of my squadron acquaintances once put it, 'I don't like making love to a Waaf. You have a sort of horrible feeling that you're playing around with one of your mates.'

Obviously her day, now reaching the fag end, had been busy and was dissipating itself in the ashes of weariness. Her dark blonde hair showed a damp rat-tail or two from under her peaked cap. Her nose was shiny. She blew a jet of air upwards from an outshot bottom lip to remove a vagrant lock from her eyes. She spoke little as she drove around the perimeter.

'Big night tonight, Skipper?' She said 'Skipper' quite respectfully, putting herself colloquially in rapport with my crew. In Ops Room, around MT Section, she would have addressed me as 'sir'. Tonight, in the dark of the MT vehicle, with the boys chattering in the back, she was one of the company.

'I mean a *long* trip?' I called her 'dear' quite as impersonally as the waitresses in the mess called us 'luv'. A sort of national in-it-together comradeship and affection that we found throughout the country.

We moved around the track and reached dispersal. The bombers stood tall, gaunt, powerfully beautiful, in black outline against the sky.

'Which one's your kite?'

'B-Baker. There where the bloke is with the torch.'

She stopped the truck and the boys started tumbling out like noisy colts out of a stock wagon. Yellow mae wests unbuttoned, untied, with tapes trailing, parachute bags, thermos flasks and snack packets, navigator bags and charts, helmets with microphone flexes dangling, wool-lined flying boots clomping over the hard-standing. Knights in their armour.

'... And gentlemen in England, now a-bed, shall think themselves accursed they were not here....' I thought of Shakespeare.

I was moving over to join my boys when she called to me, 'Skipper, you've forgotten your sandwiches.'

I came back to pick up the package. 'You might get hungry, luv.'

'Thanks a lot. Well, cheerio, girl.'

'Come back safe, Skipper.'

I looked at her. At her eyes. And impulsively, as naturally as two dumb creatures compelled by a strange animal yearning, we clung together.

I never knew her name, any more than she knew mine. We were complete strangers of but ten minutes knowing. Strangers, but lovers to the whole world by proxy, we two were all the partings that had ever parted, all the goodbyes and adieus and auf Wiedersehens that are to be whispered to the end of time, when man shall go forth and woman must wait. She has done it before. Hers is the waiting role, and she takes her part with infinite patience and stoicism.

The Waaf turned her truck in a noisy loop, squawked the horn in parting salutation, and headed back to MT Section.

Here she would log her daily mileage run, make a sentence or two of tired badinage with the duty NCO and walk to her dormitory block. She would fling her cap on her bed, look in the ubiquitous mirror, and mutter a distasteful 'Gawd!' at her tousled reflection....

...We were at the end of the runway, and I stopped the machine for final check before take off....

I called to the boys, 'Everyone set for take off?' and the call came back, 'All crew in take-off position.'

At 90 knots I pulled the Halifax off the deck and we climbed away. We were off to the Skagerrak and Kattegat and the sea routes of the German ore boats and the skies of the enemy night fighters.

I sit in contemplation. We will descend to 200 feet as we approach the Skagerrak and, flying on radio altimeter, traverse the complete circuit of the patrol area at this height. This will be exhausting work mentally for concentration is acute at this perilously low height.

And then it happens.

'Contact twenty degrees starboard at ten miles!' Stan Sharp had got it taped.

'Change course oh-nine-five, Skip', says Rube.

'Oh-nine-five it is.'

We home onto the contact, which Stan has identified as a ship of some substance. He chants the distance to run.

'Seven miles.' A pause. 'Five miles.'

'Bomb doors open!'

'Left-left Skipper. Steady. S-t-e-a-d-y ... *Bombs gone*!'

A stick of six 500-pounders plummets seawards, with air-burst pistols fused. A mighty roar above the drone of the engines, and six flashes tell us the bombs have hit the surface....

The flight home [to Stornoway].

I have taken my eight men into hostile territory in one of His Majesty's aircraft. We have harried the enemy in his own grounds and have emerged safely....

The wheels go 'boip' and two puffs of burned rubber smoke rise on the crisp morning air. B-Baker runs her allotted length of the runway, and I turn her onto the perimeter....

Somewhere in the women's dormitories, I guessed, a Waaf may be astir, even at this uncivilized hour. She'd hear the kite come in and look at a bedside clock and reckon on time elapsed that B-Baker had come home. She'd check at Ops Room when she went on duty and see if the boys had got shot up. But at least they were home....

Ships Dead Ahead!

Jacques Andrieux, who was to become a general in the French Air Force, was one of the most able of the Royal Air Force's Free French flyers in wartime, winning the DFC and bar and other decorations. The spirit which he brought to the task can be sensed in this short extract from his book Le Ciel et L'Enfer *(Presses de la Cité, 1983)*

Six Hurri-bombers were waiting for us at Predannack (in South Cornwall). I would be flying with 130 Squadron. Planned take-off time was 0934 hours for 130 Squadron and 0937 hours for 234. The whole trip would be at sea level across the Channel to Brittany and there would be complete R/T silence.

There were flak ships among the eight vessels to be attacked and we all knew what that meant. They had enormous fire power: four to six quadruple-mounted 20-mm pom-poms, not to mention six to eight 37-mm automatic guns.

Stuffy, of my ground crew, who knew just how important little details are, had redoubled his efforts on my aircraft. He knew I liked order in my cockpit. He rubbed over my canopy with special paste so that it was immaculate; not a drop of oil or any trace of finger prints....

I was flying No. 2 to the Wing Commander (Wing Commander Minden Blake from New Zealand). I liked flying with him; he was a good pilot and an excellent leader and knew how to handle a formation. He was no slap happy gong-hunter. He really looked after his pilots.... A 'real gentleman', this Blake.

Putting on my helmet I realized that my forehead was wet with sweat. My heart was hammering. My face felt hot and the veins in my neck seemed to be swelling up; my parachute buckle was pressing hard against my knotted stomach. All these symptoms disappeared the moment the Wing Commander gave the signal....

I stuck to my No. 1 like glue. My wing was inside his, my wingtip level with the roundel. The moment I noticed his wings waggling slightly as we neared the ships it would mean 'Spread out a bit, I'm attacking'.

The sea slid past under our propellers. We had been flying for 35 minutes, so the target could not be far away. The Wingco must have spotted something. Sure enough, he started to waggle his wings.

I also now caught sight of the ships. There were eight of them, split up into two formations of four with the defending flak ships fully visible. Only too clearly I could pick out the front and rear flak batteries.... It was our job, as the anti-flak squadron, to knock out these gun turrets, stuffed full of guns.

'Look out, chaps, ships dead ahead. Dead ahead, full throttle. Pull left after the attack and for God's sake break the moment you have finished firing.' The Wingco's instructions were quite clear.

Like madmen we rushed towards the enemy. I opened up too far away and then saw my shells hitting the target. I kept on firing, aiming at the bridge. The flak was coming from both flak ships at the sides of the convoy. I stayed close to my leader. Together, we flew through the protective screen of machine gun fire, level with the ships' masts. I just had time to catch a glimpse of the crews throwing themselves flat on the deck. Wham! A Hurri-bomber had scored a direct hit with a bomb. I felt like cheering but I had other things on my mind.

... Enemy shells were striking the sea all around us, sending up fountains of spray. Suddenly, there was a terrific noise in my cockpit. The stick was almost snatched out of my hand ... I could not hear anything ... I was in another world. I'd been hit by a shell.

I gained height and cautiously tried my controls. Marvellous, they seemed to be operating normally. The Sept Iles were fading into the distance behind me. The enemy ships had lost their nice formation. A plume of opaque smoke was rising from two of them....

... The Wingco was flying alongside me, giving me the thumbs up sign. Very soon the most beautiful sight for the pilot of a crippled aircraft would appear just above the nose of my aircraft: the English coastline....

The ATA

Lord Beaverbrook – Minister of Aircraft Production, Minister of Supply, Lord Privy Seal and, for a period, member of the War Cabinet in Churchill's National Coalition

The men and women of Air Transport Auxiliary were civilians in uniform who played a soldier's part in the Battle of Britain and who performed, throughout the war, a task of supreme importance to the RAF.

They brought the airplanes to the squadrons. In foul weather and fair, by night and day, they kept the ferry moving.

I came upon the ATA almost at the start of its activities. I watched it grow from a mere handful of pilots into a vast organization. And I saw how the fine courage and cheerfulness of the early days endured as numbers swelled and responsibilities multiplied.

'Be Your Most Fetching'

Francis Francis, Geneva, July 1982

Commander Francis Francis, CO of No. 1 Ferry Pool of the Air Transport Auxiliary at White Waltham, near Reading, in Berkshire, combined wealth with diverse talents.

He flew his own aircraft and (flying) boats in the thirties all over Europe and by the outbreak of war had 1500 hours in his log book. He had been selected to represent Britain in the 1928 Olympics in three different sports, but illness forced him to pull out. He turned to golf and became an England player. He was one of the founders of the Martin-Baker Aircraft Company, thereby recognizing the design genius of James Martin, inventor and manufacturer of the ejector seat which has already saved over 5000 lives. He gave fighter aircraft to sustain Britain's war effort and saw to it that others never knew. He concealed his generosity by the simple expedient of having nothing said about his benevolent acts.

Frank Francis's ordered and disciplined mind required that an aircraft be flown precisely – or not at all. With these credits in the account, he became, arguably, the ATA's most effective ferry pool commander. A few months before he died in 1982, he wrote a first-hand piece on the Service for this collection. Here is an extract from it.

'Air Transport Auxiliary was founded early in the war by Gerard D'Erlanger, banker and peacetime private flyer. It offered a specialist service because its pilots were trained to fly any aircraft on sight whether or not they had ever flown the type before. The key to this – apart, of course, from flying ability – lay in the pages of a small blue book.

This was a brilliantly-compiled edition containing every item of necessary flight information for every aircraft, British or American, flown by the Royal Air Force or the Fleet Air Arm. . . .

We were masters of our own fate. In collecting and delivering aeroplanes we could overrule station commanders and chief flying instructors. An incident which occurred in the middle of the war illustrates the point.

I had to go one day to collect one of the (then) new US B-26 Marauder bombers from Stoneyhurst, an American base in the New Forest, in southern England. It was to be delivered to the Royal Air Force's experimental and testing establishment at Boscombe Down on Salisbury Plain, which I knew very well. I was looking forward to the job because I had never seen or flown the aircraft before.

I started with a problem when I was flown over to Stoneyhurst in a

Walrus, an antiquated-looking single-engined pusher amphibian, which the pilot was afterwards to deliver to the Fleet Air Arm at Lee-on-Solent. The most charitable thing you could say of the Walrus, which was used for air-sea rescue and other things, was that if Noah had been able to stow an amphibian inside the Ark at the time of the Great Flood, this would have been his best bet. So they looked at me curiously when I got out at the American base.

I duly reported to the base commander with my papers. As I left his office I asked him whether, by chance, there was an off-duty pilot who could spare a moment to show me the taps on the B-26. It would save me time.

It seemed to stop him in his tracks. "Say," he said, "do you mean you are not familiar with this type of ship? Have you not flown it before?"

My answer did nothing to reassure him. He was adamant that I shouldn't take the Marauder away.

"All right, sir," I said. "If you don't trust me, perhaps you would telephone the Air Ministry. They will confirm my credentials."

He put down the receiver. "The guys there seem as crazy as you are, so off you go. But no pilot of mine is going to help you kill yourself. You must solve your own problems."

I spent forty-five minutes sitting in the aircraft with the blue book watched by a number of incredulous US Army Air Force pilots who had seen me step out of Noah's Ark....

The flight over to Boscombe was short and uneventful, but when I got there I saw they must have a problem. The fire engine crash crew and ambulance were all standing by at the end of the runway. I guessed a test pilot was in some difficulty with a new aeroplane. I flew away for a bit and came back into the circuit. The emergency services were still there, but this time the control tower gave me a green.

I thought it was odd, as I touched down, that the crash tender came haring along after me. As I stopped to turn off the runway, I opened the hatch. A very familiar, but astonished face looked up at me from the ground. "Good God, it's you! Whatever's going on? They called us a few moments ago from Stoneyhurst to say a madman had taken off in a B-26 and was headed for here."

Weeks later, a Marauder had to be ferried to Stoneyhurst. I checked my list of top-grade lady pilots (they were very good indeed) and I called in the most attractive of them. I told her the story.

"Look," I said, "you go and collect this aeroplane and take it over to Stoneyhurst. Make sure you look your very, very best *and most feminine* and when you land smile at everyone; use all your charm – particularly with the base commander. Be your most fetching."

It worked. After a beautifully flown circuit and landing the pilots were stunned to see this attractive young lady step elegantly down from the Marauder onto the tarmac. As she chatted and laughed with the crews her taxi aircraft landed and pulled up beside the control tower.

With a wave and a smile and a "hope to be back soon", she was quickly away, leaving, we heard afterwards, a group of incredulous American airmen wondering what had hit them.'

The Walrus and the Yank

Recounted by The Rt Hon. Lord Justice Kerr, London, February 1983

A Royal Air Force Walrus, on air-sea rescue duty, had to land at an American base near by. The CO saw this [to him] incredible sight and had himself driven round it three times in a jeep in total silence.

Then he said: 'Jesus Christ, now I've seen everything.'

The Experience to Judge

Lieutenant-Colonel F.C.L. Wyght, the *Record News*, Smith's Falls, Ontario, 12 November 1980

... [Midway through the war] ... I transferred from One Canadian Div. to the RCAF, enlisting as aircrew....

At No. 9 Elementary Flying Training School, Southam, Warwickshire I was assigned to an exceedingly fine ex-Coastal Command pilot-instructor who would have been about twenty-eight, heading for over-the-hill. Well, in those days, at that age you were working on it....

...Once on the ground, my instructor, a kindly and patient man, demonstrated he had limits. 'Now, tomorrow,' he said, 'try to do something different, like not trying to drive the propeller through the bloody turf....'

Early afternoon the next day we took off... I shot three landings: the first – perfect; the second – a little bouncy; the third – a daisy-cutter.

We took off again after the third to start yet another and as we turned onto the downwind leg a Wellington, a venerable medium bomber, went lumbering by on our starboard and a friendly tailgunner swung his four guns in salute and we dipped wings to him. About 2 miles ahead the 'Wimpy' started a turn to port, stuck its nose down, and went straight into the deck. Out of the boiling black smoke came occasional bursts of dark red flame and so burned the pyre of five aircrew. Not a word passed between us.

Just before we turned across wind over on our port, at about a 45-degree angle, came a twin-engined craft at obviously great speed and headed east for cloud cover.

From the instructor: 'Type?' From a woolly-mouthed student: 'A Junkers 88, sir. German medium fighter-bomber.'

Seconds behind the 88 came two Spitfires. From the instructor: 'Watch your height.' Student: 'Yes, sir.'

But my concern wasn't height, but, 'By the gods, if this is what flying is all about, it's here I'll hang up the skates as soon as we're down.'

So, more in a trance than in a deliberate series of movements, I swung in on final and set her down as though the sod were eggs. I pulled up by our flight shack where the instructor hopped out, busied himself in the rear cockpit for a moment while I awaited his order to taxi over for refuelling. And the order came....

'Now is as good a time as any for you to go – off you go for one circuit – and make it like the others.' With a clip on the shoulder, I taxied out. The circuit was without the high drama of the last one – the landing without fault.

Later when I made up my log book it totalled 4 hours and 35 minutes. Usually you went to 8 to 10 hours before you soloed.

It became clear in my mind that he had reasoned: 'If he doesn't go now, he may never go.' And he was dead right....

There was a clear and not-to-be-violated regulation at No. 9, and other EFTSs, that no student was to be sent solo until he had been given a confirmatory check by another senior instructor. I had had no such check. Therefore I had not soloed – No. 9 did not wish to be blistered by Training Command. But my log book carries a small red S beside that flight, put there by my thoughtful instructor, who realized just what that day meant, and cost.

Wimpy

Arthur B. Wahlroth, 405 Squadron RCAF, Mount Albert, Ontario, 1981

...There were, in all, about seventeen Marks of Wellingtons, the major change, except for those that were singled out for special duties, being the provision of larger, heavier, and higher-powered engines. (It's NOT true that the first ones were powered with elastic bands, this being impossible because the country that controlled the world's supply of natural rubber was on the other side!)

We who flew them were of the generation nurtured on the 'Strength through Spinach' philosophy of Popeye the Sailor, whose bosom pal was Mr J. Wellington Wimpy, hence the nickname of the machine.

'Bomber Command's Worst Defeat'

Gordon W. Webb, *Airforce* Vol. VI, no. 1, March, 1982

Gordon Webb flew on the Nuremberg raid in Halifax LW-M ('Pistol Packing Mama') on the night of 30/31 March 1944. He and his crew were among the lucky ones; they returned. Plenty didn't. In a specially written story, originally prepared for the 1982 issue of Icare *on the 38th anniversary of the attack, Webb called the operation 'Bomber Command's worst defeat of the war'.*

His penetrating, first-hand account of this rough experience was published, as a tribute to the crews who took part, in Airforce *(Volume VI, Number 4) in March 1982, and extracts are reproduced here by kind permission of* Icare *and Air Force Publications Ltd of Ottawa. The editor is also indebted to the Director of the Directorate of History, National Defence Headquarters, in Ottawa, for proposing this important contribution to the record of the great bombing offensive.*

...We were by all standards a capable and efficient crew. Each man knew his job thoroughly and we had been together and flown on operations long enough so that we were now considered a seasoned crew. We knew all about the percentages for and against finishing a tour. We had long since come to terms with the possibility of one night not coming back. We had each of us agreed, however, that should we 'go missing' or as it was commonly called 'go for a burton' it wouldn't be because of something we did or should have done and did not do. If they were going to get us, they would have to work at it. . . .

The night's operation was scheduled to take us through the very heart of Nazi Germany and would be, we were quite certain, a one-way trip for many.

Usually the briefing officer was a former aircrew type and knew the feeling well. This evening, pointer in hand, he signalled to have the huge curtain drawn back while announcing in the time-honoured tradition:

'Gentlemen, your target for tonight is Nuremberg!'. . . .

From Eastmore, in Yorkshire, the red line reached out across England to pass south of the Wash and over the North Sea. It crossed the Belgian coast just west of Ostend and continued southeast to a point near Charleroi. From there it made a turn due east and began a long 265-mile leg toward the old German town of Fulda. Here, another sharp change of course led the ribbon south, southeast directly to Nuremberg. It was a long, long way into enemy territory.

What this classic tableau of a 'Target for To-night!' scenario really told us was that we were going to challenge the full might of the Luftwaffe's night-fighter squadrons plus the deadly, highly sophisticated German radar and flak defensive system. And we were going to do this under weather conditions which gave the defenders, who already held a big edge, an overwhelming advantage. It must be admitted that to a man we felt that Bomber Command HQ was being uncommonly generous with our lives. . . .

It is a matter of record and perhaps worth noting at this point that over 800 heavy bombers took off from England that night. 712 actually crossed into enemy territory. Less than 600 returned. More British and Commonwealth aircrews were lost on this one raid than were lost in the entire Battle of Britain [editor's italics]. . . .*

Twenty-five minutes flying time into Belgium brought us to the vicinity of Charleroi. Here, our flight plan called for a sharp turn to port to begin the long, straight and deadly leg that led to Fulda. Unhappily, long before we reached this turning point it had become increasingly clear that the bomber force was in for deep trouble.

To begin with the moon, which was hanging low in the northern sky, was the brightest moon any enemy night fighter could have asked for. Visibility was virtually unlimited for a night sky. At our altitude cloud cover simply did not exist. What cloud there was hung low to the ground providing a perfect backdrop silhouetting the bombers for fighters prowling hungrily above.

Secondly, the forecast winds turned out to be incorrect. Our navigator, Vic, discovered this long before the Windfinder Aircraft, whose job it was to detect any shift in the wind, reported the change back to the stream. In

*Detailed research by Martin Middlebrook (*The Nuremberg Raid*, published by Allen Lane, 1973 and reprinted 1980) revealed the actual figures for the raid. These were: 781 aircraft dispatched, 55 aborted, 95 were missing, 634 returned of which 10 either crashed or crash-landed. (Ed.)

addition, the winds we discovered were far stronger than those found by the Windfinder crew.

And thirdly, we saw that aircraft at our altitude were leaving vapour trails that must invite attacks from the fighters....

From my vantage point in the pilot's seat the scene spread out before me was growing more and more horrifying the further we advanced into Germany. The night sky seemed to be full of Jerry fighters which is not surprising considering we'd been flight-planned to fly south of the Ruhr and practically over the night-fighter holding beacons of Ida and Otto.

It's been said before and I say it again now because we were there and did see much of the awful destruction visited upon Bomber Command that night; you could almost find your way to the target by navigating along the line of shattered and burning aircraft scattered across Germany. Remember, nearly 100 bombers were lost along those 265 miles between Charleroi and Fulda.* An average of one aircraft every 2½ miles. Death, sudden and violent, was everywhere. I lost count of the aircraft I saw shot down or blown up.

As we pounded south on the new course for Nuremberg we fully expected to pick up the target dead ahead and clearly marked with the large coloured flares dropped by the Pathfinder crews. To our consternation however, target-markers, searchlights and heavy flak activity became clearly visible several miles away, off our port wing. If we were on course all this uproar would be in the Schweinfurt area but supposing we weren't on course? Could we be west of Nuremberg? Were the Windfinder's winds right and ours wrong? Could the Pathfinders be so confused that they were dropping markers on the wrong target? What was going on?

Our navigator, Vic, with the assurance that a full tour of Ops had given him, had no doubts at all. We were right on track – right where we were supposed to be and that was that. Find the target, drop the bombs and let's go home....

Did we actually do this sixty-six times in two years? Well, not actually. Most Ops weren't this bad for which we could all be duly thankful, but now the moment of truth was at hand.

With Vic having been proved right and with Nuremberg and our target just ahead we began the run in. There was a lot of flak. 'Left-left-steady-steadddyy.' A passing searchlight floods the 'Greenhouse'. 'Hold it Skipper,' from Cy up in the nose. Can a Halifax only weigh 25 tons? More like 125 tons. 'Steady-steady-Bombs gone!' The last was superfluous as 'Mama' leapt about 200 feet with the release of five tons of HE. Now we fly precisely straight and level for 30 seconds, the longest 30 seconds anyone will ever know, so that we can get the required photo of the drop for the IOs† back at base....

...With bombs gone, it was time to set course for home and breakfast. We twisted, we turned, the old girl was hurting, no doubt about that. But she got us home.

We touched down at Eastmore at 0640 hours, exactly 8½ hours after take off. Bone-tired and mentally beat though we were, sleep was out of the question....

*Middlebrook's figure was 59 bombers lost on this long leg. (Ed.)
†Intelligence Officers.

It didn't take long to confirm what most of us already suspected, Bomber Command had just suffered its worst defeat of the war. Headquarters admitted that 108 four-engined aircraft had been lost* over occupied Europe or in the cold, grey waters of the North Sea. A further 50 or 60 had landed in various states of disrepair at aerodromes scattered across England. Over 800 experienced aircrew were dead, wounded, missing or prisoners of war....

Nuremberg was our crew's last operation with 6 Group Main Force in Bomber Command. We said farewell to 'Pistol Packin' Mama' and moved on to join the Pathfinder Force flying Lancasters. It's worth noting here, I think that 'Mama' did not return from her next trip to Hitler's Europe. Perhaps she missed us....

'A Pilot to all Ravaged Cities'

Chosen by David Scott-Malden

Along the pillared sheets of cloud
    the wide unceilinged skies across
    we sweep on slim swift wings, and pass
among the stars no stain of blood;

For borrowed is our flight, and earth
    reclaims from all the skies her dead.
    We fly, but when our path is sped,
there is no present shame of death.

Brave? Ay, brave, if men must write
    words to enhance the shift of war;
    but mind how broad and sunlit are
the silent spaces where we fight.

For in our battles men have seen
    renewed the pomp of lance and plume
    and said: "How bravely do they come!"
They do not know how cold and clean

Are the unfurrowed fields we cross,
    the swirling walls we strive among;
    fields where the dead may not stay long,
walls that divide to let us pass.

There is no city of the moon
    where worn women bend and weep
    and press dead mouths and cannot sleep
nor heed the faltering eye of dawn;

There are no cloud-lit walls, destroyed,
    and in their wrack pale children's eyes
    that from bent bodies whitely gaze
in their frail final attitude.

*The loss of bombers publicly announced after the raid was, in fact, 94. Source: Air Historical Branch. (Ed.)

For those I weep, who see thrown down
man's love in riven house and street
who sort the tangled skeins of hate
amid man's agony to man.

We have our triumphs and our fears,
but at our going they are gone;
there is no blot against the sun,
across the inviolate sky no scars.

We fly apart. The wide world's wars
are mute and distant as the grave.
The earth contains our store of love,
Our hate cleaves to the silent stars.

Hugh Popham (Lieutenant, RNVR)

The Norwegian House of Christie

Major-General Johan Christie (Retd), Royal Norwegian Air Force, The Christie Archives, Oslo, 1982

The wartime deeds of the old Norwegian family of Christie (a forebear was President of the country's first parliament) are secured in the operational records of the Royal Air Force.

Werner Christie became, first, a squadron commander in the Norwegian Spitfire wing and then, towards the end, leader of successive Mustang (P-51) wings as they ranged deep into the heart of Germany alongside the marauding forces of the US 8th Army Air Force. His brother, Johan, on the other hand, was wedded to Bomber Command where he finished up as a flight commander of 35 Squadron in the Pathfinder Force with whom he flew 48 marking operations. Later, each achieved the rank of major-general in the peacetime Royal Norwegian Air Force.

Both brothers were highly decorated, winning, among other things, a DSO and a DFC apiece, while their sister, Katrine, just to top things off, rose to the summit of the RNoAF's Women's Service. A rare composite record for a family.

Johan Christie disciplined himself to keep a regular, wartime notebook. From it he set down his views on the selection, build up and needs of a first-rate bomber crew in a striking, analytical document from which an extract is reproduced below:

The essence of what I learnt was very simple, and although I think it more important for bomber crews than for others (because the issue at stake for them is higher), the conclusions reached would convey success in most jobs in life. They can be summarized shortly:

1. Be much more systematic (Be over-systematic)
2. Work much harder

Most bomber crews are very young (average age about 22–23 years). This may be an advantage in many respects such as in the ability to endure physical and nervous strain; but a man of that age is often too 'happy-go-lucky' for such a serious job. A captain of an aircraft (sometimes he may be only 20) is the supreme commander of a big ship and six other highly skilled men, often working under terrific difficulties.... That things go as well as they do is a great tribute to the guts and spirit of the youth of the Allied nations in general and the British Commonwealth in particular.

I have often wondered whether the policy of the Royal Air Force regarding age has been correct, particularly as regards heavy bomber pilots. I believe that provided a man is physically and spiritually 100 per cent fit, it would be an advantage if he were between 30 and 35 (in special cases even as old as 40) rather than, as now, between 20 and 25. But it may be that the number of useful men of that age, especially those who have the necessary mental outlook, is rather small. Spirit is more important than physique.

The advantages of higher ages are:
1. More developed characters
2. More suitable to lead
3. Probably more systematic
4. More liable to realize the dangers and, therefore, probably harder working

The disadvantages are:
1. Probably slower reactions
2. Definitely far slower in picking up new 'gen' (information)
3. More liable to suffer from strain (eyes, nerves, physique)
4. Probably more liable to lose morale (because of realization of dangers)

Aircrew should not be married. If they are they should be stationed as far as possible away from their families. (Best overseas.)

By observing the average aircrew types in the Royal Air Force and comparing them with their counterparts in other Services, it is striking how seldom one finds the 'tough' ones among them. Generally, they are small or medium-size; they do not look as if they would be liable to distinguish themselves in sporting championships (where physique counts). They tend to be, nervously, more highly strung than the average man. Naturally enough, they nearly always have a gambling (i.e. risk-taking) streak in them. Also, they are nearly always ambitious.

Norwegians, generally, seem to do fairly well. On the whole, they are harder-working than most other nations. They are very ambitious. They seem to take strain and danger at least as well as any other....

The Salt of the Ground

R.A.N. Douglas, London, 1982

Ronald Douglas was an Australian success story in an RAAF Squadron where success was hardly won. He rose from (very) young sergeant pilot to squadron leader, and one of 460 Squadron's most experienced bomber captains, in the course of a sustained operational run. With 14 out of a possible 16 attacks on Berlin to his credit during the roughest months of the offensive, he has his own views on the worth of these raids.

Now Agent-General for Western Australia in London, after managing the Shell Company in Perth, he is the first non-political appointment to this office. Against such a background, Douglas's opinions carry special weight.

'...I operated against a number of the major targets during that period (1943–44) including many attacks on "the big city" – Berlin. I am aware that these raids have been criticized as a senseless waste of aircrew and airplanes; an unnecessary damaging of the fine cities of Germany, and not particularly rewarding to the war effort. To this day I remain implacable in my view that it was a most important segment of the air war. It drew, for its effort, in the main, on the strength of untried and mostly inexperienced airmen having their first sensation of war in a most hostile environment, reacting to the concentrated training they had received, and accepting the role for which they had been chosen.

'A lot has been said and written about their courage, but in my case – and it seems to me it applied to many other young men – it was the relationship with the older men of the ground crew that helped us most.... There developed a father-son, elder brother-younger brother, friendly uncle, bond between us.

'The unspoken care these men offered to returning aircrew; their constant good humour in the face of the appalling conditions in which they worked on the aircraft; their proprietary interest in the individual crew – and to those crews who had the good fortune to live in the sergeants' mess; the ongoing relationship with the senior NCO's who controlled them; the seat being held in a poker game and given up even when on a winning streak; the insistence on the extra drink (to make you sleep); and, above all, the hand on the shoulder and the gruff – "You all right, son?".... These are my lasting memories.

'Despite every effort I make to remember the strength required to fly a Lancaster with two engines out on one side; the fear and the sweat trickling in the flying boots over the target; the magnificence of the sky on a clear night full of target indicators, searchlights and tracers; the acrid smell through the oxygen system after you had been hit.... The recollection remains indistinct and harder to fix than the face, and the voices of my ground crew. These men who, because of their age, had come to war from many and varied experiences all had one thing in common – the memory of the unemployment queues of the Depression. I really believe they had a greater concern for the future than I did....

'Why Berlin?... To me, there were always two very good reasons for planning the offensive against the city. Berlin nestled a long way behind the borders of German-occupied territory.... A proud capital from which to issue the devastating orders of the Third Reich to its forces on land, sea and air.

'As a young man, it seemed to me inherent in the thinking of all servicemen and women in our defensive situation, that we should show our enemies we were capable of striking back.

'London had been bombed and burnt, as had many other cities. There was an affinity, then, between the shock of this and the major attacks on Berlin. I have never questioned that the raids were intended to do anything more than carry war to the heart of the German people.

'My secondary point is that it seemed fundamental to the thinking of all major nations, irrespective of the changing fortunes of war, that the capital city implied strength, authority, organization and direction. I believe this could certainly be said of London because whatever went into decentralization and deployment of industry and administration, the major war effort was run from there....

'I am sure the same thing applied to Berlin. For this reason alone, the merciless bombing of the German capital was, to me, warranted....'

'The Ladder of St Augustine'

The heights by great men reached and kept
Were not attained by sudden flight
But they, while their companions slept,
Were toiling upward in the night.
H.W. Longfellow

Turreted Victories

Reproduced from Allied Airforces' Reunion Newsletter, 1952, by courtesy of Wartime Pilots' and Observers' Association, Winnipeg, Manitoba. Chosen by Don Aylard, Association President

In a Bomber Command attack during the last week of May 1944, on the military camp at Bourg Leopold, Sgt P. Engbrecht, mid-upper gunner for F/L/Capt. O.J.G. Keyes (USAAF) of 424 RCAF (Tiger) Squadron, shot down an FW 190 and an ME 110 while beating off fourteen attacks. The 190 blew up in the air, and the 110 exploded as it hit the deck.

During the raid on Versailles-Matelots on 10 June, Engbrecht repeated his double of the previous month, aided by the rear gunner, Sgt C. Gillanders. Their victims were an ME 109 and an ME 110. In recognition of these accomplishments, and for the general excellence of his record, Engbrecht was awarded the CGM and was also commissioned.

On the 12 August the Engbrecht-Gillanders duo scored again, with two more added to their total, on a mission to Brunswick. The enemy aircraft to hit the deck this time were an ME 410 and an ME 109. Gillanders was awarded the DFM.

Late in August this potent combination made the headlines yet again when they downed an ME 410 and an ME 109 during an attack on German positions at Falaise. Both of these aircraft burst into flames. Four nights later Gillanders claimed a probable.

The 'Great Escape' – Stalag Luft III, 1944

1 MURDER

The Polish Air Force Association, *Destiny Can Wait* (William Heinemann, 1949)

One of the organizers of this escape was Wing Commander M. Brzezowski, O.C. No. 301 (Pomeranian) Squadron, who had been taken prisoner on 2 July, 1942, when shot down near Groningen, in Holland. He has written the following account of this brutal and gloomy crime:

Some 80 miles west of the Polish–German frontier is a small place called by the Germans, Sagan, where they prepared a prisoner of war camp (Stalag Luft III) in which were put some 2000 RAF officers, including a number of Poles. In international conventions concerning prisoners of war it is forbidden to extort undertakings not to escape from camp; on the contrary, it is recognized that all combatants have a duty to try to escape from captivity. So, despite the immense difficulties and risks involved, the prisoners at Sagan soon conceived the idea of organizing a mass escape.

In March 1944, one tunnel was finished. The work had taken a year to complete under most difficult conditions. The workers were enthusiasts for freedom, but they were severely handicapped by the physical weakness of the undernourished. Lots were drawn to decide who was to leave in the first batch of two hundred. The winners were supplied with civilian clothes tailored in the camp and with appropriate documents and maps, all expertly prepared. Each had a different route for his escape, so that the difficulties of the pursuers would be multiplied and travelling in large groups avoided.... Most of the routes were toward the Baltic, France or Czechoslovakia.

On the night of the 24–25 March 1944, the inhabitants of Barrack 104, where the tunnel started, had to evacuate to make room for the lottery winners. The escape began.

The failure of the electrical current and thus the intense darkness in the tunnel made progress slower than had been expected. Only seventy-eight of the two hundred ... got away. Then, as luck would have it, some of them were seen by guards in the wood near the exit, the secret of which was betrayed by footprints in the snow.

The alarm was given....

A few days later we heard that almost all who had got away had been recaptured. Rumour had it that they had fallen into the hands of the Gestapo and would not return to the camp.

Several weeks later a batch of dispirited officers returned with the news that those who had been captured had been taken to Gorlice for cross-examination by the Gestapo. While there they themselves had seen other fugitives.... All of them were quite fit and expecting to return to the camp.

When they did not reappear, the senior officer went to the German commandant to make inquiries. The German replied by showing him a list of forty-one of the escaped officers who, he asserted, had been killed while trying a second time to get away. Not long after another list was forthcoming, making fifty-three in all.

All those who returned to the camp were agreed that at Gorlice there had been no idea of attempting a further escape. The story that there had been such an attempt did not ring true – much less so since all fifty-three were shot dead while escaping. Not one was wounded. Not one got away....

Two years after, in 1947, when the murderers of the fifty-three officers were being tried for this crime, it was ascertained that the victims were first interrogated by the Gestapo at Gorlice. They were then taken in batches to the Wroclaw motor road, where they were shot down in cold blood; those who still showed signs of life were finished off by revolver shots.

2 SOME GOT AWAY

Captain Per Bergsland, Oslo, 1982

Per Bergsland escaped to England in 1940 when the Germans overran Norway. He trained in Canada and during the attack on Dieppe in 1942 was shot down and picked up in the Channel by a German warship. He was sent to Stalag Luft III where he adopted an alias – Peter Rockland, a British citizen.

'In 1943 the German commandant circulated an order saying that if a POW escaped wearing a German uniform and was caught, he would be liable to be executed.

Three days afterwards I walked out of the main gate wearing a German uniform made in the camp. A German sergeant, named Glemnitz, met me outside the gates. After he had passed me, and was about 200 yards away, it must have dawned on him who I was. He ran after me and stuck a gun in my back.

I was brought before the commandant who told me I would be sent to Leipzig where he expected the court would sentence me to death. There, I was put in a cell in solitary confinement and was given two pieces of bread without butter and a glass of water per day. After two months I was very sick and nearly died. I had a severe fever with a very high temperature. I assumed they had decided not to send me to the court, but, instead, to kill me by slow starvation. However, I got through it and was sent back to Sagan early in 1944.

Shortly afterwards came the "Great Escape" and I was one of the escapees. I decided to be a Norwegian Nazi working in Germany. Another Norwegian, Jens Müller, had the same idea and agreed to escape with me.

We got out of the tunnel together as Nos. 42 and 43 and walked immediately to the nearest station where we bought tickets to Stettin in the Baltic. We changed at Küstrin where the military police checked us. Our papers were so well faked that the police weren't at all suspicious.

In Stettin, we walked about the streets looking for Swedish sailors. We found two who took us to the harbour, but they left with their ship before they could get us on board. We were asked at the gates what we had been doing in the harbour. I said we were electricians and named the ship we had supposedly been working on. The guard checked the list of ships in the harbour and let us out without looking at our papers which weren't then valid.

The next day we found two more Swedish sailors and they took us on board their ship, putting us down in the anchor room. They said we mustn't be introduced to the captain because he would hand us over to the Germans.
The ship was searched with dogs before it sailed. As the dogs couldn't come down to the anchor room, the German guard searched it with a torch. Jens Müller was lying on top of me with his head over my shoulder. We had planks and canvas covering us. Our faces were completely black.
The guard prodded all over the heap and stuck one of his fingers into my eye, but he did not realize two people were lying there.
We landed at Gothenburg, got off the ship, jumped the harbour fence and went straight to the British consul. "You should have gone to the police," he said, "don't you realize there's a war on?" He was quite serious.
We burst out laughing, we couldn't help it!'

I.P. Tonder, Bagnols-en-Forêt, Frejus, France, 1982

Ivo Tonder, a Czech, who could speak fluent German, was flying with 312 Squadron when, on 3 June 1942, he baled out of his Spitfire after a contest with FW 190s over the Cherbourg peninsula, and was captured. In due course, he ended up in Stalag Luft III at Sagan. He recalls subsequent events.

'I made several unsuccessful attempts at escaping; but my real chance came when I joined the party digging the tunnel, "Harry". Eventually, when the time came to get out, I received a "free" place among the twenty who did not have to draw lots. I paired up with Johnny Stower who was already a most experienced escaper. He had recently been recaptured swimming the Rhône to Switzerland, having successfully crossed through Czechoslovakia....'

Having got out of the tunnel, Tonder and Stower, like Bergsland and Müller, made for Stettin. However, once there, they did not enjoy the Norges' luck; no Swedish ship was due to sail. They elected, therefore, to head south for Mladá Boleslav where Stower, after his previous escape, had a good contact. The journey embraced a change at Görlitz.

'While I was buying the tickets, a young man in SS uniform stood directly behind me, looking at me suspiciously. Taking the initiative, I turned round and showed him my papers which described me as a Czech engineer working at the Focke-Wulf factory. I said I was going to Mladá Boleslav with a Spanish friend for a holiday. The young SS member then became very friendly, helped me to buy the tickets and put us on the right train.'

At Žitava, a party of Kriminal Polizei boarded the train and at once started checking passengers' papers. Tonder's and Stower's credentials appeared satisfactory, but just as the Polizei were leaving the compartment one of them turned back. He had noticed that Stower's trousers were the same colour (Australian Air Force blue) as those of another escaper already held in the local prison. The whole party then returned to make a thorough search.

'This time, we were for it. Our clothes had no lining and our suitcase contained chocolate. We were quickly arrested and taken to Liberec prison where we met four of our friends who had been arrested before us. Later that day they left – allegedly to return to Sagan.

We were interrogated and put in a cell. After a few days, the Gestapo arrived and marched Stower away. I protested loudly, thinking that he was being returned to Sagan while I, because I was a Czech, was being kept behind....

I was taken to Prague and there put in solitary confinement in Pankrác prison. The prison barber slipped me a note to say that, immediately after my escape from Stalag Luft III, my mother, brother, his new wife, whom I had not met, and even her parents, were all taken and were being held somewhere in Pankrác prison....

Bauer was the name of the man who was interrogating me. Indirectly he probably saved my life.'

Other Czech Air Force officers had also been sent to Prague for interrogation before being returned to Sagan. Because of this, one of them, Colonel Busina, quite by chance walked past Tonder's cell one day while he was being shaved. He was very surprised as there had been no news of Tonder. This enabled Busina to report his whereabouts to the Senior British Officer at Stalag Luft III on his return. As a result of this information, a complaint was made to the Red Cross. Following it, Tonder and another Czech, Flight Lieutenant Dvořak, who had also been held in Prague, were moved to the prison camp at Barth.

'On our way there the guards told us the dreadful news of the execution of our comrades. We could hardly believe it. Our four friends from Liberec and Johnny Stower, my partner, were among the victims. We were absolutely shattered.

After a week in the camp at Barth, Dvořak and I were rearrested and put in the cooler. The next stop was Leipzig where we were tried for treason and, despite the services of an excellent lawyer, sentenced to death. The sentence was to be carried out within ninety days. With that we were sent to Colditz.

By now, however, it was 1945 and the war was in its final stages. The Germans had other preoccupations. On 16 April, my birthday, the United States' forces liberated the camp. We had been spared the firing squad....'

Czech Secret

Squadron Leader M.A. Liskutin, ex-No. 312 (Czech) Squadron: Extract from manuscript of *Distant Horizons*, Fareham, Hampshire, 1983

'During one of my firing practices, using two guns with 100 rounds each, I achieved 76 holes in the "flag" target.... The result was unbelievably good. It created a stir in our Wing Headquarters, being by far the best ever achieved.... For me it represented the essential reassurance that my air-firing technique was good ... that I could shoot well.

My mediocre performance in shooting down the Do-217 was due to excessive tension ... in the heat of battle. I have identified the problem... To be successful in combat, the pilot (myself in particular) must remain calm, fly accurately and press the button at 250 yards distance, plus or minus 50 yards and be sufficiently lucky not to be shot down at the same time! My own eyesight at the time was at its best and I could fly accurately. In any future combat my problem was to remain calm. Coolness in the heat of battle, remaining calm in the presence of death, reducing tension.... Herein lay the secret....'

Clandestine Courage

Roger Malengreau, Brussels, 1982

There was no greater courage shown in wartime than by a small group of Belgian patriots who, for more than two years, were feeding secret weather information to the Royal Air Force in the United Kingdom from occupied territory. For most, the risks would have been unbearable, but for these stalwarts it was a matter of duty.

Roger Malengreau, who escaped to England after the fall of France in 1940 and then fought with 609 Squadron after spells with 87 and 56, has recorded a note of this astonishing service which was provided under the nose of the enemy. As a member of his country's diplomatic corps until his retirement, and as a former ambassador, Malengreau is well-placed to provide the facts.

'The unit which operated this secret radio network for the Royal Air Force from Belgium between 1942 and 1944 was called "Beagle", after the code name of its leader, Albert Toussaint.

'Toussaint, who had escaped via Dunkirk in 1940, got tired of waiting with the Belgian brigade which was being formed in Malvern and volunteered for the "special forces". As he was a meteorologist, who had been trained at the Royal Observatoire in Brussels, it was suggested to him that he might establish a secret field meteorological service in Belgium. The RAF was badly in need of regular "met." reports giving scientific data which was outside the scope of the information provided by the regular flights of reconnaissance aircraft.

'After a short training period, Toussaint was dropped by parachute on 23 August 1942, with two radio operators. Although they landed some 60 miles from the target area, they managed to reach their destination without losing their precious equipment.

'The first station, Beagle 1, was set up with the help of Professor Max Cosyns of Brussels University who, with Professor Picard, had flown the first stratospheric balloon in 1931. It started transmitting from Rienne, in the Ardennes, just north of the French frontier, in September 1942.

'A few months later a second station, Beagle II, was operating from Berneau, halfway between Maastricht and Liège. The person sheltering the radio transmitter was a lady of seventy-two who had won the Military Cross in the First World War.

'The third station, Beagle III, was established at Wingene, in west Flanders, in the house of the mayor of the village. Later it was moved to a point near Cluysen. The fourth station, Beagle IV, which became operational at the beginning of 1943, was located in a farm between Leuven and

Malines. The network then covered the whole of Belgium and thus the main flightpath into Germany.

'Some forty agents were trained as met. officers and radio operators at the original base at Rienne. The Germans were on the alert the whole time, and were often puzzled and suspicious, but, in spite of all the risks of detection, the station managed to supply up to seven reports a day for much of the time it was operating. This was the direct result of the perfect coordination between the agents carrying the responsiblity and also between the network and the UK.

'Some seventy-five messages were also sent giving information about German dispositions and radar installations.

'This clandestine meteorological network – the only one to exist in occupied Europe – was closed down towards the end of 1944 when Belgium was liberated. Albert Toussaint, who died in 1982, was awarded the DSO for his courage and leadership.

'From the photograph [opposite p. 193], taken in 1943, it is possible to sense the pride of this little group in being able to serve the Royal Air Force and the Allied cause – even if, for obvious reasons, they could never wear our uniform.'

'D-Day'

Jean Accart, *Silence, On Vole* (Arthaud, Paris and Grenoble, 1946)

Jean Accart, ex-345 (Free French) Squadron, Groupe de Chasse Berry, took on the alias 'Francis Bernard' when, in 1943, he escaped from France, over the Pyrenees and into Spain and North Africa, to England. He did this because he was known to the Germans and he was concerned for the safety of his wife and five children who were still in France. It was as Commandant Francis Bernard, not as Jean Accart, that the Royal Air Force knew him. Accart, who rose to the rank of general, wrote this story of D-Day in June 1944, while it was fresh in his mind; the English translation is his.

0430 hours on 6 June 1944, over the Channel. . . . The twelve aircraft fly in three columns in close formation so as not to lose contact in the night which is still dark. There is nothing to see; but there is no doubt about the forces that are on the move, for the headphones are alive with the calls of the squadrons which are already giving cover to the fleet or heading for the landing beaches.

A little way out from the English coast, the flashes of explosions fascinate the pilots and lead them to the target zone. As they approach, they can see the evanescent red flames spreading along the coast of Normandy, once so peaceful. The explosions of falling shells and bombs mingle with flashes from the firing of the German batteries, which fade one by one, as daybreak reveals the mightiest armada in history. In a few minutes thousands of ships of all shapes and sizes emerge from the shadow, lit only by a moonbeam which has pierced the clouds. The warships, moored broadside on as if on an exercise, silence the most aggressive gun positions. The landing craft are ready to go into action.

Against a gradually lightening sky, the fleeting shadows of the fighters become sharper as they sweep over their sector in stacked groups crowding in between the water and the clouds. We make out the powerful silhouettes of the Thunderbolts which pass above us, dipping a little to check our identity, and of the suspicious Lightnings which come and sniff at our tails. It is a miracle that all these squadrons can manoeuvre in so small an area without colliding.

And now day is born.... The flames are no longer so visible, but there are rising columns of smoke, and the coast is signposted with wisps of red and black. The heavy ships are still marked out by intermittent tongues of flame. The landing craft move towards the beaches while the land in front of them is torn by immense flaming caterpillars which follow the track of the last bombers.

It is H hour.... All at once a great mass of warlike machines of every sort move to attack Fortress Europe, and their fire overwhelms the batteries which have not yet been subdued. Will they all turn back in disorder after a few seconds ... after a few minutes? It is an agonizing moment. But the flooding tide, which rises ceaselessly, drowns the defence; the Atlantic Wall is shattered – definitely. The pilots return confidently to base buoyed up by the sight of the vast number of ships moving in from every side.

The initial impression is of highly organized might ... of calm and confident power ... of an overwhelming tide. This day of the great seaborne invasion is crowned by the amazing vision of the attacking airborne divisions.

In the early morning mist, and precisely at the appointed time and place, columns of towing planes and gliders appear and move onwards in a procession more than 45 miles long. As far as eye can see the lines of heavy aircraft and huge gliders fly low over the Channel, covered by swarms of fighters weaving over them like watchful sheepdogs. The French coast appears and becomes clear as the gliders pass over it precisely on time amid the puffs of a few bursts of flak.

One after another the gliders cast off, spiral down and land lightly on their designated field, assembling with masterly skill on a pocket handkerchief. Above them the fighters provide an impenetrable defence – and impenetrable it is, for the problem is to avoid collisions. Never before had we possessed such absolute domination of the skies.

Then we turned for home, still escorting the towing planes which, flying faster now without their charges, headed north only a few miles away from those still making their outward journey in this amazing Channel shuttle.

Our old Spit V Bs landed at nightfall after our fourth D-Day sortie.

On this invasion day our pilots lived intensely from dawn to dusk. They could hardly believe it to be true, so long had they waited for it. But it cost the Berry group, which completed 100 operational hours' flying, a heavy price. We had lost one of our dear members, Lieutenant Joubert des Ouches, who perished over the sea, a few miles from the French coast.

'I know my sacrifice will not be in vain,' he wrote before taking off on his last mission.

Let us hope, with him, that this D-Day, which we greeted with such solemn joy, may lead to a victory which shall be both early and fruitful for mankind.

Scott of 123 Wing

Desmond Scott, the New Zealander, was twenty-five and one of the youngest group captains in the Allied air forces, when he was given command of 123 Wing to lead into Europe. He was very forthright. He said what was in his mind and exactly as he saw it and he didn't mind too much who heard it. An engagingly earthy humour, coloured by the use of breath-taking similes, usually delivered him from trouble. He was a rock-solid competitor when the way was uphill. But he felt loss and hardship intensely for he was sensitive about his fellow-men – which is one reason why he can write as he does.

Scott was an unusually practical officer who was a recognized master of the art of air-to-ground attack and close support of the armies. His Wing of four rocket-firing Typhoon squadrons – 183, 198, 609 and 164 – was, perhaps, the most aggressively successful formation in a very nasty business. From Normandy, on through the Low Countries and into Germany, the enemy's armour suffered grievously before its arms. Its leader waited nearly forty years before he put his reflections down on paper. Time has not diminished his pungent convictions.

Desmond Scott, *Typhoon Pilot*, (Leo Cooper/ Secker & Warburg, 1982)

Much has been written about the thrusts, counterthrusts and controversies of those early days in Normandy. But as an airman, I must here speak up on behalf of all members of 123 Wing – indeed of every branch of the Allied Air Force. Their role has been consistently understated.

Back in 1942, the build-up of the Desert Air Force under Air Marshal Coningham had lifted General Montgomery above the level of his unfortunate predecessors and placed him on the pedestal of public popularity. The press headlines were always 'Monty Strikes Again', 'Monty's Victory', 'Monty's Army'. The Desert Air. Force, Tedder or Coningham were seldom mentioned.

Montgomery appeared to revel in this idolization and carried his opinions across the beaches and into the bomb-scarred fields of Normandy. His declaration – 'I would sooner use a thousand shells than lose one single life' – was well received by his troops, and in their eyes he became a god of war who could do no wrong.

But the view that aircraft and their pilots were expendable, mere weapons of convenience, rang a discordant note in the ears of all of those who flew....

Without air supremacy the invasion could never have been attempted, and undoubtedly our command of the air was the most important single factor in the success of the invasion. Without Coastal Command the seas would still have been alive with U-boats. Without Bomber Command and the Eighth Air Force the great industrial cities of Germany would still have been churning out weapons in ever-increasing quantities.

Our strategic bombing had had a profound psychological effect on Germany's civilian population. The Tactical Air Force, rocket-firing Typhoons and fighter bombers, maintained a successful interdiction of the

Seine bridges and ensured victory in the land battles before they had even begun, crippling the enemy's armour as it lay waiting in the hedgerows.

It was the day of the eagle, yet it was General Montgomery who accepted the accolade....

Field-Marshal Rommel summed up the situation in Normandy, when, on 12 June (D-Day +6) he sent this teleprinter signal to Field-Marshal Keitel, Chief of the Wehrmacht Supreme Command:

> The enemy is strengthening himself visibly on land, under cover of very strong aircraft formations. Our own operations are rendered extraordinarily difficult, and in part impossible to carry out. The enemy has complete command of the air over the battle zone, and up to about 100 kilometres behind the front, and cuts off by day almost all traffic on the roads or byways, or in open country. Troops and staffs have to hide by day, in areas which afford some cover.... Neither our flak, nor the Luftwaffe seem capable of putting a stop to this crippling and destructive operation of the enemy's aircraft.

One morning, when the Normandy sky was still tinged with red, our Typhoons roared off the dusty airstrip and swung low over the pockmarked landscape east of Bayeux.... No sooner had we crossed the railway to the northwest of Falaise than I caught sight of the object of our early-morning mission. The road was crammed with enemy vehicles – tanks, trucks, half-tracks, even horsedrawn wagons and ambulances, nose to tail, all pressing forward in a frantic bid to reach cover before the skies once more became alive with the winged death of the 2nd Tactical Air Force.

As I sped to the head of this mile-long column, hundreds of German troops began spilling out into the road to sprint for the open fields and hedgerows.... There was no escape. Typhoons were already attacking in deadly swoops at the other end of the column, and within seconds the whole stretch of road was bursting and blazing under streams of rocket and cannon fire. Ammunition wagons exploded like multi-coloured volcanoes. A large long-barrelled tank standing in a field just off the road was hit by a rocket and overturned into a ditch. Several teams of horses stampeded and careered wildly across the fields, dragging their broken wagons behind them. Others fell in tangled, kicking heaps, or were caught up in the fences and hedges.

It was an awesome sight: flames, smoke, burning rockets and showers of coloured tracer – an army in retreat, trapped and without air protection. The once proud ranks of Hitler's Third Reich were being massacred from the Normandy skies by the relentless and devastating fire power of our rocket-firing Typhoons.

During the first week in August the Germans prepared to launch a counter-attack in the area of Mortain, aiming to force their way to the coast at Avranches and cut off the Americans under General Patton – who had reached Le Mans.

It had been a bold move on the part of the enemy, but its defeat spelt the beginning of the end of the German occupation of Normandy. It also proved conclusively that major ground offensives can be defeated by the use of tactical air power alone. Von Kluge wrote after the war: 'The armoured operation was completely wrecked exclusively by the Allied Air Forces, supported by a highly trained ground wireless organization.' 123 Wing rocket-firing Typhoons alone attacked 87 tanks on the afternoon of 7 August.

The Falaise pocket – as it was to be known – became the chopping block and the graveyard of Hitler's Seventh Army. The divisions sent from the north to reinforce it were systematically torn to shreds . . . [as the] ... low-attack squadrons ripped into their armour, and the greatest German defeat since Stalingrad began....

For the Germans [Falaise] was a catastrophe. Many had earlier escaped the trap and fled for the Seine, but they left behind 1300 tanks, 1500 field guns, 20,000 vehicles and 400,000 troops, half of whom were taken prisoner....

Russian Reflections

While the ball was now rolling decisively the Allies' way in the West, on the Eastern front the Russians continued their stiff, even unyielding, attitude to renewed overtures for cooperation by the British. It was difficult, sometimes, to accept that the Soviet Union and her Western Allies were on the same side. The examples were there to see.

NIL RESPONSE

Air Commodore H.A. Probert, RAF (Retd), *Air Clues*, July 1981

...After ... [the earlier] ... sharp rebuffs no further attempts were made to base RAF squadrons in Russia, but as the Allies closed in on Germany in 1944 two situations arose in which the use of Soviet facilities was highly desirable.

The first was in August when the Polish Home Army rebelled in Warsaw and the Russians refused to help them, criticizing the revolt as the work of adventurers. Moreoever, while they could not object to Allied aircraft dropping arms in the Warsaw area, they would not agree to them landing afterwards in Soviet-controlled territory. RAF aid was therefore limited to 189 exceptionally hazardous sorties (many flown by Polish crews) from Mediterranean bases, involving 1750-mile journeys largely over enemy-held territory; 91 of the sorties were successful and 31 aircraft were lost.* When, in September, the Russians eventually agreed to cooperate it was too late.

Secondly, once the Russians were in occupation of eastern Poland, there would be great advantages if damaged bombers attacking Central and East German targets could make for Soviet controlled airfields, and the British Ambassador took this up with Molotov, the Soviet Foreign Minister, on 3 September. While Molotov agreed in principle, no assistance was offered on the matters that needed resolution if the arrangement were to work in practice, namely identification procedures, airfield locations, and night-flying facilities.

Repeated pleas elicited no response, with the result that throughout the final winter of the Bomber Offensive the emergency facilities that could have made so much difference to crippled aircraft were effectively denied. Only

*SAAF squadrons, flying within the RAF's 205 Group, itself commanded by the South African Brigadier J.T. Durrant, were also heavily engaged in the Warsaw operations. (Ed.)

when it no longer really mattered, on 7 March 1945, did the Russians offer three airfields for this purpose to the British and Americans. Though far from well situated, these were better than nothing, and 30 Mission tried to exploit this success by requesting sites for several Gee stations; the Russians were, however, easily able to stall on this until the need was past.

The record speaks for itself. Occasionally relations were good, especially when direct links could be established between fighting men, but this happened largely in the early stages, when the Russians needed assistance at almost any cost. As the war proceeded, the Russian attitude varied according to the political temperature of the moment and their latest estimate of their Allies' military fortunes, and the more they became confident of their ability to handle the Germans on their own the less were they disposed to cooperate. Their general attitude was to take advantage of whatever aid was on offer, provided no strings were attached, and to give virtually nothing in exchange; in particular there must be no contamination of the Soviet forces or of the nation at large. Consequently as few British personnel as possible must be allowed on Russian soil and they must be kept away from anywhere significant.

Moreover, direct contact with the air staffs with whom the Air Mission should have been able to do its business was for all practical purposes forbidden. The close military cooperation so obviously desirable in the interests of defeating the Germans was nearly always overriden by the political determination to keep the foreigner at arm's length, leaving few who had had close dealings with the Russians under any illusions about the future.

FRUSTRATED LIAISON

Group Captain P.H. Hugo, Victoria West, Cape Province, 1982

From 21 November 1944, to 13 December 1944, I was attached as Liaison Officer to the Allied Military Mission with Marshal Tolbukhin's 2nd Ukrainian Army. This army formed the left flank of the Russian forces, operating mainly south of the Danube. It was at that time advancing slowly from Bulgaria towards Hungary, with Army Headquarters in Sofia.

The weather was very bad and air activity was at a minimum. Neither the Russian nor the German air force carried out any tactical or strategic bombing. What bombing there was was confined to very short distance attacks mainly by Russian Stormoviks and German Ju 87s on local strong points and gun positions, these aircraft being used as mobile artillery. There was no sign of fighters being used as fighter escort, but during strikes against targets by the bombers, they gave area cover and at other times patrolled on their side of the lines or made very shallow incursions into enemy territory.

From what I could find out the air effort by the Russian air force was concentrated on the central and northern sectors of the front. The southern armies were expected to 'take it on the chin' and rely on artillery instead of the air.

The Russian aircraft mostly in evidence were Stormoviks for ground attack and various marks of Yaks, Migs and Laggs as fighters. For the use it was put to, the Stormovik appeared adequate; it was strong, heavily armoured and dependable and although the bomb-load was not heavy it could be accurately delivered.

The fighters were not impressive and could not be compared with their American, British or German counterparts. They appeared to be strongly built but of inferior design. I, personally, would have preferred to tackle the GAF in Hurricanes.

Occasionally Airacobras (P-39s) were seen, and these seemed to be popular with the Russian pilots. However, most Russian aircraft appeared dilapidated and it was no rarity to see instruments missing. The attitude appeared to be that provided the engine worked and the airframe could convey the pilot and armament that was all that was required. I never saw any of the Guards Air Regiments where the standards were reported to be much higher.

I saw no evidence of the use of radar or even ground-to-air radio control. Pilots were briefed on the ground and expected to operate in accordance with their briefing. I had very few opportunities of speaking with Russian pilots as it had to be done through an interpreter with a political commissar always listening in. So they were understandably cagey and non-committal. Their flying ability appeared average; but, judging from the amount of wrecked aircraft to be seen, the wastage due to flak, the weather and the GAF, was high. I even began to believe the astronomical claims by GAF fighter pilots!

A part of my duties was to signal possible targets to Air Force Headquarters in Italy for possible air attack by American fighter bombers. It was soon evident that Russian-nominated targets could not be accepted as they expected limited strongpoints and even individual gun positions, to be attacked by our aircraft flying up to 350 miles over enemy-occupied territory. Their target description and map coordinates, and even front-line positions, were hopelessly inaccurate. I therefore had to go and verify the targets requested. This caused such long delays that, by the time AHQ, Italy, received the information, it was too late to act on.

In general, the Russians were suspicious of us, mistrustful of our intentions, and prepared to go to any lengths to frustrate and obstruct us. They wanted a Russian team stationed at AHQ, Italy, to pass target information to the Allies, who would accept their requests without reservation. No cooperation that entailed any risk of American or British personnel obtaining any information about them was acceptable to them.

We were under constant surveillance – quite openly – and had to have Russian interpreters, who were, in turn, spied upon by political commissars, and needed special permission to go anywhere (for our own safety, they claimed). Life became quite unbearable.

Air Commodore Murray, my immediate superior, and I were stopped at a road blockade late one night on our way back to our villa. We were kept for two hours in freezing weather, while I was detained for the regulation two hours in the 2nd Army guard room on my way to work. At that point I decided I had had enough.

Air Commodore Murray and I returned to Italy where our explanations were not cordially received. I was then posted back to the UK to the Central Fighter Establishment, and what became of the rest of the Military Mission I have no idea.

CLOSED MAPS

Squadron Leader T. Bennett

More characteristic evidence of the Soviet Union's determination to keep any formal cooperation* with the Royal Air Force at arm's length came in September 1944, during the first of the three wonderfully determined attacks on the German battleship, *Tirpitz*, nestling menacingly up in the North Cape area of Norway. 617 Squadron, led each time by the redoubtable Willie Tait (Wing Commander J.B. Tait), who was in overall charge of the operations, and 9 Squadron, were the two units engaged.

'Ben' Bennett (Squadron Leader T. Bennett), navigation leader of 617 at the time and one of Bomber Command's most tested navigators, has recorded the detail of the initial mission.

The outward flight on 11 September, which lasted 10 hours and 35 minutes, was from Woodhall Spa, in Lincolnshire, via North Uist in the Shetlands, across Norway, Sweden and eventually to Yagodnik, '*a grass, Elementary Flying Training School airfield* [editor's italics]† some 20 miles southeast of Archangel.'

It was from Yagodnik that the attack was launched on 15 September by the force of Lancasters and it was to this airfield that the aircraft returned 7 hours and 15 minutes later.

'Willie, in the van, had a view of the *Tirpitz* superstructure (the ship was lying in Kaa Fjord, an offshoot of Alten Fjord) and his bomb-aimer obtained a hit on the ship that rendered her non-effective from the moment the 'Tallboy' (12,000-lb bomb) struck – as was admitted after the war when the German reports were studied.'

But the outward leg that day and particularly the run-in to the target:

'... presented some unforeseen difficulties. The RAF maps of the route were not accurate.... They did not present a true picture of the geographic features en route which made it difficult for such features to be used for accurate wind velocity calculations, although allowing the track to the target to be reasonably maintained. *The Russians refused all requests for accurate maps from their own sources* [editor's italics].'

On the flight back to Woodhall Spa the next day, the Lancasters were routed:

'across Russian and Finnish territory, through the Skagerrak and thence across the North Sea. The leg across Russia and Finland was flown at about 1000 feet *with no navigational assistance except the occasional pinpoints on major lakes and the Finnish coast* [editor's italics]. Thankfully, we were able to pick up the city of Stockholm in all its "lit up" beauty.... Our flying time was 9 hours 40 making an overall total of operational time for the whole round trip of some 27½ hours....'

*Russian aircrews were friendly enough; it was the presence of the political commissars which fouled things up.

†A grass training airfield was the best the Russians were prepared to offer! (Ed.)

Sadly, on the homeward flight, one of the crews of 617, captained by Flying Officer Frank Levy, a 27-year-old South African, flew into a mountain top at Nesbyen, in the Hallingdal Valley, in south-central Norway, some 180 kilometres northwest of Oslo. All the occupants of the Lancaster – there were ten of them* – were killed. The crash was not due to enemy action; a navigational misjudgement, or even a compass error, may have been the cause. The events that followed deserve to be recorded.

'On the morning following the crash [17 September] the [German] occupation troops forced the citizens of Nesbyen to bury the crew in a shallow mass grave close to the site of the accident. Soon after this a wooden cross mysteriously appeared by the grave. Ten nails were hammered into the cross, one for each of the crew members believed to be on board. Shortly after that a poem was nailed to the cross by one of the men who was first on the scene and had been deeply moved by this evidence that a foreign nation was still fighting for Norway's freedom. . . .'

The name of the poet, an inhabitant of Nesbyen, is Jorgen Syversrudengen. For the perceptive and touching English translation of the poem the editor is indebted to David Scott-Malden.

> A bleak grave on the hillside, marked out in land unbless'd;
> A plain unvarnished wooden cross; no names of those at rest;
> Only a set of rusty nails, driven deep into the cross,
> Tell how many died here and commemorate their loss.
>
> This was where it ended for you – the pain, the fear,
> Just one among so many of the tragedies of war;
> And in that hour no witness, no timely friend was there
> To hear your last words spoken, carry them to those who care.
>
> But now your mission's story, the glory and the pride,
> Is told to all your loved ones, unaware of how you died
> And through the pall of sorrow that lies on those who mourn,
> For sister, father, mother, the bright memories return.
>
> Rest in peace then, stranger, who for home and country's sake
> Must sleep in foreign soil, never to awake.
> The cold North wind shall whisper, over your lonely grave,
> A final benediction for the life you freely gave.

Thankful we must be that the spirit of the people of Nesbyen should find its expression in later, determined action.

'At the end of the war, roles became reversed. The Norwegians forced the Germans to re-bury the bodies in the community churchyard where the graves have been carefully looked after ever since.'

Sources: Squadron Leader T. Bennett, the 617 Squadron archives, The Defence Department, British Embassy, Oslo, and the Commonwealth War Graves Commission.

*Established by the Commonwealth War Graves Commission (Index No. Nor 47). (Ed.)

'THEY'LL SHOOT HIM, OF COURSE!'

John Iverach, Winnipeg, Manitoba, 1982

John Iverach had as many operational hours flying as any navigator in the Allied air forces. In four – repeat four – tours of flying duty his roles varied from 25-hour Catalina flights from the UK to Russia, conveying military missions or other alumni, to clandestine sorties in a captured Heinkel 115 picking up or depositing agents in enemy waters or territory, and four-engined Liberator missions to the Far East and beyond. 240, 353 and 232 Squadrons were his wartime home from home:

'Our return trip [from Archangel] got somewhat delayed. One of our passengers was the Russian, General Golikov, who was making his second trip to Britain with his military mission. The press referred to him as "General Golliwog" ... but later he was rated among the ten best generals of the world, mainly because of his success around Stalingrad.... I asked him one day through our interpreter what were his views of Christianity.... The interpreter, whose name as I recall it was Irene Vladimir, was reluctant to ask, but finally did. Golikov explained in a volley of Russian which, of course, none of us understood. When we finally wormed out of Irene his reply, the subject was dropped forthwith. "The General," she said hesitantly, "says that anyone who believes in God should be shot!"

We had flown for about an hour out over the White Sea, with our escort of three Russian fighters, when we spotted what looked like the same three fighters heading towards us, about 1000 feet above. As we watched, one of them peeled out of the formation and dived towards us. I was standing between the two pilots, looking over their shoulders. Bradshaw said "Looks like he wants to play!" "Play, hell!" I yelled. "He's heading right into the cockpit!" Instinctively I ducked.... There was a God-awful crash and the old Cat went careering down towards the sea. Just a few feet above the surface the two pilots, using all their combined strength, got it back under control. Sluggishly, gradually, they managed to turn back towards Archangel. The fighter, in the meantime, had also spun down out of control but had likewise recovered just above sea-level, and now General Golikov, rudely shaken, came storming forward, jabbering in Russian. The interpreter explained "The General says he has never seen such a terrible display of flying, and will see to it that you pilots are severely disciplined!" We couldn't convince the Russians that we had been rammed by one of their own fighters.

When we landed back at Archangel and inspected the damage we realized just how lucky we had been. The fighter ... had struck the top of our starboard wing, taking off all the aerial wires that ran between the tail and the wings, and tearing off about two-thirds of our tail-fin and rudder....

The General refused to believe that one of his own fighters was the culprit. Eventually, after Bradshaw had miraculously landed the damaged Cat, we persuaded the General to let us go to the nearby airfield and inspect the fighter. Sure enough, one of the first aircraft we looked at had a damaged port-wingtip and some telltale aerial wires still hanging on it.

We asked the interpreter what they would do to the pilot. "Why," she said in a rather surprised tone, "they'll shoot him, of course!"'

Pistol for Captivity

Johannes Wiese, correspondence with the editor, 15 September 1982, Kirchzarten, West Germany

Oberst Johannes Wiese succeeded Johannes Steinhoff (General Johannes Steinhoff) as the commander of Jagdgeschwader 77 in 1944. Awarded the Oak Leaves, few had had more experience of day fighter operations against the Soviet Air Force on the Eastern front.

'We did not worry ourselves unduly about the stories we heard about the Russians' treatment of prisoners of war.

The Russians whom I had shot down and who had then parachuted into German-occupied territory behaved variously when captured. One officer, I recall, was picked up not far from the airfield he had just bombed. He was unharmed and stood waiting in the open until our Volkswagen came along to pick him up.

When hailed – in his own language, and in a friendly way – he calmly put a pistol to his head and died instantly.

Russian propaganda had put it about that we were all torturers, castrators, and murderers.... As if we, who risked capture ourselves, would sign our death warrant by such behaviour!'

'A Sound Thinker'

Jim Goodson (Colonel James A. Goodson), former American Eagle Squadron member, became one of the US Eighth Fighter Command's exceptional leaders. Pete Peterson (Major General Chesley G. Peterson), commander of the Fourth Fighter Group and a discerning judge, has stamped his authority upon him. 'Jim was, to my mind, one of my finest pilots and leaders, especially in the long-range escort business, both on P-47s and P-51s. He is a great ace – 28 kills or so I believe – but, above all, he is a *sound thinker....*'*

Vern Haugland in *The Eagles' War* (Jason Aronson, New York, 1982) adds to the detail:

'Goodson eventually was to become the premier Eagle ace of the 4th Group, outscored in that elite organization only by two men who had not served in the Eagle Squadrons, John T. Godfrey and Ralph K. Hofer. Blakeslee in 1981 had occasion to write to Eagle Squadron Association President, E.D. "Jesse" Taylor, in support of Goodson's belated application for ESA membership: "It would be an honour for the Association to have a member of Jim's calibre. If I sound biased, it's because I am. Jim saved my bacon dramatically."'

Now Goodson has himself gone on the record (*Tumult in the Clouds*, William Kimber, 1983) with his own experienced observations on three topics:

*General Peterson – correspondence with the editor, 22 September 1982.

EASTERN VERSUS WESTERN VALUES

'... The Luftwaffe made a very clear distinction between the Russian front and the Western front. "Macky" Steinhoff* said of fighting the Russians: "It was like shooting ducks." It was JG 26, first under Galland, then under Schoefel and then "Pips" Priller, that defended the Western front from the beginning to the end of the war, and even the aces of the Russian front rated Galland's 109 victories and Priller's 101 in the same category as "Bubi" Hartmann's 352 and Barkhorn's 275 on the Russian front.'

AIR SUPERIORITY OVER THE SOVIET UNION

USAF's Historic Russian Shuttle

'The generals of the Eighth Air Force needed all the understanding they could muster in backing Blakeslee,† but he never let them down... Their real genius was in encouraging their wing, group and squadron leaders to take the initiative and play it as they saw it. The German fighter leaders had brilliance, experience and incredible bravery, but they were defeated by the lack of understanding and backing of Göring and Hitler.

Probably the event that underlined the final, complete air superiority of the long-range fighter, and represented the peak of Blakeslee's combat career was the England–Russia–Italy shuttle. The plan was [for the Fourth] to escort bombers‡ from England 1600 miles across Europe to Poltava in Russia.... [This] would be the longest leg, but it should only take 7½ hours, and we had already been airborne (in the Mustang (P-51)) for 8 hours before.... Estimated time of arrival was 7.35 p.m. At 7.35 exactly the Russian flares floated up from the base. Blakeslee had hit it on the nose!

From Russia, the bombers would take off for Italy, bombing an oil refinery in southern Poland on the way, escorted by the Fourth as far as the Yugoslav coast where Mustangs from the Fifteenth Army Air Force, based in Italy, could take over. From Italy, the bombers and their escort would fly back to England, bombing railroad yards in France en route....

But if the Eighth Army Air Force had shown the Germans that they had air superiority over Germany, the Luftwaffe showed [at Poltava] that they [the US Eighth] didn't have it over Russia.... On the second night, Luftwaffe bombers attacked Poltava where most of the B-17s had landed. *Although it was never reported, 43 bombers were destroyed. A few Ju 88s bombed and strafed the bases where the Mustangs were parked, and put 15 of them out of action*§ [editor's italics].'

*General Johannes Steinhoff.

†'Don Blakeslee ... the greatest fighter leader of the war.' General Peterson in correspondence with the editor, 22 September 1982.

‡There were to be 104 B-17s on the mission. Concurrently, the USAAF would be mounting a 1000-bomber raid on Berlin. Of the Russian shuttle, Blakeslee said at the briefing: 'This whole thing is for show.'

§So much for the quality of the Russians' airfield defence, and for the effectiveness of their radar screen. On the other hand, the feeling was abroad that the Soviet Union was deliberately indifferent to the safety of the American aircraft, thus providing a further example of their 'arm's length' cooperation with the Westen Allies. (Ed.)

The Eighth Air Force's daylight offensive versus (by implication) the onslaught of the Royal Air Force's Bomber Command by night

'The big Berlin (daylight) raids had started for us in early 1944, and in a little over a month, we'd lost Freeburger, Brandenburg, Fiedler, Saunders, Richards, Lehman, Hobert, Davis, Hustwit, Barnes, Goetz, Vilhinger, Herter, Skilton, Siefert and a few others I can't remember. Looking at the names now, they seem bare as skeletons, but to us, each one was a warm, laughing, close "Ol' buddy", so much alive we couldn't believe he wasn't. We were lucky. There weren't bodies or funerals, they just weren't around any more.

Maybe that's why we were still willing and eager to climb into the cockpit ... take off by twos, wheel into formation and climb on out on the now familiar trail: out over Orfordness, over the North Sea, crossing into Holland over the Hook, dodging the flak on the coast, moving past the flat Dutch countryside, stretching out below, with the sun flashing on the many glass greenhouses, then through the flak of the Ruhr, out over the open, green fields of Westphalia, past the reliable checkpoint, the perfect, round Dümmer Lake, and on past Hanover, Braunschweig, Magdeburg, and finally the vast stretch of built-up area, "Big B" itself.

I don't know that we realized it at the time, but the Luftwaffe fighters were giving their all to defend the cities of their homeland. After that summer of 1944, the Luftwaffe would never be the same. As 1940 had been the Battle of Britain, 1944 was the Battle of Germany. Galland, who was then still General of the German Fighter Command, wrote in a report to the German Air Ministry: "Between January and April 1944, our daytime fighters lost over 1000 pilots. They included our best Staffel, Gruppe and Geschwader commanders. Each incursion of the enemy is costing us some fifty aircrew. The time has come when our weapon is in sight of collapse."

Albert Speer, [Hitler's] Minister for Armaments and War Production, agreed that this was the period during which Germany lost the war. In his interrogation under oath, (18 July, 1945) he said: "The Allied air attacks remained without decisive success until early 1944.... *The Americans' attacks which followed a definite system of assault on industrial targets, were by far the most dangerous. It was, in fact, these attacks which caused the breakdown of the German armaments industry. The attacks on the chemical industry would have sufficed, without the impact of purely military events, to render Germany defenceless* [editor's italics]."'

America's Belief

Laddie Lucas, *Five Up: A Chronicle of Five Lives* (Sidgwick & Jackson, 1978)

... We were seeing, almost daily, evidence of the gathering strength of the daylight attacks of the US Eighth Army Air Force as they began to penetrate farther and farther into the heart of the German Fatherland....

Apart from seeing it all at first hand from the limited escorting or diversionary missions which the wing was able to fly in support, there was the additional proof of shot-up Thunderbolts (P-47s) and ... damaged Fortresses (B-17s) and Liberators (B-24s), which, at their last gasp, had found sanctuary in the haven of Coltishall as they lobbed down after struggling back across the sea. Each example bore testimony to the extent and severity of the combat which all were now having to endure....

The US Eighth Fighter and Bomber Commands in late 1943 and 1944 performed the impossible. They did what we, in the Royal Air Force, had told them they could never do – what we, with our equipment, could never have done. What's more, the Luftwaffe generals – and indeed, the colonels who were actively mixing it in combat with the Fortresses and Liberators – had advised Hitler and Göring that it couldn't be done.

They flew, in broad daylight, escorted bomber missions into every sector of Germany. There wasn't a target in the Third Reich – even in the most eastern part of the country – which was safe from their attentions. And all this was accomplished in the teeth of the strongest and most lethally-equipped defensive fighter force which had ever been assembled....

To the British mind the extent of the achievement remains, even with the passage of time, incomprehensible. It stands as a paragon of America's belief in the art of the possible. And of her refreshing conviction that if convention says no, the answer most probably is yes.

The German Reaction

Johannes Steinhoff, *The Straits of Messina* (Andre Deutsch, 1971)

In my mind's eye I clearly see the faces of the young men for whom I was responsible. I see them as they were when they reported for the first time 'at the front', self-confident and psychologically prepared for the battle (of which they had only a hazy conception) by slogans such as 'the final victory will be ours,' or 'the German soldier does not cede a foot of ground.'

They soon lost the superficial veneer acquired through ideological indoctrination in the Hitler Youth. The jargon of the Third Reich disappeared from their vocabulary and they became human beings who, out of self-preservation, drew ever closer to one another.

Once the newcomers had known what it was to lie in a slit trench during an air raid and to fly a sortie against the Flying Fortresses, they quickly adapted themselves to the flippant style of the old hands, those who, with scathing irony, made light of everything (though inherently they had become mature and serious-minded men who now lived only from one day to the next). During those days I can remember few cases in which our communal life was upset by anyone's insufferable behaviour, or lack of adaptability....

Time Expired

There was no more decisive difference between the Luftwaffe and the Allied Air Forces than in their respective attitudes to the time aircrew should spend on operational duty. Under the Allies' rules, an operational tour for bomber crews rested upon the attainment of a specified number of missions; in the Royal Air Force, the figure was normally thirty. This was generally followed by a rest of six months.

With fighters, the period took account both of time and the hazards and stresses of a particular theatre. According to the circumstances, a fighter pilot's tour might run for a year or eighteen months – or significantly less if the going was very rough. It was the responsibility of a commander – squadron, wing or station commander, as the case might be – to determine when a man had had enough. It was often a finely balanced decision which demanded experience and judgement.

The Luftwaffe High Command's approach was altogether harder, dictated to some extent, as the war went on, by the Germans' chronic shortage of experienced operational pilots. Two of the Luftwaffe's well-tried and successful commanders have put the position straight. First, Wolfgang Falck, one of the German Air Force's foremost night-fighter leaders, a man of feeling and understanding:

Wolfgang Falck, Tirol, Austria, 1982

'We did not have a special "tour" system, neither did we have limitations or programmes relating to the number of missions flown. We had to fly until we were so worn out that our superior would recognize that we were no longer fit to operate. The physical, as well as the psychological, strength of pilots differed greatly....

Depending upon their mentality, I would send pilots off home, or on vacation, or skiing etc., for short spells. Then they would return.

Sometimes I had my problems with pilots who did not want to go on vacation, fearing that their score of victories would be passed by their comrades. This called for firmness.

We also had cases where a whole Gruppe was withdrawn for a time, usually when there had been heavy losses and it was necessary to restore morale. New personnel and equipment could then be introduced while the older pilots could take a vacation....'

This view is substantiated by Oberst Johannes Wiese, who also knew the pressures.

Johannes Wiese, Tirol, Austria, September 1982

'In striking contrast with British and American pilots, there was never, at any period, a fixed tour of duty in our Service. Whoever was fortunate enough to be trained as a pilot and sent to an operational staffel just flew and flew and flew until he was killed – or was lucky enough to survive.

In a very few and exceptional cases, more senior officers were transferred from flying duties to positions of responsibility elsewhere. Among the earliest of these was Werner Mölders, and then, later, Adolf Galland, Gordon M. Gollob, Adolf Dickfield and others.

"Worn-out" pilots – that is to say those who needed a rest on physical or psychological grounds were occasionally (but never in any particular rota) sent to the aircrew recuperation centre at Tegernsee for a short period of good food and care. But then it was back again to the grind...

The Luftwaffe's command system demanded the fullest dedication from every individual. How little this was appreciated by the Reichsmarschall (Hermann Göring), and how frequently he sought to denigrate it, is now well known. . . .'

Nightrider

Oberstleutnant Georg Hermann Greiner, Hintergarten/Schwarzwald, West Germany, 1982

The important advances which had been made – on both sides – with airborne radar in the last year or so of the war brought a new precision to night fighting which the earlier exponents of the art could hardly have contemplated.

Oberstleutnant Georg Hermann Greiner, who knew his way better than most about the night sky over Germany and the occupied countries, was to keep sharp in his mind one of his first experiences with SN2, Germany's sophisticated and fatally effective airborne homing device:

'I was positioned over Leeuwarden [northern Holland] by the [ground] controller and then given a vector towards an enemy aircraft. At the same time Feldwebel Kissing, my radar operator, had already picked up the target and was giving me the distances from it, measured by SN2. These reduced with such rapidity that the logical thought – that the two aircraft were approaching head-on – did not even occur to me.

Then, suddenly, a Mosquito shot like a ghost past my Me 110G. I could pick out the outline of the pilot's head and shoulders quite clearly, and I feel sure he must have seen me, too.

After I had recovered from the shock (felt, I suspect, on both sides), I was left with the feeling of incredulity – amazement – at the extraordinary precision of this device. . . . That, and the regret that I hadn't had time to salute.'

Brilliant Aristocrat

Alexander Brosch, Neu-Ulm (Burlafingen), West Germany, 1982

'Interestingly, the leading German night fighters never caught the public's imagination as the day fighters did. This was probably because the night fighters had their greatest moments when German armies were already in retreat, and war was coming ever closer home. . . . The time for victories was past.'

There was, however, a winning young German aristocrat who made an enduring mark with the Luftwaffe in the night skies over Germany, occupied Europe and the Eastern front and who did, indeed, capture the public's gaze. Heinrich Prinz zu Sayn Wittgenstein Sayn's story is conspicuous for being far removed from the general run. Here, based on his research of family papers and other available material, Alexander Brosch tells of this officer's brilliant and unremitting service.

The editor warmly acknowledges Herr Brosch's notable contribution to the record, just as he does the original suggestion of an old friend, Carl August Freiherr von Thuna, who stimulated this account and prompted Heinrich Sayn Wittgenstein's nephew, Alexander Fürst zu Sayn Wittgenstein Sayn, to encourage its preparation. The editor expresses thanks to Alexander Sayn-Wittgenstein for this help.

'Heinrich Prinz zu Sayn Wittgenstein Sayn's background and upbringing were more compatible with a career in the army than the air force. He was born in Copenhagen on 14 August 1916, the second of three sons of Gustav Alexander Prinz, the German diplomat, and his wife, (née) Baroness von Friesen. When, in 1919, his father left the foreign service, he took his family to Switzerland.

'After school at Neubeuern, in Upper Bavaria, and then at Breisgau, Heinrich Sayn Wittgenstein arrived at Freiburg, became a member of the Hitler Youth and was made a team leader in 1935. He was also active with his local National Service unit and, in April 1936, he reported for duty with the 17th Cavalry regiment in Bamberg where, ten years earlier, the future would-be assassin of Hitler, Count von Stauffenberg, had also entered the army.

'It was while he was serving with his regiment that the Prince decided to volunteer for the Luftwaffe. He was accepted by the Brunswick flying school in October 1937, and commissioned nine months later. After finishing his training, he was posted to various air force stations and his career advanced quickly.

'Starting as an observer, Sayn Wittgenstein soon remustered as a bomber pilot and, after taking part in the drive through the Low Countries and France in May and June 1940, he then became engaged in the attacks on England. As a Heinkel 111 pilot, he flew in the raids on Biggin Hill and other airfields in the southeast.

'With Hitler's attack on Russia, the Prince was moved in June 1941, to the Eastern front where he operated until November with army group "North", flying some 150 sorties against railway installations, transport columns, harbour facilities and shipping. He spent this period, first, in Kampfgeschwader 1 and then in 51.

'After eighteen months of bomber operations, Sayn Wittgenstein decided to throw in his lot with the night fighters. His mother held that he took this decision because he "recognized that bombing the civil population would cause too many casualties". It was his view that: "Night fighting is the most rewarding role of all, and also the most difficult. There I have everything, the flying, the stalking of the quarry and the attack."

'Four months after his transfer to Nachtjagdgeschwader 2, he was leading his Staffel on the night of 8 and 9 May 1942, when he scored his first success on the Western front. His score mounted rapidly in the operations over Holland and Germany as the Allies' night offensive built up. After his 22nd combat victory he was awarded the Knight's Cross of the Iron Cross.

'Promotion accompanied success. He became Kommandeur of IV Gruppe in Nachtjagdgeschwader 5 and, later, of 1 Gruppe in NJG 100; but the strain of the unrelenting struggle had begun to take its toll and in February and March 1943, he was in hospital being treated for a stomach ailment. By April, he was back again on operations – at Insterburg, in East Prussia – tackling the Soviets' night attacks on the Eastern front. After a couple of months in the East, there was a short sojourn in southwest France, protecting the U-boat bases on the Atlantic coast, before being moved north to Holland.

'Sayn Wittgenstein's stay in the West was short-lived and in July and August 1943, he was back again on the Eastern front where, on 25 July, he gained his highest score – 7 victories in a single night.

'After that, it was the Western front until the end. As the Kommandeur of 11 Gruppe of NJG 3, the Prince was based in Schleswig from August to November and, in September 1943, after gaining his 59th kill at night, he was called to the Führer's headquarters to receive the Knight's Cross with Oak Leaves.

'This extraordinary operational run found its reward in January 1944 when he was promoted Kommodore of the elite NJG 2 at Deelen, in Holland. Here, he and his squadrons faced the full weight of the Royal Air Force's massive night onslaught on Germany. In what was to be the last letter he would write to his mother, Heinrich Sayn Wittgenstein left no doubt about his continuing spirit. "This is really good hunting country," he wrote on 15 January.

'He had never had a time like it. It was by far his most successful spell since he turned his back on bombing and took up the chase at night. Between 1 and 21 January, in three peak weeks, he shot down 15 enemy aircraft and brought his grand total up to 83 – higher than any other German night fighter had reached at that point in the war, and 2 in front of his great rival, Lent, who had 81.

'The end came on the night of 21 January. Ostheimer, Sayn Wittgenstein's radio operator in the Junkers 88 on that last sortie, survived to provide the detail. "We were in position to attack an aircraft, and the pilot was about to fire, when there was a terrible flash and explosion in the aeroplane. The port wing immediately caught fire and the machine started diving steeply. When I saw the canopy fly off, I knew I must get out. I pulled off my oxygen mask and helmet and was thrown out by the centrifugal force; my parachute opened and fifteen minutes or so later I landed near Schonhausen.

'"As far as I could make out, we seemed to have been hit from below; but I've no idea whether it was from our own flak or fire from an enemy fighter."

'Prince Heinrich was found dead in a wood near Stendal.

'Hitler and Göring both sent messages to the family and the loss was announced in the official communiqué. Sayn Wittgenstein received the posthumous award of the Oak Leaves with Swords and was buried in the Air Force cemetery at Deelen, not far from his Geschwader's headquarters.

'In 1948, the remains were transferred to Ysselsteyn, in Holland, the resting place of more than 30,000 German soldiers.'

Ghoulies and Ghosties

The Germans introduced a new dimension in aerial warfare in Europe with the advent, in late 1944 and early 1945, of the Messerschmitt 262 and 163. The introduction of these two aircraft put Germany well ahead of the Allies in the jet and rocket propulsion stakes. The editor, who was commanding a Mosquito squadron in France at the time, had a first-hand taste of it.

Returning one dark, moonless night from a patrol up and down the Berlin–Lübeck autobahn and railway, he and his navigator saw a tiny, ominous-looking light swinging in on the port quarter in a very fast curving attack from 7 o'clock on the same level. It could only have been a jet; no propeller-driven aeroplane would have been going at that speed. The little glow in the sky was from the burners.

After taking suitable evasive action to ensure a safe return to base, the crew reported to Intelligence: 'The enemy is using the Me 262 at night.'

'He's not,' they said. 'In daylight, yes; but not at night.'

2 Group confirmed the contradiction; so did Headquarters, 2nd Tactical Air Force. 'Rubbish,' said the crew.

But now we know very well that they were operating this beautiful aircraft at night. Oberstleutnant Hans-Joachim Jabs, a foremost exponent of the nocturnal art, whose NJG 1 was operating the Heinkel 219, the Me 110 G and the Ju 88 in the four Gruppen, has offered an interesting postscript.

Oberstleutnant Hans-Joachim Jabs, Ludenscheid, West Germany, 1982

'A staffel of Me 262s, under the command of Oberleutnant Kurt Welter, was being *operated at night from the Lübeck–Hamburg autobahn, near Rheinfeld* [editor's italics].... The usual bases were too much frequented and receiving too much attention for this kind of activity. So the autobahn was the answer.

Friends of my family – my wife and two sons had just moved from Magdeburg to Rheinfeld – owned a large property beside the Hamburg–Lübeck motorway. This included a farm. Oberleutnant Welter had his operations room in the farm buildings and the officers and the pilots were also billeted there.

The 262s were stationed in the covered stands beside the motorway. Take offs and landings were effected with relatively simple technical and navigational aids. The aircraft were only flown operationally at night.

Oberleutnant Welter was very successful and was awarded the Oak Leaves in 1945. Unfortunately, he died in an accident after the war.

This is the story of the night-fighter jets – Operation (or Kommando) Welter – at the air base, Autobahn Rheinfeld.'

The Heavens Were a Different Place

Lieutenant-General H.J. Martin and Colonel Neil D. Orpen, *Eagles Victorious* (Purnell, Cape Town). Reproduced by courtesy of the War Histories Advisory Committee

By the late summer of 1944, the Me 262s under the skilled lead of Germany's Major Walter Nowotny, based in the Munich area, were already proving to be a fierce thorn in the side of the Allies' high-flying reconnaissance aircraft

from San Severo in Italy. Hitherto, the Mosquitoes, operating alone and unarmed over southern Germany, had had things much their own way; but no longer.

In August, 60 Squadron of the South African Air Force, in the names of Captain Pienaar and Lieutenant Lockart-Ross, had fought a dogged and determined battle with the jet. Now, as the old year was dying, another experienced crew from the same squadron, Captain C.H.H. Barry, with Lieutenant G.R. Jefferys as navigator, escorted this time by P-51 Mustangs, got a taste of the phenomenal ascendant speed of Willie Messerschmitt's new prize.

'... Over Augsburg ... they sighted the vapour trails of an unidentified aircraft.... They all set off after it, but it lost height, gained speed and easily evaded them. About fourteen minutes later, as the Mosquito was approaching Gunzburg–Leipheim airfield at about 7300 metres, one of the Mustangs sent out a warning of an approaching aircraft.... It was closing on [Barry] from dead astern.

Coming in at high speed, the enemy opened fire at 900 metres. Barry put the Mosquito's nose down violently. Still firing, but without scoring any hits, the grey Me 262 overshot and the escorting Mustangs tried to turn into it. They could not complete more than a 90-degree turn before the jet flashed past.

After attempting to complete his photographic run over the airfield, Barry ... was warned by the escort leader: 'Break right!' ... [The Me 262], firing all the time, began to turn into the formation.... [It] overshot at a true airspeed of about 800–880 km/h....

As [it] went into a gradual climbing turn to starboard, the fighter escort and the Mosquito chased after it, but were outpaced....'

All at once the heavens were a different place.

The Real-Life Dreams of a Photo-Reconnaissance Pilot

Air Vice-Marshal F.L. Dodd, Tarland, Aboyne, Aberdeenshire, 1982

Squadron Leader Frank Dodd (Air Vice-Marshal F.L. Dodd) brought an advanced aeronautical knowledge to his work as a young PR pilot of No. 544 (PRU) Squadron in the last years of the war. His own skills were well complemented by those of his navigator, Flight Sergeant Eric Hill. Between them, they made up one of PRU's most effective pairs, and their results showed it.

Dodd, now the administrator of the MacRobert Trust – embracing the lovely estates of Douneside and Alastrean on Deeside (see pages 245–6) – writes fictionally about his PR role in wartime; but this cannot disguise the authority which experience brought him.

'The company secretary settled comfortably in his window seat of the Jumbo Jet, winging himself and his wife to the Mediterranean. The company meetings were over, the minutes despatched and now he could enjoy the week with a clear conscience.

"There goes the Seine down there on the left," he murmured.... The lunch had been good and he was soon well away. Or was he?

It seemed real enough.... The Seine was certainly there; but in place of the two jet engines in the port wing, two enormous paddle-blade props were now threatening to saw through the wooden cockpit.... He was back in his PR Mossie* heading south across France with the diplomatic bag in the nose for Churchill's meeting with Roosevelt in Athens on Boxing Day, 1944. It was time for a top-of-climb, 30,000 feet, track check.

"Eric," he called to his navigator, "the Seine's down there to the left. Is that where the Seine should be? Are we on track?"

"ShSpain?" The retort was unexpected. "ShSpain should be over to the right a bit." Silence followed.

Ah, well, the sergeants' mess had had one tremendous party last night.... Better let him sleep it off.... after all, we had been caught on the hop – bottom of the "ops" roster, and not expecting to be on for a day or two over Christmas....

A glance, now, at the navigator brought instant panic. He was slumped unconscious in his seat. His finger nails were black – a sure-fire symptom of oxygen lack, or anoxia, as the doctors had it. It was the insidious killer, unlike the painful, but normally non-fatal "bends" which resulted from flying for long periods in unpressurized aircraft.

The oxygen supply was now turned up to emergency – but no response. What to do? Return to base in the hope he might be revived? Press on at lower altitude and risk running out of fuel? The checks were gone through again. Then discovery! And blessed relief! The cause was found. An oxygen tube connection had come adrift....

Recovery was far quicker than there was a right to expect. Astonishingly, Eric sat to attention in his seat and, in the confines of a Mosquito cockpit, ripped off one of the smartest salutes and cheeriest "Good morning, sir," it had been his pilot's lot to receive in twelve months of crewing together.

Eric soon appreciated the true situation and the flight resumed a more normal pattern.

"Good morning, sir, tea or coffee?" The stewardess's friendly, smiling face disturbed the reverie.... The dream had been too near the truth for comfort. "Bends", cockpit misting up, oil film over the camera lenses, interceptions in trails and out of them, unsettling bursts of ack-ack ... these were all part of a normal day's ops.... But anoxia – this was quite different. How many lone PR Spitfire pilots, posted "missing believed killed", might have succumbed to its insidious effects?

Now the Alps, with their early covering of winter's snow, slid by on the left, to fan another flash-back to July 1944, and the reconnaissance of the fjords of northern Norway, hunting for the *Tirpitz*.... There had been 10/10ths cloud all the way up from Narvik, radio silence and no radar aids – only D/R† navigation – as the crew groped a way through to the peaks of the Lofoten Islands.

*Photo-reconnaissance Mosquito.
†Dead reckoning navigation.

Unforecast winds could – and did – negative the D/R plan, anyway. But luck prevailed as the aircraft broke cloud in a Lofoten valley, with the cloud-topped mountains on either side. Fortune smiled again forty-five minutes later, when, following a low approach up Alten Fjord, the Tirpitz *loomed up, its lair revealed and duly recorded.... How lucky, indeed, could one be!* [editor's italics]...

The memories came racing back.... PR had always been an interesting role. It could be exciting – as witness that engagement (it was one of the very first) with a pair of very persistent Me 262s over Magdeburg, and the Mossie's ability to out-turn them – and it could be frustrating and yet wonderfully satisfying.

There was the great sense of one-upmanship in operating over the heartland of Germany in a clear blue sky, covering all targets, evading interception, and on the same day seeing the rushed PI photos at Ewelme and learning of the intelligence value of the target coverage.

This was the compensation for the flights – fortunately relatively few – when everything appeared to have gone well only to hear the fateful words of the photographers as they opened the camera hatches on return.

"Lenses misted up with oil."'

The Inequality of It

Photographic reconnaissance was now suffering ill-afforded casualties. After Wing Commander Lord Malcolm Douglas-Hamilton had led 540 (PRU) Squadron during 1943 and 1944 and survived thirty-three deep-penetration sorties ('... throughout,' said his DFC citation, '[this officer] demonstrated to his subordinates his willingness to undertake the most difficult and dangerous missions...'), the next, and youngest, of the four brothers, Squadron Leader Lord David Douglas-Hamilton, fell a victim of the enemy on 2 August 1944.

A flight commander of 540 Squadron, David was nursing his damaged Mosquito back to Benson, PRU's base in Oxfordshire, on one faltering engine. The aircraft had been hard hit by flak during a long excursion into enemy territory and the navigator was critically wounded. As they approached the circuit of the airfield the remaining engine quit and the aircraft plunged to earth, killing pilot and navigator.

David Douglas-Hamilton died undecorated. His was as good an example as any of the inequality which existed in the dispensation of awards for gallantry, for so much was owed to luck.

He had led 603 Squadron with signal spirit and success through the height of the battle for Malta in 1942. That alone should have ensured recognition, but some brush with Keith Park, the AOC, right at the end, almost certainly denied him an award. Park was the best tactical commander of the air war, but there were times when he allowed personal prejudice to taint his judgement. He had shown it before in the Battle of Britain with another similarly deserving squadron commander who had rubbed him up. It was a weakness in the New Zealander's otherwise formidable armoury.

Taking together Malta and his thirty-one daylight photographic missions

– all unescorted and unarmed, and many flown deep into Germany at the most hazardous period of the air war – none could deny that Douglas-Hamilton's record was undoubted. Men received DSOs for less. But life will always be unequal.

Tony Holland (Flight Lieutenant A.C.W. Holland), himself a twice-decorated officer, and a comrade-at-arms of David's in Malta in 1942, knew him and his deeds, as well as anyone.

A.R.W. Holland, Head's Nook, Carlisle, Cumbria, 1982

'David, as the leader of 603, was a calm pillar of strength at the head of the squadron.... With more understanding of humanity than most, he combined enthusiasm with an unrivalled patriotism and a thorough knowledge of, and flair for, the job at hand.... His concern for those serving with him was paramount.

If conditions were favourable in combat, he never hesitated to attack, and he had the knack of positioning his formation so that all might have an opportunity to fire. On 10 May 1942, during a 35-minute sortie, he led me and others on a hectic scramble,* intercepting a massive Luftwaffe attack on Grand Harbour. We arrived over Valletta at about 18,000 feet as the first wave of Stukas, and their escort, were starting their dive and the fierce harbour barrage was opening up. David led us straight through the barrage, giving everyone a chance to score hits on the enemy....

In thinking of him now, I recall some verse sent to my father by the Chaplain of RAF Station, St Eval, in Cornwall, when my own brother had been killed flying on operations with 233 Squadron of Coastal Command.

Remember me when I am gone away,
Gone far away into the silent land;
When you can no more hold me by the hand,
Nor I, half turn to go, yet turning stay.

Remember me when, no more day by day,
You tell me of our future that you planned:
Only remember me; you understand
It will be late to counsel then or pray.

Yet, if you should forget me for a while
And afterwards remember, do not grieve:
For if the darkness and corruption leave
Some vestige of the thoughts that once I had,
Better by far you should forget and smile
Than that you should remember and be sad.

Remember
Christina Rossetti

*Interception.

Expression

Desmond Scott, *Typhoon Pilot* (Leo Cooper/ Secker & Warburg, 1982)

'Unlike many officers of air rank, I believe Air Marshal Keith Park would have made a fine bishop.'

'... Gliders, losing a wing, would spin down like sycamore seeds. Altogether 11,466 aircraft were engaged in the Rhine crossing, of which 108 were shot down. But the doorway to the Third Reich was thrust open and burning at its hinges.'

Colonel Gregory 'Pappy' Boyington, USMC (Retd), *Baa Baa Black Sheep* (Wilson Press Inc., USA, 1958)

'... If this story were to have a moral, then I would say: "Just name a hero and I'll prove he is a bum."'

Pacific Mirror

And since you know you cannot see yourself
So well as by reflection, I, your glass,
Will modestly discover to yourself
That of yourself which you yet know not of

Shakespeare, *Julius Caesar*: Act 1, Scene ii

Throughout 1944, the United States' Navy and Marine Corps waged the Pacific war with a ruthless resolution and sweep which few who weren't there can possibly comprehend. Outside America, Australia and New Zealand, the extent of the operations, in human terms, is, even now, barely understood.

When Lieut.-Commander F.H.E. Hopkins (Admiral Sir Frank Hopkins), then commanding 830 Squadron of the Fleet Air Arm in Malta, was posted to the Naval Staff in Washington in July 1942, he had been on operations continuously since May 1940. He had had his full share of the fighting. And knew the form.

After a spell in the US, Hopkins was sent to the Pacific as the British Naval Air Observer. There, he was embarked in various USN carriers and saw, at uncomfortably close range, the great sea actions as they unfolded. It was his task to send back to Admiralty, in London, eye-witness accounts of the engagements supported by the detail of the conduct of the carrier operations and the control of the astonishing US air activity. The purpose was to ensure that the British Carrier Force, which was then preparing to join the Americans in the Pacific, would know what would be expected of it when it arrived. No British officer obtained a more intimate insight into all these affairs.

Here, in a specially written, factual and deliberately low-key appreciation, Sir Frank provides an acute picture of the magnitude of the wonderful United States' effort. Americans may be surprised to see mirrored for them this impartial and authoritative British assessment of their Navy's maritime capability.

Admiral Sir Frank Hopkins, Kingswear, Dartmouth, Devon, 1982

'The most remarkable thing about the Pacific war was the huge scale on which everything concerned with it was conducted.... Very great distances, many hundreds of ships, thousands of aircraft and astronomical losses by the Japanese.

For example, during the United States' assault and capture of the Marianas and Saipan in June 1944, the Japanese deployed 9 aircraft carriers with about 500 aeroplanes embarked. The Americans had 15 large carriers, with 5 small ones in support, and nearly 2000 aircraft. They also had 600 other ships.

In the operations, the Japanese lost 3 carriers, and 3 were seriously damaged and put out of action, together with 465 aircraft. Their admiral reported the day after the battle that he had only 35 aircraft left out of his original 500.

The United States lost 23 aircraft, shot down, while 4 ships sustained slight damage. The US Navy flyers christened the day the "Mariana Turkey Shoot".

The battle of Leyte Gulf on 20 October 1944, was another instance of the sheer magnitude of the Pacific operations.

Here the Japanese had 9 battleships, 2 of which were fitted with a flight deck and could carry a few aircraft, 4 large carriers, 21 cruisers, 40 destroyers – but only 115 aircraft. These were all the aeroplanes they could muster after their recent heavy losses. The Americans deployed 12 battleships, 46 aircraft carriers, 28 cruisers, 235 destroyers and a total complement of some 1590 aircraft. In addition they had about a further 900 other ships – troop transports, logistics vessels and so on.

The Japanese made a determined effort to prevent the landing on Leyte, but, at the end of the day, they had lost all 4 carriers ... and, additionally, 3 battleships, 20 cruisers and 11 destroyers. All their carrier aircraft were shot down or lost with the carriers.

For two weeks before the landings on Leyte, the United States' fast Carrier Task Force had delivered heavy air attacks on the fifty-seven airfields in the Philippines and also on airfields in Okinawa, Formosa and Hong Kong. The Task Force consisted of 16 large carriers, with 1110 aircraft embarked plus battleships, cruisers and destroyers in support.

The Americans usually flew 1200 to 1400 sorties a day over shore targets and maintained a permanent combat air patrol of 96 Hellcats over the Task Force all day.

On 15 October, Tokyo Radio announced that Japanese forces had gained "a glorious victory for the Emperor". They claimed to have sunk "11 aircraft carriers, 1 battleship, 3 cruisers and 1 destroyer" and to have "severely damaged a further 8 carriers, 2 battleships and 4 cruisers." The United States' fleet, they said, was now retiring "helter-skelter".

All of us in the Task Force heard this announcement direct from Tokyo.

Admiral Nimitz, the Commander-in-Chief, Pacific, replied with a message broadcast on worldwide radio: "Admiral Halsey reports he is now retiring towards the enemy following the salvage of all the ships of the Third Fleet sunk by Tokyo radio."

The actual losses sustained by both sides, up to 15 October, i.e. five days before the Battle of Leyte Gulf, were as follows:

United States	No ships lost, 2 cruisers damaged and 76 aircraft (out of 1110) lost
Japan	41 ships – mostly merchant ships – sunk, 525 aircraft either shot down or destroyed on the ground (Japanese figures)

The result of all this was that on D-Day – 20 October – there was virtually no Japanese air opposition to the landings on Leyte.

It may be said that it was easy for the Americans, with their great superiority both in quantity and quality of pilots and equipment, to achieve these results. The fact is, however, that they were just as good when the situation was reversed.

For example, the Battle of Midway Island showed how well the Americans could do when the cards were stacked against them.

The Japanese had decided in June 1942 to capture Midway, some 1500 miles northwest of Pearl Harbor, as the first step towards taking the Hawaiian Islands. Their force for the operation consisted of 6 battleships, 4 large aircraft carriers and 1 small one, 15 cruisers, 43 destroyers and a lot of transports. 310 aircraft were embarked.

The United States had no battleships to call on as all had been sunk in the attack on Pearl Harbor. Their force, therefore, consisted of 3 aircraft carriers, one of which had been badly damaged in the Coral Sea and was only partly repaired, 8 cruisers and 17 destroyers. 253 aircraft were embarked in the carriers while a further 53 navy and marine aircraft were available on Midway. At this time the US Grumman F4F fighter was much inferior in performance to the Japanese Zero.

The outcome of this critical battle was that the Americans sank all 4 of Japan's large carriers, 280 aircraft were either shot down or sank with the carriers, while the Japanese lost the cream of their front-line pilots. In the face of these losses, the enemy abandoned the operation and returned home.

The United States' losses were 1 carrier sunk – the previously damaged *Yorktown*, 1 destroyer sunk and 130 aircraft shot down. Of the 48 torpedo bombers involved in the operation, 43 were shot down by the Japanese.

This great victory, the turning point of the Pacific war, was won by naval aircraft inferior in number and performance to those of the enemy. It is a great tribute to the superb courage and excellent training of the USN aviators.

'At the End of the Day Friendship is All . . .'

Though the years roll by
If friends remain
No year is lost
But each a gain

Harold Balfour
1983

PART SEVEN

Aerial Dividend

When the rot sets in, it travels fast – not least in an air force. Recrimination, blame and intrigue always accompany failure and defeat. Internal strife can be just as lethal a weapon as any possessed by an enemy.

As the Allies' eastern and western armies closed in on the Third Reich, and the hard winter of 1945 gave way to spring, dissension in the Luftwaffe was rife. Göring's disparagement of the Fighter Arm of the Service had rebounded upon him. The Reichsmarschall was now a deflated dirigible and regarded with contempt by many of his more senior operational commanders.

Galland, outspoken and forthright to the last, had been replaced as General der Jagdflieger by Gordon Mac Gollob (a Scottish godparent was said to have been a Gay Gordon), an officer trusted by few, but nominated by Himmler of the SS. Gollob lasted a couple of months from early February until 7 April when he applied for sick leave. Göring ignored the application so Gollob wrote out his own medical chit and 'joined one of the greatest migrations in history, the trek of high-ranking Luftwaffe officers to south German hospitals in the last days of the war'.

In such an ethos, with the German Air Force now back on its heels against the ropes, it could only be a matter of time before the towel was chucked into the ring. But before the fight was over there were still some nasty punches to trade.

In the Pacific, the Japanese, now greatly weakened, were still slugging it out with the enemy, and in some cases 'honourably' committing suicide, as the Imperial Navy and Army withdrew to the homeland under cover of a Setting Sun. As in the West, there were still some horrible engagements to be fought. But the American Eagle, dominant and aggressive, was striking again and again at its prey. Then the Bomb fell on Hiroshima and Nagasaki and all was over. World War II, after six cataclysmic years, was at an end.

1945, then, was the year when the aerial dividend was paid to Allied shareholders – the year when those aircrew who had survived the holocaust could contemplate the hard-won profits by which the pay-out had been earned.

The 'Enjoyment' of War

Group Captain W.G.G. Duncan Smith, *Spitfire into Battle* (John Murray, 1981; Hamlyn Paperbacks, 1983)

The fact is I enjoyed the war. This may seem strange and shocking, but it became my driving force because it had a purpose and I accepted the risks, the excitement of combat, survival and retrospect in that context. We all did. A dangerous game played with no rule-book, it rested on the individual – individual responsibility in leadership with faith and courage, so that the sense of belonging to a great Service in its hour of triumph lifted the spirit above the horror and bestiality of war.

Naval Maestro

Those who served with Bobby Bradshaw (Lieut.-Commander R.E. Bradshaw) in the Fleet Air Arm in wartime speak of him as a latter-day marvel, a maestro, a man apart – the Royal Navy's answer to the Royal Air Force's Adrian Warburton.

Donald Judd, himself one of the FAA's operational successes, and a City solicitor in peacetime, knew Bradshaw as well as anyone.

'Bobby Bradshaw was, without doubt, one of the finest operational pilots thrown up by the Fleet Air Arm in World War II as well as being an outstanding personality with a very strong individuality of his own. He became a legend far beyond the confines of the FAA.

'His wartime career speaks for itself. He was awarded a short service commission as a midshipman in the FAA – RN (A) or 'A' boy as they were known – in 1939. He joined the newly-formed 826 Naval Air Squadron in 1940 at Bircham Newton, in Norfolk. The squadron spent the rest of the year carrying out light bombing and minelaying operations in their 100 m.p.h. Albacores along the Dutch, Belgian and French coasts. In October, they joined HMS *Formidable* bound for the eastern Mediterranean, where they carried out the attack on the Italian Fleet in the Battle of Cape Matapan, among other successful operations. But when *Formidable* was badly damaged on 26 May 1941, off Crete, after an attack on Scarpanto airfield, the squadron was posted to the Western Desert in June 1941. That was when I first met Bobby Bradshaw as I joined the squadron at that time.

'He was by then a sub-lieutenant, but in spite of his youth and junior rank, he was clearly the personality of a squadron which did not lack characters. He was to spend the next two years operating with 826 up and down the Western Desert, engaging mostly in night dive-bombing, ending up in Tunisia as the Commanding Officer. So, while the average tour of duty for aircrew was about a year, Bobby spent three years with 826, rising from midshipman to lieutenant-commander.

'He then converted to Avengers and joined 852 Squadron in the United States as CO and served on HMS *Nabob* in the Home Fleet for shipping strikes and minelaying off Norway. He thus spent four and a half years almost continuously on operations, being awarded the DSC and two bars, and being mentioned three times in Despatches.

'This record would be enough for any man. But behind it, was an extraordinary person. A touch of the aristocrat and the born leader, combined with an individuality and a happy-go-lucky disposition, made Bradshaw a personality in any surroundings. He was always colourful, often crazy and generally a larger-than-life extrovert. But with all this, he had a singular modesty of his own.

'He had no regard whatever for authority, rules or convention. Fortunately, like T.E. Lawrence, the Desert was made for him.... No formal naval discipline, a casualness in dress, and night operations that gave scope for individual initiative, it all suited him down to the ground. He had one governing ambition – to cause as much damage to the enemy as possible. I remember on at least two occasions, when we were near enough to Rommel's forces to do two operations each night, Bobby did three – without the consent of the CO, of course.

'In many ways, he was wasted on the Albacore and, later, the Avenger, both carrying three aircrew. He should have been a fighter pilot which would have given full vent to his dash and individual flair. One of his navigators, Jeff Powell, once told me that it was all rather like the aircraft having two pilots – Bobby flying the front half and himself the rear!

'Sadly, having come through the war unscathed, Bobby Bradshaw died with his wife in a car crash in the 1950s. A short life, but one which was full of colour, incident and achievement. It is remembered by many.'

'Bogey at Fifteen Miles'

The Oxford Book of Twentieth-Century English Verse: chosen by Philip Larkin (Oxford University Press, 1973)

The usual exquisite boredom of patrols.
    The endless orbiting at twenty grand;
the growing pins and needles in the arse,
    the endlessness of time, the crawling clock,
and nothing to relieve monotony.
    The necessary plague of oxygen,
the routine checks of fuel, pressures, temperatures,
    oil, coolant, guns—
(as if there's any chance of using them!)—
    browned off with looking round, above, behind,
conducive to no rational thought or act
    but the conditioned reflexes of fear.

The ships below, sculling like water-beetles on a pond
    in white-tracked zig-zags, escorts on the flank.
And the careful, modulated tones
    of the directing officer: 'Steer One Fife Eight!
Bogey at fifteen miles.' Turn onto course—
    another bloody Sunderland, for sure.
And still half-an-hour to go. 'Blue two—
    open to five. And keep a good watch out.'

For eighteen months one has awaited this,
    this consummation justifying all,
the means the end, the end the means. Just this,

to have the bastard steady in one's sight—
a spot more deflection—now. And the dulled drumming
of the wing cannon, the ruled tracer lines,
and smoke flowering from the engines,
and the dive away into the neutral sea.
The wild oblivion of excitement,
mastery beyond all thought or doubt or argument.

Last time—an 88 it was—above the fleet;
we picked him up at dusk at thirteen grand,
chased him as he dived into the dusk:
somebody hit him, saw the bits come off.
Then the last light went. We lost him
to the eastward, going like hell
right down on the drink. And can those pockers go.
We only got a probable for him. He might
have made it home. Not the clean, concise
explosion of the bastard that you jump,
two quick sure bursts—and there an end.

But here, the mainland a dim shadow
in the haze, there's not much chance.
'Blue two: we'll start going down, I think'—
blow out your ears and spiral down. Thank God,
the bar should be open when we land.

Hugh Popham

Modern Vikings

Air Marshal Sir Douglas Morris, Northiam, Rye, Sussex, 1982

Taken collectively, the composite record of the Allies who flew with the Royal Air Force in wartime was unique. That it was possible to blend together in a matter of months all the different factions and nationalities, with their language and temperament problems, was due primarily to three factors.

First, the spirit and ability of the Allied airmen themselves, their innate preparedness to accept a good lead, and their will to fight to the death for their homelands, alas not always achieved.

Second, the organization and leadership – and sheer professionalism (often disguised) – which the Royal Air Force was able to offer those who had accepted to fight under its banner. And, third, the Allies' total commitment to the cause to which they rallied.

The Norwegian Wing, made up of 331 and 332 Squadrons, was conspicuous among its counterparts. Its (North Weald) sector and later, (132) wing commander from 1942, right through until February, 1945, was 'Zulu' Morris (Air Marshal Sir Douglas Morris), a tough and experienced leader, whose South African upbringing matched the exacting backgrounds of these modern Vikings. None is now better equipped to penetrate their character.

'... The most striking characteristic of the Norge pilots was their intense desire to engage the enemy. All of them had made incredible journeys, in some cases almost half way round the world, to get away from the Germans and ultimately to join their fellows in Canada or America, there to begin their training. These experiences had toughened young men, beyond their years, into formidable opponents when their turn came to face the enemy.

'Although there were no outstanding, high-scoring "aces", the general level of flying was very high and, although over one hundred Norwegian pilots were lost from such a small force, the balance was greatly in their favour. Mention should, perhaps, be made of a rather exceptional young Dane, Kai Birksted.

'"Birk" had managed to escape from Denmark and join up with the Norge squadrons. By his great tactical understanding of air combat and his natural powers of leadership, he rose to become the first Allied officer to lead into battle a Wing that included British squadrons. He was decorated with the British DSO, and DFC, in addition to Norwegian honours – awards that were earned both for the meticulous way in which he planned his operations, and for the concern had for his less experienced pilots.

'... Although, on the ground, a rather shy, quiet and reserved character, in the air he was a determined and confident leader, a wonderful example to young pilots. His successful leadership of the Norge Wing was widely recognized. This was further demonstrated when, on being rested from operations, he was employed on planning fighter actions, particularly those involving support for the US 8th Army Air Force's daylight, bomber raids into Germany. Here again he showed a remarkably mature understanding of the problems.

'The Norges were, possibly, the easiest of all the Allies to handle. Their sense of humour – or sense of the ridiculous – and their reaction to any form of pomposity, was easily understood and accepted. They all spoke more or less perfect English, and when British and Norge pilots were together it was only by the different insignias of our uniforms that strangers could tell us apart.

'I was always fascinated listening to Norge pilots in conversation when relaxing off duty. On mundane matters the discussion would probably be conducted in Norwegian, but when, inevitably, it turned to the latest combat with the Luftwaffe the pilots broke excitedly into English – and RAF phraseology, the universal language of the air.

'Their upbringing has made the Norwegians naturally self-reliant and experts at "making-do". This was most apparent during the moves of 132 Wing on the Continent, from Normandy to Holland, when we were often located on ex-German airfields. Because winter was approaching I was asked whether the wooden huts left by the Germans could be dismantled and taken forward with us. Having been assured by my chief engineer and transport officers that nothing would be left behind and our ability to move would not be impaired, I readily agreed.

'From then on, wherever the Wing went, we took our accommodation with us. Thus, were we able to stand the cold of winter which would have been impossible in tents. When we arrived at a new site the huts would be erected and inside each hut huge stone fireplaces would be built to provide the snugness. When we moved on the huts would be dismantled leaving the fireplaces standing, naked and neglected, a monument to our temporary occupation.

'Another example of their ingenuity was shown at an ex-German airfield we occupied at Bergen-op-Zoom, just north of Antwerp. There, the camp hot water and sewerage systems had been completely destroyed by the Germans or the advancing Allied forces. Within a few days everything had been restored to normal. We could use the lavatories and enjoy the supreme luxury of a hot bath!

'However, not all the actions of the Norges were as helpful, and some were quite infuriating. While at Bergen-op-Zoom we were brought to a high state of readiness as Intelligence considered that the Germans might make an attack across the Rhine to support their Ardennes offensive by cutting off the key Allied port of Antwerp. Everyone was at a high and nervous state of alert.

'On Christmas Eve, we held a Carol Service, well attended by all ranks, and conducted by our very popular Norwegian padré. Imagine the upheaval when, at about midnight, a fusillade of rifle fire was heard and immediately everyone assumed that the long-awaited attack had begun. When, eventually, it was found to be no more than a few Norges, who had celebrated too well, engaging the full moon, my short-lived anger can be understood.

'There were, of course, the sad moments – especially when we lost good men. One which hit me very badly was the loss of Wing Commander (Lieut-Colonel) Rolf Berg, RNoAF, who flew over 450 fighter sorties against the Luftwaffe. In the course of his operations he was awarded the British DSO and DFC, in addition to the highest Norwegian honours.

'Rolf was the Wing Commander Flying when I took over 132 Wing and we rapidly built up a very close affinity. I flew as his No. 2 on many occasions when we went off together on opportunity ground strafing missions. We enjoyed many relaxing moments in camp or in Lille, Brussels and Antwerp.

'In January 1945, I realized that Rolf had done enough and, despite his protests, decided it was time he was rested. His replacement duly arrived and on the day before Rolf officially handed over, he came to me and asked if he could lead the Wing for the last time. It seemed a fairly straightforward mission – a ground strafe of a small airfield in North Holland – and I reluctantly agreed.

'When the Wing returned, I was gravely concerned when the Spitfire, with Rolf's identifying initials, RAB, did not land ahead of the others. My worst fears were realized when it was confirmed that he had been shot down by ground fire.

'The sequel was equally sad. Some months later I had to fly up to Leeuwarden to attend a large Victory Parade of the Canadian Army. As I flew round before landing, there, lying almost accusingly on its side and clearly visible, was the silver-coloured fuselage of a Spitfire with the letters, RAB, staring at me.

'Years later, flying from England to Norway, I put down at the civil airfield of Leeuwarden. While talking to the airfield controller I found there was a small cemetery in the town where Allied airmen were buried. The town Burgomeister, who knew the history, drove me to the little graveyard. There I found Rolf's grave and there I paid my final farewell to a very courageous fighter and very loyal friend.'

'Hurricane, 1940'

Just twisted scrap thrown on a dump
Strips of wing and a Merlin sump
    Old Fighter plane
    Your flight is done
Your landings made and Victories won

Gun barrels scorched and motors tired
Your masters fought as men inspired
    Old Fighter plane
    They trusted you
Who faithfully served the Gallant Few

Casually now they fly around
Jet propelled at speed of sound
    New Fighter planes
    Fierce in your power
Spare thought for those who had their hour

Harold Balfour

The Numbers Game

Among all the participants, the claims by fighter pilots and gunners of enemy aircraft destroyed in combat proved to be, on the evidence collected *after the war*, often misleading and greatly exaggerated. There were diverse reasons for the discrepancies ... duplication, speed, excitement, competition, two-upmanship, imagination, euphoria, mistakes, the desire to excel ... many factors, human and otherwise, were involved. First-degree dishonesty was the exception, not the rule.

A UNITED STATES VIEW

Vern Haugland, *The Eagles' War* (Jason Aronson, New York, 1982)

Dixie Alexander* says that the granting of credit for enemy aircraft destroyed was relatively severe early in the war, and often lax in the latter stages of the air offensive.

*R.L. 'Dixie' Alexander flew with the Royal Air Force in the American Eagle Squadron(s) and then with the US 8th Army Air Force in the daylight offensive against Germany. There weren't many tricks he didn't know.

'The RAF was tough. You either had good pictures – and just strikes were not enough – or you had to have confirmation by a couple of persons. Later in the war, claims became a laugh.

I broke in several of the later real big aces, and without mentioning names, they never showed me too much. They were not good shots and were not the best in the thought department – two very necessary items. Some of these huge scores were run up on confirmation such as A confirming two for B, and B confirming three for A. This went on and on, with the same people working together.

After 168 sorties of various types in combined operations in the two armies, I finally wound up with 6 destroyed and 1 probable. Don Blakeslee, who was just about the best, had 12 in about 450 missions. There were many more of the same....

I remember a show on 1 August 1942, when gunners on the bombers were credited with 17 German aircraft destroyed. The truth was that we were bounced on the way out from the target by three 190s. They came down from about 30,000 feet and were going full bore when they made their pass. They went over on their backs, belching diesel fuel, and on to Germany. We couldn't have caught them with anything less than a jet.

I saw no hits, and we didn't even bother with them. Anyhow, the bomber boys all fired, and claimed 17. There were 17 air medals handed out for this one.'

Alexander makes some critical points about unit effectiveness in terms of the so-called kill ratio.

'The British kill ratio in the war was about 3½ enemy aircraft destroyed to 1 lost. The record of the 4th Fighter Group, Blakeslee's outfit, showed a ratio of 4½ to 1. This was the oldest, most experienced fighter group in the ETO,* studded with experienced and accomplished leaders. It doesn't make good sense to believe that the 56th Group, a younger, less experienced outfit flying in the same skies with the same type of aircraft, could achieve a record of 8 to 1.

All the other outfits are somewhere in between in the scoring. The figure given by the 56th, to bring up its ratio, has got to be fictitious, and no doubt has a great deal to do with the fact that they harboured such aces as Francis Gabreski, Dave Schilling and Hub Zemke within their group.'

A BRITISH VIEW

Group Captain W.G.G. Duncan Smith, *Spitfire into Battle* (John Murray, 1981; Hamlyn Paperbacks, 1983)

'... The victories credited to the German "aces" Hartmann (352), Neumann (302), Nowotny (258) and Marseille (158) appear fantastic by any standards.

There is no doubt the Germans destroyed in the air and on the ground hundreds of near-obsolete Soviet aircraft during the summer and winter of 1941.

*European Theatre of Operations.

... In assessing individual combat claims, it has been suggested that leaders of German fighter formations were credited with enemy aircraft claimed as destroyed by pilots in the formation and these were included in the leader's running total...

Discussing these matters with some German fighter pilots after the war got me nowhere because they hotly denied, not surprisingly, that combat claims were ever fiddled. Be that as it may, there is the interesting if perplexing case of Hans-Joachim Marseille. Fighting over the Western Desert he is reported to have shot down 61 RAF aircraft in the period 1–7 September 1942, with 17 British aircraft downed in a single day (1 September). RAF records show that there were no losses that day in the areas the claims were made.* Marseille was killed on 30 September 1942. His total score by then was 158; 150 of them claimed over the Western Desert. It appears that his claims were accepted at face value and forwarded to the German High Command by the Jagdgeschwader concerned.

British pilots had to have confirmation of any kills. This was usually provided by other squadron pilots, by assessment of cine-gun camera film or an eye-witness on the ground. If none of these requirements were met it became the responsibility of squadron and sector Intelligence Officers to assess pilots' claims from the details of combat reports and determine to which category the claim belonged, i.e. destroyed, probably destroyed or damaged. No claims were admissible for any aircraft destroyed on the ground as these fell into the category of ground targets....

If comparisons must be made, it is interesting that the score credited to the leading "ace" from each of fourteen Allied nations (excluding the Soviet Union) totals 352 which averages out at 25·14 aircraft destroyed. Pat Pattle of South Africa, 'Johnnie' Johnson of Britain and Richard Bong of the USA – three quite outstanding fighter pilots – destroyed 129 enemy aircraft between them.

Can there be such a wide differential between the relative performances of top German and Allied fighter pilots even allowing for the huge number of aircraft German pilots engaged over the Eastern front and Germany itself in the months following the Allied landings in Normandy? And what about the leading Soviet "ace", Colonel Ivan Kozhedub, with his modest claim of 62 German aircraft destroyed?

Even if the best Soviet pilots were not up to the standard of the best Germans, which I doubt, surely Soviet fighter pilots flying some superb aircraft by that time could not have been as ineffectual as the figures suggest?'

'Editor's note: Few, if any, Allied fighter leaders had more operational sorties in their log books than Duncan Smith. He was 'at it' with Fighter Command both during and after the Battle of Britain, with the Desert Air Force in Sicily and Italy and he was still going strong at the end of the war. Like Dixie Alexander, he knew the operational form.

It is only fair, again, to make a further point. When comparisons are made

*In fairness, it must be said that this statement has been challenged by other sources. '... In the areas [where] the claims were made' could be interpreted as a number of separate and tightly circumscribed localities. (Ed.)

of contemporary kills registered by the Royal Air Force with the numbers of aircraft actually lost by the Germans, according to their own records, significant discrepancies are often revealed. Similar comparisons between USAAF claims and actual, recorded losses by the Luftwaffe, expose the same kind of differences.

THE GERMAN VIEW

Set alongside the two foregoing opinions are three similarly authoritative, German observations. The pilots concerned were all exceptionally experienced.

Sources: Johannes Wiese and Wolfgang Falck – correspondence, 1982. Gunther Rall's quote was represented by Werner Schroer, 1982, and taken from *Fighter Aces of the Luftwaffe*, Col. R.F. Toliver (Retd) and Trevor Constable (Aero Publishers Inc. of California).

Johannes Wiese

No pilot could claim a victory without strict corroboration. Other pilots, flying at the same time, were usually best-placed witnesses and would report the position of a crashed aircraft by R/T to the ground station. In the hurly-burly of combat, double reporting may have been inevitable, but specific details were always required before a victory could be confirmed – aircraft type, height and sector, amount of ammunition expended and the precise time of the 'kill'. If these details could not be established without doubt then the claim was downgraded to 'probable' or 'possible' – and these credits did not count towards the award of decorations.

On the score of the large differences between the totals of the leading German pilots and those of other nations, these clearly arose from the great operational differences which existed between the various combatants. The Luftwaffe's circumstances were often quite different from those of the opposing forces in the matter of providing opportunities in combat. The comparisons are not on a like-with-like basis.

To my personal knowledge, there was never an instance of a flight leader being credited with victories achieved by others in his section, squadron or Gruppe. If this had been the case my own score would have reached a thousand. One must be quite clear about this. The success of a Staffel or Gruppe reflected well upon its leader and he gained personal kudos from the victories gained by his unit. Decorations, however, were awarded strictly according to one's own personal kill-rate.

Wolfgang Falck

It is correct to say that a 'kill' had to be confirmed by a witness before it could be credited to a pilot. This was an essential prerequisite for obtaining confirmation and acceptance of a victory.

However, it is not correct to suggest that a Staffel or Gruppe received fractional credits for aircraft destroyed or that these were added to a leader's personal total. In every case a victory was added to the personal score of the pilot who achieved it; it was never credited to someone else because he happened to be the leader of the formation. There were no exceptions to this.

Gunther Rall
The wartime combat reports of the Luftwaffe fighter pilot were highly detailed. Every evening you had this business to go through. Witness, air witness, ground witness, your account of the combat, the type of enemy aircraft, the kind of ammunition you fired, the armament of your aircraft, and how many rounds of ammunition. These reports were a nuisance to us,* but when I was on the staff of Galland I saw how valuable they could sometimes be.

We found that Marseille needed an average of only fifteen bullets per kill – which is tremendous. No other fighter remotely approached him in this respect. Marseille was the real type – an excellent pilot and a brilliant marksman. I think he was the best shot in the Luftwaffe.

Mirror, Mirror...

At the height of Bomber Command's offensive against the Third Reich, and at the moment when Don Bennett's target-marking techniques were coming to fruition, the Luftwaffe air staff (Luftwaffenfuhrungstab) circulated a secret intelligence assessment (No. 61008 Secret Ic/Foreign Air Forces West A/Evaluation West) entitled 'British Pathfinder Operations as at March

*Gunther Rall's view is understandable. The Luftwaffe's documentation requirements were far more comprehensive than those of the Allied air forces. The essential papers were: a Victory Report (Abschussmeldung), an Action Report (Gefechtsbericht), a Witness Report (Zeugenbericht) and a CO's Report (Stellunghahmen). These had to be forwarded through the appropriate channels to the Air Ministry (Reichsluftfahrtsministerium). In due time the claim was returned, allowed or disallowed. In the case of confirmation, the pilot received a certificate stating that the victory was officially recognized. (Johannes Wiese's comment: 'At the end of the war I had collected 130 certificates – all now lost!')

Rall himself would have been entitled to 275 certificates as the third highest scorer on the German side. How he managed his administration at the peak of the fighting on the Eastern front – or, better still, how Marseille handled his 17 returns on the evening of 1 September 1942 (see Duncan Smith, A British View, p. 348) in his tent in the Western Desert, makes a most interesting reflection.

Had an Australian pilot in 249 Squadron been asked to sit down in the evening and deal with that kind of paperwork at the end of a long day at the height of the fighting for Malta in 1942, the editor wouldn't have given much for the squadron commander's chances on the first sortie at first light the next morning.

For the victors, if not the vanquished, the endless controversy over the Numbers Game may best be resolved by recurring to Douglas Bader's throwaway quote on the discrepancies in the claims of both sides in the Battle of Britain. 'Who cares, anyway? We won, didn't we?' Nevertheless, several undeniable facts remain.

Pilots in the Luftwaffe went on and on. They had no set operational tours and no prolonged rests; they went on until they were killed or, for one reason or another, were incapable of flying against the enemy. The survivors were operational from 1939 to 1945, virtually, without a break. The opportunities to make 'kills' were endless. Beyond that, their aerial targets were infinitely more numerous and, on the Russian front, often made of wood and, therefore, more inflammable. Apart from isolated periods, the Allies' targets were in short supply and, what there were, were well constructed and elusive. Pilots were off operations – continuous operations – for, perhaps, a total of eighteen months or two years out of nearly six years of war; sometimes it was much more.

The two positions were thus quite different. There was, therefore, no general and fair basis for comparing the level of victories on either side. (Ed.)

1944'. There are three comments to make upon it:

1. The operations staff of 8 Group, Bomber Command (the PFF Group) would have been hard put to it to produce a more detailed thesis on their own organization and operations.
2. Whatever its sources, the Luftwaffe's knowledge of the intricacies of PFF's operations was penetrating, comprehensive and substantially accurate. It leaves little doubt about the effectiveness of the Germans' intelligence arrangements – at any rate in this sphere.
3. As a staff paper, it was a model, balanced, well marshalled and tightly written. By any test, it was a first-rate piece of staff work.

Space allows only a few short extracts from the memorandum's sixteen pages.

(i) The Contents summary shows the spread of the document. Under each head and subhead there is a detailed supporting back-up.

BRITISH PATHFINDER OPERATIONS

CONTENTS

Preface

A. *Development*

B. *Organization and Equipment*
- I. Organization and aircraft types
- II. Personnel

C. *Pathfinder Operations*
- I. General
- II. Markers
- III. Execution of Pathfinder Operations
  - i) Dividing of the Pathfinder crews
  - ii) Route markers
  - iii) Target markers
  - iv) Release of markers
  - v) Navigation

D. *Conclusions*

(ii) The opening paragraph of the Preface answers the critics of PFF:

'*The success of a large-scale night raid by the RAF is, in increasing measure, dependent on the conscientious flying of the Pathfinder crews* [editor's italics]. The frictionless functioning of the attack is only possible when the turning points on the inward and outward courses, as well as the target itself, are properly marked. Lately, these attacks have been compressed into about four minutes for each wave averaging 120–150 aircraft.'

(iii) After detailing the composition of PFF:

8th Bomber Group at present consists of
Five 'Lancaster' Squadrons
One 'Halifax' Squadron

Four 'Mosquito' Squadrons (including two special bomber squadrons with 'Bumerang' [Oboe] equipment)

One 'Mosquito' Met. Flight.

For further information concerning the organization of these units, see 'Blue Book Series', Book 1:
The British Heavy Bomber Squadrons –
the navigational and special equipment carried by PFF aircraft is set out in depth under (a) 'Four-engined aircraft (Lancaster and Halifax)' and (b) 'Twin-engined aircraft (Mosquito)'. The student is also directed to a study of the 'normal navigational aids' with special reference to 'Blue Book Series', Book 7: *British Navigation Systems.*

(iv) The Personnel section supplies a useful narrative about PFF crews:

'The crews are no longer composed mainly of volunteers as was formerly the case. Owing to the great demand and the heavy losses, crews are either posted to Pathfinder units immediately after completing their training, or are transferred from ordinary bomber squadrons. As in the past, however, special promotion and the Golden Eagle badge are big inducements to the crews.

At first Pathfinder crews had to commit themselves to 60 operational flights, but because, due to this high number, there were not sufficient volunteers, the figure was decreased to 45.

After transfer to a Pathfinder Squadron, a certain probationary period is undergone. The crews are not appointed Pathfinders and awarded the Golden Eagle until they have proved themselves capable of fulfilling the requirements by flying several operations (about 14) over Germany. Before the award of the Golden Eagle each member of the crew has to pass a special examination to show that he is fully capable of performing two functions on board, for example gunner and mechanic, or mechanic and bomb-aimer etc.'

The training methods, and the areas over which training flights took place, are accurately described. The last paragraph ends decisively:

'If, on several occasions, the [flight] schedule is not [strictly] adhered to, the crew is transferred to an ordinary bomber squadron.'

(v) The general description of Pathfinder operations gives credit to the fertile and innovative character of PFF direction:

'New methods of target location and marking, as well as extensive deceptive and diversionary measures against the German defences, are evident in almost every operation.'

Three additional paragraphs show the advance which was achieved with PFF methods and organization. Praise is given where due:

'Whereas the attacks of the British heavy bombers during the years 1942–43 lasted over an hour, the duration of the attack has been progressively shortened so that today a raid of 800–900 aircraft is compressed into twenty minutes at the most. (According to captured enemy information, the plan for the raid on Berlin on February 15/16 1944 called for about 900 aircraft in five waves of four minutes each.)

The realization of these aims (concentration of attack and saturation of defences) was made possible by the conscientious work of the Pathfinder group and by the high training standard (especially regarding navigation) of the crews.

The markers over the approach and withdrawal courses serve as navigational aids for all aircraft and above all they help them to keep to the exact schedule of times and positions along the briefed course. Over the target, the markers of the Pathfinders enable all aircraft to bomb accurately without loss of time.'

(vi) Eight pages now follow on Markers – Target Markers (ground and sky), Route Markers, Allocation of Crews to Specific Marking Duties (each duty decribed in detail), with a pointed reference to navigation:

'The basis for all Pathfinder navigation is dead reckoning, and all other systems are only aids to check and supplement this. H2S equipment is valueless without dead reckoning because the ground is not shown on the cathode ray tube screen as it is on a map.

To facilitate the location of the target, an auxiliary target, which experience shows to give a clear picture on the cathode ray tube, is given during the briefing. This auxiliary target should be as close to the actual target as possible, in order to eliminate all sources of error. Cities, large lakes, or sometimes even the coastline features are used as auxiliary targets.

The course and the time of flight from the auxiliary target to the actual target are calculated in advance, taking the wind into consideration. The H2S operator then knows that the main target will appear on the screen a given number of seconds after the auxiliary target has been identified.'

(vii) The special duties assigned to PFF's Mosquitoes – six are stated in detail – disclose the importance the Luftwaffe attached to the activities of this elite force. The paper concludes with an unanswerable vindication of PFF and the demolition of its critics:

1. Strong criticism from amongst their own units was at first levelled against the British Pathfinder operations, but *they were able to prevail because of the successes achieved during the years 1943/44.*

2. *The original assumption that the majority of bomber crews would be less careful* in their navigation once they became used to the help of the Pathfinders, and that therefore the total efficiency and success of raids would diminish, *has hitherto not been confirmed.*

The navigational training and equipment of the ordinary British bomber crews has also been improved.

3. *The operational tactics of the Pathfinders cannot be considered as complete even today. There are in particular continual changes of all markers and marking systems.*

4. The trend of development will be towards making possible *on one and the same night two or more large raids on the present scale*, each with the usual Pathfinder accompaniment. (Editor's italics.)

Who said the Germans didn't know what we were doing?

Pathfinder

The following extract is taken from Charles Woodbine Parish, DFC, RAF, 12 April 1915–22 April 1943, Pathfinder Force' *written by the subject's father, C.W. Parish, and circulated privately among the family and close friends. This beautifully bound little book, a moving tribute to a devoted son, was printed by the Medici Society Ltd, MCMXLIII.*

'In the next few days hopes ran high that he might have baled out as he had done over the North Sea in September 1940,* but with each passing week that brought no news these hopes slowly fell until early in June the slender thread that held them snapped on news from one of his crew.

This was a young Canadian, Sergeant Smith, RCAF, aged twenty-one, his engineer on nineteen raids, who came to see me at Bateman's ... with the news that he alone of the crew survived, that he had escaped from enemy territory ... [and] that Charles had perished at his post.

In that dark hour my cup of bitterness overflowed and I echoed David's cry: "Oh my son Absalom, my son, my son Absalom! would God I had died for thee, O Absalom, my son, my son!"'

Analysis of Defeat

Johannes Steinhoff, *The Straits of Messina* (Andre Deutsch, 1971)

There could be, by now, but one outcome to the struggle and it had long since been predestined. In retrospect, the most critical and authoritative appreciation of the denouement of the German Air Force has been written by General Johannes Steinhoff – 'Macky' Steinhoff, who fought virtually right through the conflict, from the Battle of Britain, Stalingrad, North Africa and Sicily, through to the eventual formation of JV 44, the distinguished Me 262 unit, with General Galland at its head, and a handful of the Luftwaffe's alumni in support – Barkhorn, Hohagen, Schnell, Krupinski, Lutzow and the rest. It read like a World Cup squad which had come into the competition too late.

Steinhoff has gone straight to the kernel of the matter.

'The Luftwaffe that entered the war was an incomplete weapon and, when that war had to be conducted against great powers on several fronts, the high command and its instrument were very far from adequate to the task that confronted them.

The first manual of aerial warfare (LdV 16) was produced under the first Chief of Air Staff. It was concerned mainly with air attack, air defence being relegated to second place.

*Charles Parish, the sole survivor of the crew of a Wellington, came down in the sea in September 1940, returning from Germany. Seven miles short of the English coast, he swam – yes, swam – in the darkness, reaching the shore with little life left in him. The army heard his cries and saved him. A week later, he was back again on the Berlin run. Parish completed 54 operations, many as a Pathfinder captain, before he died. (Ed.)

Thus in 1939 the military conceptions that governed the Luftwaffe failed to take into account the possibilities and limitations of our own forces. Air warfare as an "independent" factor – the destruction of the enemy's vital centres, aptly named "strategic air warfare" by the Allies – was never consequentially planned by the Luftwaffe General Staff. Air offensives were, instead, to be conducted by means of medium-range bomber forces whose size and composition would enable them, at best, temporarily to disrupt the enemy potential, but not to destroy it.

"Ural-Bomber" was the name given to the project for a big bomber which, however, never came into production. There can be no doubt that the economic resources of the Third Reich would never have permitted the construction of an air arm adequate for the conduct of both air defence and strategic air warfare with big bombers.

At the outbreak of war, the defence of the Reich had been grossly neglected. The intention – an over-optimistic one – was that active air defence should devolve almost exclusively on the anti-aircraft artillery. There was no overall organization for the conduct of defensive warfare since air raids on the Reich were not anticipated.

When the British began their night raids – shortly to be followed by American daylight attacks – it was too late to repair these omissions.

Whereas Göring, like Richthofen before him, had himself been a serving pilot and hence must have been aware that air warfare had come to be an independent factor in war, many Luftwaffe generals and high-ranking officers were of military or naval origin and very few of them succeeded in understanding the different laws governing air warfare. They did their best by qualifying as pilots or observers, but since few of them had "grown up" with this new weapon, their notions of air warfare derived from the narrower concepts of land warfare.

Moreover, during the course of the war, the High Command grew increasingly out of touch with the fighting component of the air arm.

To the junior officer the High Command's mistakes and wrong decisions became obvious at about the time the Battle of Britain was drawing to a close. That battle had inflicted on the fighter arm – then only five years old – losses in men and material which, relatively speaking, could not be made good. The extension of the air war and the beginning of the Allied bombing offensive against the Reich eliminated any possibility there might have been of conducting, with material superiority, either an offensive war in the air or an effective aerial defence. The reaction increasingly took the form, not of careful calculation and organization, but of improvisation and over-precipitate action with insufficient means. The great technological advances in the field of jet propulsion and rocket techniques could not change the course of events.

The High Command began to compensate for its mistakes and omissions with a "psychological war effort". The demands on the courage and endurance of the fighting forces were stepped up to the point of ruthlessness. The self-immolation of the German soldier was expected to succeed where equipment had failed.

At the very point when a realistic assessment of military prospects would have readily revealed the inevitability of defeat, calculation and foresight were cast to the winds in favour of that pathetic unknown quantity, sacrifice and heroism, which has played so disastrous a role in German military history.'

Cry for Mercy

Nicola Malizia, Rimini, Italy, 8 August 1982

Nicola Malizia is a well-informed Italian air historian. He has written extensively on the Regia Aeronautica's operations in World War II. Born on 3 July 1930, he was at an acutely impressionable age when, in the early 1940s, his country's air force, often in company with the Luftwaffe, was flying its missions over the Mediterranean, Malta and the North and East African theatres.

Malizia, son of a railwayman, was living then in a small village in southern Italy. There, in his boyhood, he recalls seeing Axis troops and supplies, and the formidable arrays of aircraft, making their way to Sicilian ports and bases. He remembers, too, the later impact of the Allies' strategic and tactical bombing of the mainland. It is upon these recollections that his researches – of 'war veterans' and documents – have been built.

If some of the historian's controversial conclusions are unlikely to be accepted by the former opposition's 'veterans', who were actually there, no matter. The editor happily offers a place to a sometimes critical and fractious Italian view.

'For the first two years of the war, the morale of the Italian flyers was very high. Victory then, seemed certain. Successes, announced almost daily on the radio, confirmed this and made the aircrews even keener.

Huge concentrations of aircraft and armaments were being assembled in Sicily for the conquest of Malta. The Luftwaffe was based at the most strategically placed and important airfields – Catania, Comiso, Trapani "Milo", Gela, Gerbini and Syracuse. Their presence convinced the Italians that British resistance on the island of Malta would be crushed. German aircraft – Stukas, Heinkels, Junkers 88s, Messerschmitts – were, in the early days, superior to the Italians' Fiat CR 42s, Macchi 200s, SM 79s, Reggiane 2000s and Fiat BR 20s....

Relations between the Italians and Germans were extremely cordial. There was much goodwill and mutual admiration, something that is still remembered by World War II veterans.

After some time, however, when in spite of incessant attacks, it was clear that Malta was no nearer subjugation, the Italians began to realize that the Germans were little better than themselves! The capacity of the British island to defend itself amazed everyone. Not even the strong German detachments, with their numbers and their equipment, seemed to shake its implacable resolve.

The failure to neutralize Malta was disastrous for our aircraft convoys ferrying troops and supplies across the Mediterranean to the forces in Libya.

During 1941, they were mown down along the Italy–Tripoli route. It came to be called "Death Alley"....

On offensive operations, the Luftwaffe bombers preferred to be escorted by Italian fighters. In the earlier raids, the attacking forces heavily outnumbered the British.

It was in the spring and summer of 1942 that the fighting over Malta reached its climax. The most brilliant Italian fighter unit at the time was the 51° Stormo Caccia Terrestre (CT), equipped with the Macchi 202, whose emblem was a black cat with three green mice. This was commanded by Lieut.-Colonel Aldo Remondino. Stationed at Gela, it had a strength by the end of May 1942, of 71 operational aircraft. Although, by this time, the Regia Aeronautica's pilots were conscious of the hard tasks they faced every day when attacking Malta, morale was still high, and belief in victory remained strong.

The German Me 109 squadrons flew a much looser, more open formation than their Italian counterparts. Although they were not particularly afraid of the Royal Air Force's Spitfires, they never seemed to be noticeably aggressive when repelling attacks on the bombers they were covering.... Like the Me 109, the Macchi 202 was considered to be a first-class aeroplane, although weak on armament. The pilots regarded it to be a good match for the Spitfire VB.

However, the Italians had a high opinion of the Spitfire and felt that the Royal Air Force possessed some of the very best fighter pilots of the war. They did not, however, regard them as being any braver, or more heroic, than themselves, bearing in mind that, in the earlier days, the Regia Aeronautica's aircraft were inferior.

The Italian airfields in Sicily were seldom bombed in 1941 and the earlier part of 1942 and life went on in a reasonably carefree way. Although the Luftwaffe aircrews occupied separate messes, the two allies often visited one another and, as guests, ate together.

But strange things sometimes happened. One day 360 Squadriglia of 155° Gruppo CT (51° Stormo CT) had to be stood down from operations. Its self-appointed NCO cook, one Sergeant Luigi Vulcano, bought some deliciously fresh fish from the market in Gala. When he got back to the mess he found he had no olive oil to cook it with. Instead, he used the engine oil which fuelled the Macchi 202 engines. No one in the squadron knew he had done this.

The next morning, all the pilots reported sick with diarrhoea and stomach pains, and were off duty for the day. It didn't go with a swing.

As the fighting over Malta reached a crescendo and, later, when the Allied bombing began to intensify, it was then that the pilots and aircrews began to feel the long separation from their homes and loved ones. They were becoming anxious about the effects of the bombing upon their own families.

The oustanding Macchi 202 pilot of 20° Gruppo CT (51° Stormo) was Captain Furio Doglio Niclot, commander of 151 Squadriglia. He was a courageous and able leader and his aircraft always carried a distinctive white < on the fuselage which he regarded as a sign of defiance against the English. The RAF had been trying for some time to shoot him down, but without success. One day, after running out of ammunition in a combat, he fought off a Spitfire which was attacking a lone JU 88 which had been damaged over Malta and had become separated from the rest of the formation as it returned to Comiso. Instead of landing at his own base, the pilot of the 88 followed Niclot back to Gela. He wanted to thank his saviour personally and give him a hug.

Eventually, Niclot was shot down – on 28 July 1942, over Malta. It is almost certain that his conqueror was the Canadian, Sergeant George Beurling, who, on the same day, had accounted for Sergeant-Major Lalliero Gelli in a battle over Gozo.

Another of the Regia Aeronautica's exceptional pilots was Niclot's friend, Ennio Tarantola, who was shot down three times during the war – once, when piloting a JU 87 off the coast of North Africa, a second time as he returned after a fight over Malta and, again, over northern Italy towards the end of the fighting. He survived and now lives near Rimini.

While the wartime propaganda stirred up the Italian and British people to dislike and despise one another, the airmen of the two countries felt differently. Certainly early on in the war there was mutual respect for each other and on the North African front there were many instances of chivalrous treatment, of information being passed to relatives of men who had been shot down.

But this seldom applied over Malta. Several Italian pilots and crews who were shot down over or near the island, and were taken prisoner, told how they had been tortured and maltreated. Moreover, after the invasion of Sicily and Italy, there were authenticated stories of brutal acts known to have been committed by British pilots including the machine-gunning of Italians as they parachuted to safety.

Earlier, on 10 March, 1941, there had been the specific instance of Lieutenant Luigi De Pol, of 412 Squadriglia Autonoma CT being shot down in his Fiat CR 42 and then being gunned by a British aircraft (perhaps a South African pilot) as he descended by parachute on the East African front.

Again, on Palm Sunday, 1941, also on the East African front, Lieutenant Bruno Caldonazzi was similarly treated at the hands of a South African pilot flying a Hurricane I.

Other comparable incidents occurred over the Mediterranean, Malta and Egypt. Even our rescue service which did such good work picking up pilots and crews of all nationalities, did not escape. The Cant-Z 506Bs of 286 Squadriglia (94° Gruppo – 31° Stormo RM) were painted white with big red crosses, yet British pilots regularly ignored these mercy markings and shot the aircraft down without hesitating. The Luftwaffe also operated a rescue service with Dorniers, based on Syracuse, and reported similar experiences.*

*These are hard strictures. None who fought through those rough and taut Mediterranean days of 1942 would deny that this kind of thing may have happened. In the editor's experience, however, it was the exception in the otherwise clean and aggressive fighting. He saw but two instances at first hand in six months of the Malta battle. One was when Douggie Leggo, a fine Rhodesian pilot of No. 249 Squadron, was floating down in his parachute after being forced to bale out. A lone Messerschmitt 109, diving unseen out of the sun, fired at his canopy and it streamed earthwards.

No discipline will quell the blind fury of a squadron which has witnessed such inhumanity. A few days later the score was squared when a pilot from Leggo's squadron gunned down the crew of a JU 88 as they bobbed about in the sea in their dinghy, 10 miles or so southwest of Delimara Point, waiting to be picked up by a Dornier 24 of the Luftwaffe's air/sea rescue service. In war, one bad turn will always beget another. (Ed.)

Kesselring, the German, was a great commander, very much admired and respected by the Italians, both as C-in-C in Sicily and Italy, and, later, in 1944, when he was holding the famous Gothic Line which was such a thorn in the Allies' side. Generally, the senior Luftwaffe commanders in the Mediterranean, Sicily and Italy, were highly regarded.

Had the Axis forces succeeded in capturing the whole of Egypt, with Alexandria and Cairo, after reaching El Alamein, Malta would have fallen. The island would have been isolated and starved of supplies and food. An assault might well have succeeded in the spring of 1942, but it was never tried.*

Had there been a really serious and determined blockade by the Italian Navy right at the start of the conflict, the island wouldn't have held out even for a year. But the navy was not able to bring itself to fight a war – whether because of incapacity or betrayal, one doesn't know. The larger units could never face up to the reality of a full-scale naval battle; instead, they allowed themselves to sit about like ducks on a pond during the famous night attack at Taranto on 10 November 1940.

In consequence, there were heavy losses among our convoys and supply ships bound for Libya. One still has the suspicion that in the senior ranks of the Supermarina in Rome there were traitors who were directly, or indirectly, serving the British.

Finally, no attempt has been made here to evaluate the respective abilities of the pilots and crews of the various air forces. The Regia Aeronautica never sought to build pilots into "aces"... It is fruitless to try to make comparisons; the differences in equipment and operational conditions and circumstances were too great.

But, for the RAF and the USAAF pilots to claim the right to say – as they did later – that they were the best and the bravest of the Second World War, is both false and arrogant. There is far too much for the Regia Aeronautica and the Luftwaffe to say in mitigation....'

The Attractions of Rosie

Wing Commander P.D. Cooke,† Harare, Zimbabwe, August 1982

Captain Jack Malloch, the Rhodesian, who fought with 237 Squadron, was shot down on 22 February 1945, in the late stages of the Italian campaign. After attacking a small arms factory at Alessandria, some 50 miles north of Genoa, his Spitfire was hit by light ack-ack fire while strafing ground targets on the return.

Malloch baled out in the mountains and broke his ankle on landing. However, the local Italians, who hated the Germans, befriended him, got a doctor to put his ankle in plaster and sent him deeper into the mountains. The squadron heard – on 6 March – that he was safe and that the Italian

*Hitler vetoed it. He felt he could never rely on the Italian Navy. (Ed.)

†Peter Cooke's two elder brothers (they were twins) were both killed in World War II serving with the Royal Air Force. One, who served with 106 Squadron and then 44 (Rhodesia) Squadron, became the first Rhodesian to be awarded both the DFC and the DFM. The other was an air gunner who failed to return from the second 1000-bomber raid.

partisans hoped to return him in due course through the enemy lines.

Eighteen days later, on 24 March, Malloch's friend Peter Cooke (Wing Commander P.D. Cooke), another member of the same Rhodesian squadron, was hit by flak while attacking a supply train crossing the River Po, near Piacenza. When his engine stopped low over the mountains, and he couldn't bale out, Cooke made a miraculous belly-landing in a clearing in a pine forest on the mountainside. He enlisted the aid of friendly Italians before the Germans could find him. After a hair-raising ride on the pillion of a stolen German Army BMW motorcycle, and two days of walking and climbing, he was taken to a farmhouse in a small mountain village where, he was told, another Spitfire pilot was living. He knew at once, from the description, that it must be Malloch.

'Jack was out visiting an escaped prisoner of war in the next village when I got there. I went to meet him coming back. Imagine the surprise! There he was, all spruced up in his blue battle dress, very smart and clean, hobbling along with his leg in plaster, closely supported on either side by two very pretty young Italian girls.

'Seeing Jack with a bird on each arm was a startling sight. All the time I had known him he would have nothing to do with girls. And he didn't approve of the behaviour of the rest of us: he thought we were a bad lot when it came to the other sex. Yet, here he was now....

'These two girls, Rosie, the elder, who was just twenty-one, slim, elegant, with a very good figure, and Maria, who was eighteen, and full of fun, mischief and back-chat, came from a well-to-do family. They had been sent to live with relatives in this village to get away from the bombing of Piacenza which was an important junction....

'Jack and I shared an enormous double bed in the large farmhouse. We had the best bedroom, but on the ground floor were cows, goats, chickens and pigs – and a really dreadful smell. Still, the animals helped to keep the place warm. Above us, was a South African sergeant, named Barry. He had been parachuted in, complete with equipment, to set up an escape route for aircrew and escaped prisoners. His other job was to aid the Partiganas – the guerilla forces....

'When Barry sent off a message about Jack's injury, the reply came back saying that if we could find somewhere for a small aircraft to land they would consider flying us out. By now, an American P-47 pilot had joined us, so we all thought this a very attractive proposition....

'We found a valley where the stream ran more or less straight for 600 yards. With a labour force of local women, children and old men, we got some sort of a landing strip ready. Barry sent back the position, dimensions and direction of approach for landing. We were told to stand by for a date and time for the attempt.

'However, by this time, we were getting worried about Jack. He could get around quite well with his crutch, but was becoming very dependent on Rosie. She had to be there whenever he went out for exercise or visited another house; and she would sit for hours teaching him Italian. We began to worry that he mightn't want to go when the aircraft came for him....

'After several false starts, we got him down to the strip a fourth time on the back of a mule just as a German Fieseler Storch, with its swastikas over-painted with RAF roundels, and escorted by four USAAF P-51 Mustangs, approached. The Italian pilot had no trouble landing. He kept the engine running while we bundled Jack aboard. As there was already a passenger in the aircraft – a wounded Italian pilot, who had been picked up from another clandestine strip – there wasn't room for anyone else. The Storch was off again in less than three minutes. It disappeared down the mountain valleys while we dispersed back up to Baccapaglia.

'Jack was landed at Florence, sent to hospital for X-rays and treatment, and eventually rejoined the squadron.

'Bert, the American pilot, and I now relaxed, confidently expecting the Storch to return in a day or two. Alas, word came back via Barry that, as we were both sound in wind and limb we could walk back. That journey is another story, but I finally made it back to the squadron by the end of April....

'With the ending of hostilities, Jack and I went up to Piacenza to visit Rosie and Maria. After hitch-hiking for many dusty hours, perched uncomfortably on the top of great big fuel tankers, we eventually found the house. Jack had made a careful note of its location and address. The two girls were at home and the family made us very welcome.

'Jack and Rosie went down to the cellar for wine to celebrate. When, after some time, they came back Jack had traces of lipstick on him....

'We returned to Rhodesia about September, and, after a while at Cranbourne, were demobilised.'

Editor's note: Jack Malloch, a much-respected figure in Zimbabwe in the post-war years, was killed in an accident in March 1982, while flying a very special version of the Spitfire Mk XXIII.

Ramming to Win?

R.L. 'Dixie' Alexander, USAAF, quoted by Vern Haugland in *The Eagles' War* (Jason Aaronson, New York, 1982)

'We were told that in the early stages of the lend-lease program, the US had sent a number of P-39s and P-400s to the British. The planes had arrived at Blackpool, and were still there in crates. As desperately as the British needed help – and the lend-lease was welcome – they turned thumbs down on the P-39s at any price.

The Russians came and had a look, the story goes, and found little in the P-39s that they wanted. One Russian colonel was supposed to have commented that the planes did have metal props and thus could be considered "good for ramming". Hardly the best way to win a war!'

Russian 'Trust'

Air Vice-Marshal D.C.T. Bennett, *Pathfinder* (Frederick Muller, 1958)

I saw a good deal of the Russians towards the end of the war, because they made frequent efforts to visit my headquarters, obviously with the intention of obtaining some of our secrets. Finally the Air Ministry gave in and granted them permission, and two generals and six colonels arrived at my headquarters.... They were all smiles and courtesy and were most charming and likeable persons.

Their distrust of us and their obviously disguised views of us as an enemy were, however, quite apparent. One of the occasions of their visit was to show them a demonstration of the latest Pathfinder markers of all sorts and sizes – which we laid on one night. When we told them that this was to take place and that they would fly with me in a Lancaster, there was immediate confusion. Our interpreter finally explained to me that they were not permitted to fly in Royal Air Force aircraft owing to the dangers involved. The interpreter had apparently overheard the conversation in which it had been made quite clear that they did not trust us to fly them, as it was obviously a trick in order to kill them all.

The interpreter explained this to me, so I told him to point out that I myself was flying with them and I hardly intended to risk my own life in any way and that therefore they were quite safe. This apparently helped, but they still had to ring up their embassy in London to ask permission. In due course this permission was given, provided I was pilot of one aircraft and only one general was with me, whilst the other general and some of the colonels flew in another Lancaster. This was arranged, and the demonstration took place. I was amused to see the relief when in fact we landed back safely without killing any of them.

It's Not What You Say; It's The Way That You...

His Honour Judge R.A. Percy, Alnwick, Northumberland, 1983

'May, 1945.... A Dodge 15-cwt truck, with personnel and vital stores aboard, was waiting near Meiktila, in Burma, for the crew of a DC-3 (Dakota). They were to be taken to the temporary landing strip where the aircraft was standing.

It was so quiet and calm in the early dawn that it was hard to believe the retreating Japanese Army had been resisting here so recently and so fiercely before its rapid return to Rangoon in disarray. Only the stench of its unburied dead, and the bodies left in the adjacent lake, reminded us of the conflict. The lake was the only available water supply and the Nippons had resorted to human pollution to deny its use to the advancing 14th Indian Army.

The reliability of the truck was, however, causing problems. The passage from India, Kalewa and thence across the Chindwin, the Irrawaddy and into the central plains of Burma had put it to the test. The Indian driver/mechanic had had his difficulties. This was clearly a matter of concern to our worthy Royal Air Force pilot.

A Welshman and a flying officer, it was not disclosed how long he had been operating in this theatre. He had picked up a few words of Urdu –

enough to establish some very basic communication when mixed liberally with English. He relied on the belief that if the resultant mixture was shouted loud enough it was *bound* to be understood.

"Ghari, tikh hai – now?" (literally: "the truck, all right, is – now?") he asked of the driver, as he climbed aboard.

"Ji han, sahib! – Yes, sir," replied the Indian, reinforcing his disguised confidence with a wide, cheerful grin.

After lurching forward a short distance, amid clouds of dust, which had us all coughing, the Dodge ground to a halt. The engine had cut again. Out jumped the Indian. Twenty minutes' tinkering under the bonnet brought no gain.

The sun was now rising in the sky and the heat was beginning to gather. Accustomed to being serviced by excellent RAF ground crews, it was more than the Welsh pilot could stand. The prospect of being behind schedule on his flight to a point west of Mandalay, where the Japanese were encircled, trapped, yet still fighting fiercely, was too much.

The unhappy driver, met with a flow of expletives as he stood back from the dead engine, was finally demolished.

"Call yourself a f—g mechanic! You couldn't mend a bloody bicycle – hai!"'

Operation Oatmeal

The dropping and picking up of agents by moonlight in Japanese-held Burma and Malaya was a specialized and hazardous feature of the air war in the Far East. The times and distances flown by the Special Duties Squadrons in their Catalina flying boats and Liberators turned these operations into tests of physical and mental endurance as well as retentive courage.

The clandestine missions with the Catalinas, mounted sometimes from Trincomalee (China Bay) in Ceylon, were flown eastwards at maximum range across the southern Bay of Bengal and the Indian Ocean to the far coast of Malaya. Loaded with extra fuel tanks, even at the expense of armaments, these aircraft, cruising at 90 knots, would complete the round trip of 3000 miles in as many as 30 hours' flying or more. Towards the end of the war, if trouble was met on the way in to the landing, captains had authority to fly on eastwards to the Philippines, then occupied by the Americans. It was an uneasy option.

These operations, which were cocooned in secrecy, were planned in minute detail. So much had to be taken into account – tide and sea conditions, optimum angle of the moon for landing on the water – and, therefore, acute timing; speed and direction of surface winds; alighting distance from the shore, opposition and contingencies etc; all had to be patiently, and expertly, weighed. The sorties demanded excess resolution. Few, even now, comprehend their complexities.

Operation Oatmeal, flown on 28 and 29 October 1944, by two Catalina flying boats – aircraft letter 'S', FP 152, and aircraft letter 'P', FP 225 – of 628 Squadron makes a good example of the nature of these missions. Aircraft 'S' was captained by Flight Lieutenant Peter McKeand with Flying Officer Clive

Russell Vick as 2nd pilot, while Aircraft 'P' had Warrant Officer (later Squadron Leader) Leslie Brookes in charge, with Flight Sergeant Arthur Waddington as No. 2.

The target area for the sea landing was to be at a point just south of Khota Bharu on the far side of the Malayan peninsula. The 'passengers' were four Malayans (two were allotted to each aircraft) who would be put ashore in rubber boats which the crews would inflate immediately on touching down.

The timing and route meant that the aircraft would have to pass in daylight through the Japanese-occupied Nicobar Islands. From wave-top height, the crossing of the Kra Isthmus would be made at 6000 to 7000 feet in darkness before turning south and dropping down to sea level again for the run down the east coast to the landing area.

However, when the target locality was reached, it was at once obvious that a landing was out of the question. Exposed to the open sea, and with a long swell rolling in and breaking in a white spray on the shore, the inflated boats could not have lived in those conditions. What then to do? Leslie Brookes, at the helm of aircraft 'P' recalls it all.

'... I saw "S" turn away and fly towards the Perhentian Islands which were close by. There was no radio communication between the aircraft and my orders were simply to follow "S" and land as close to him as I could. Peter McKeand made a lovely landing in the lee of the islands and I followed him down as soon as I saw him settle in the water.

... I was, of course, watching "S" closely and to my surprise he started taxiing away from the islands. It became clear he was trying to reach the original target area which was now some 6 or 8 miles away. As we left the shelter, the sea conditions became worse and soon the Catalina was plunging with the sea spray blowing back and clouding the windscreen.

After about 3 miles, I suddenly saw "S" swing sharply round with a roar of the engines. After coming straight towards me for a short distance, the aircraft was throttled back and swung into wind where she lay motionless except for the pitching and rolling in the heavy seas. I knew something was up, so I also throttled back and waited, watching Peter closely.

The trouble soon became clear. Out of the slight mist appeared a ship on a southerly heading. At first it was eerily silent; then, as it came closer, I could hear the throb of diesels.

It was only a few hundred yards away and must have seen us but it took no action and disappeared to the south. I believe it was an enemy patrol boat, but perhaps the crew presumed we were Japanese flying boats. Whatever the facts, or the identity of the ship, the operation was now compromised. It would have been most unwise to land our passengers.

Aircraft "S" now turned round and set course again for our landing area by the Perhentians, closely followed by me. Our progress was much smoother as we were taxiing down wind and we soon reached our original alighting point where we took off in turn and set course for home independently.

The flight back to base was uneventful although we were all feeling the effects of lack of sleep and constant vibration, coupled with the noise of the engines. We made a splendid landfall at Ceylon, thanks to our navigator,

George Drummond (Flight Sergeant G. Drummond, RAAF), and eventually touched down inside the harbour at China Bay. We had been airborne for 29 hours, 20 minutes.

The trip had been a bit of a disappointment in spite of our hard work, but we were glad to be back and after a couple of beers we went to bed.

The following day we were startled by the news that Force 136 wanted us to do the job again. I must say I was worried by this. Surely the enemy would put two and two together and have a nice little reception party waiting for us. However, we were assured that all would be well....

'... There isn't really much to say about the second Oatmeal flight which took place on 31 October. All went smoothly, and this time we had planned to land in the lee of the Perhentian Islands where we knew the sea conditions would be good. The crew rapidly inflated the rubber boat and we watched our passengers paddle expertly ashore before we took off and returned to Ceylon. This time we were in the air for 31 hours, 20 minutes.'

Russell Vick, 2nd pilot in the leading Catalina, aircraft 'S', completes the narrative.

C.C. Russell Vick, 'The Russell Vick Papers', Sevenoaks, Kent, 1983

'We learned later, to our great disappointment, that the three Malayans (one* had suffered such severe air sickness on the flight that he was thought to be unfit to go ashore) had been captured. Lieut. Ibrahim Ismail, then about twenty-two, who had flown with us, and the two civilians, who had landed from the other aircraft, were quickly told by contacts that there was no hope of escaping detection.

The enemy knew of the previous landing. The smoke floats we had dropped to indicate the wind direction for landing had been found and, therefore, the intention had been given away. The inhabitants had been told by the Japanese to report anything suspicious. Hiding information would result in instant execution. The only way was to feign sympathy with the Japanese, to profess to hate the British and to offer to work wholeheartedly with the enemy.

Ibrahim Ismail, who, years later, was to become Chief of the Malaysian Armed Forces' Staff, played this role so astutely that the Japanese were completely taken in to the subsequent benefit of the Allies.

Within a short time of landing and convincing the Japanese of his "bona-fides" he had made contact with the base at Trincomalee who had then transmitted to him the pass phrase "Have you met Miriam?" When they got no response, they knew the Malayans had been captured.'

Honouring Suicide

Nihol Yuuhikai, *An Shounen Hikouhei* (Hara Shobou, 1967)

As the fortunes of war swung inexorably against the Japanese in the Pacific, the hopelessness of their ultimate plight was underscored by the readiness of the pilots to fly their suicide missions against enemy targets.

*He was subsequently parachuted into Japanese-held territory on a separate operation. (Ed.)

Corporal Toshirou Ohmura wrote this last letter before dying in an attack on the British fleet on 26 July 1945. He had entered a flying school in October 1942 and graduated as an operational pilot in July 1944.

Corporal Ohmura's aircraft dived into the sea just short of the carrier he was attacking. His citation stated that he had been promoted posthumously to the rank of Second Lieutenant.

Dear Father,

It is a long time since I last wrote to you, but I hope you will be relieved to know I am quite well. Every time I hear news of the homeland my heart aches heavily.

Here, in Malay, the conditions have worsened. I believe it is now our turn to go. As a suicide attack unit it is our duty to sink enemy carriers and battleships. Now we only wait for the time to arrive. I am pleased that I was born at the right time to serve my country.

I, Toshirou, have done nothing for my parents until now; but I hope you will see it as an act of sacrifice by a son when I attack the enemy with my aircraft.

I am proud to be a citizen of the Japanese Empire. As we shall all die sooner or later, when my ashes are returned will you lay a wreath of flowers on the casket?

As a man, this is my belief. If the enemy lands on our homeland, no distinction will be made between parents and children, so I fight for my country. I shall meet you at the Yasukuni Shrine.* It is a great honour for me, aged eighteen, to die as a suicide-attack pilot. I know that my spirit will live for ever as I am to do my duty for my country.... I shall die smiling....

It is late afternoon now in Malay and the sun is setting among the palms. Although my mother is far away in our home town, I remember her face.

I hope you are also well and that, in time, you will visit the Yasukuni Shrine to meet me.

Your son,
Toshirou

Postscript – To My sister, Tokie.

I am glad, now, to write to my sister. As your elder brother, I am sorry that I haven't been able to do anything for you until now. You are fifteen and are old enough to understand my will. I hope you will look after mother when I am gone. It will be your noble duty to guide our younger sister along the right path. I am sure you recognize the need to live an honourable life.

Your elder brother,
Toshirou

Up Ahead

Lieutenant-Colonel Lloyd F. Childers, USMC (Retd). Reproduced by courtesy of John B. Lundstrom, Milwaukee, Wisconsin, 1983

Now and then, a piece of reflective writing emerges to embellish the deeds of

*This shrine was built to honour the Japanese war dead. It was believed that the dead warriors gathered here so that the families felt, when they visited the shrine, they would be meeting their loved ones.

an engagement long since past. On 3 December 1974, John B. Lundstrom, the noted US aviation historian, received from Lloyd Childers (Lieutenant-Colonel Lloyd F. Childers, USMC (Retd), radioman 3rd class in the Torpedo Squadron Three (VT-3) of the Yorktown *Air Group, his personal account of the attack on the Japanese carriers on 4 June 1942, during the historic battle of Midway Island.*

Lundstrom had written to Childers to clarify his research. For their honesty and transparent genuineness these short extracts from the response merit inclusion in the Pacific record.

'... I decided the splashes in the water were long-range gunfire from the Japs, because it was only a short time before I spotted smoke on the horizon at about the 1.30 or 2 o'clock position. I pointed out this info to Corl (Mach Harry L. Corl – pilot) on the intercom; Corl somehow got the attention of the skipper (Lt. Cdr Lance E. Massey) or the Chief Radioman (Leo E. Perry ACRM) in the back seat and pointed in the direction of the smoke. The skipper immediately turned starboard and we were on our way. Altho' I am not sure, I think that we dropped altitude about 500 feet or so at this time, but not down on the water – yet.

Since I was facing aft, I had some difficulty watching dead ahead. Suddenly, Corl yelled, "Up ahead! Up ahead!" I thought, "What the hell is up ahead." Then I saw, looking around the port side, a Zero coming at us head-on. I jerked my gun into position, and as the Zero passed between us in a vertical bank, I pressed the trigger. Nothing happened, because my gun was still on safety. I was furious because I was so sure that I could have put a few rounds in him at that point blank range....

That was the beginning of the melee with about 30 Zeros going crazy in the most undisciplined, uncoordinated attack that could be imagined. I got a couple of no deflection shots on pilots who were in a helluva hurry to shoot us down. I fired about 200 rounds or two cans of ammo as we moved in on the target.

I observed the F4Fs above us mixing it with the Zeros. At one point, when I was not shooting at a Zero, I saw one coming almost straight down, not smoking, smacking the water within a hundred yards of us. So, I knew the F4Fs were not losing every encounter, even tho' badly outnumbered. I saw other aircraft fall out of the sky, but I could not tell whether ours or theirs. I recall thinking, "My God, this is just like watching a movie."

Altho' it required some stretching by me, I tried to see what we were headed into. One look showed me three carriers, cruisers and destroyers going top speed and turning. It was an awesome sight that frightened me badly. By this time the AA fire was coming at us in multiple black puffs. Corl yelled in his high-pitched voice, "Look at the skipper!" I looked to the port just in time to see the skipper's plane hit the water in flames, caused by a direct AA hit I assumed. As we got closer to the Japs, I happened to look over the side at the deck of a cruiser which we were passing over at a hundred feet or so. That also scared the hell out of me.

I was not aware or did not feel the torpedo drop, probably because Corl was turning and trying to jinx a little bit. A few days later I asked him when he dropped. He said that when he realized that we seemed to be the only TBD still flying and that we didn't have a chance of carrying the torpedo to normal drop distance, he dropped at greater than normal range.

At about this time probably, I couldn't figure out what he was trying to do and the flak was really bad; so I yelled into the intercom, "Let's get the hell out of here!" It is possible that my yell helped him to make his decision. Corl was a fine pilot who always exercised good headwork in our flights together as a crew.

As we turned our tail to the Japs, we entered an entirely different phase of the battle. The Zeros were waiting for us, working in pairs with good discipline and great precision. They seemed to be trying to keep one plane in firing position at all times, alternating from side to side.... On one high side run, two 7·7 slugs hit me in the left thigh. They didn't hurt much; scared me badly; and the bullets were hot.

I recovered quickly by talking to myself, saying, "You damned sissy, you're not badly hurt!" In a later run from forward, I was holding the gun butt at arm's length with my right leg stretched out for bracing when another 7·7 bullet hit me above the right ankle. It hurt so badly that I would have jumped out of my seat if not strapped on. A few minutes later, my gun jammed. Altho' I had a few tools handy, I was unable to clear the jam while watching a pair of Zeros make one run each.

As they came at us again, I pulled my ·45 cal pistol, fired four rounds at the first one as he came close by and three rounds at the second Zero. Those were the last runs on T-3, which was shot up badly with the engine misfiring regularly for the 2½-hour return flight. We were joined by Esders (Rad Elec Wilhem G. Esders) and Brazier (Robert B. Brazier ARMZc) to exchange hand code. Esders asked me if our ZB homer was working and I told him no. Since his engine was in good shape, he left us behind quickly.*

We landed alongside the head destroyer in the *Enterprise* group, after we flew past the damaged *Yorktown*.'

Editor's Note: VT-3 operated 12 Douglas-TBDs and 1 Devastator torpedo aircraft in the attack. Apart from the two that ditched on return, all the remainder were shot down.

Saburo Sakai – Japan's Greatest....

Henry Sakaida with John B. Lundstrom, 'Saburo Sakai Over Guadalcanal', taken from *Fighter Pilots In Aerial Combat* (Blake Publishing, California, 1983)

... Sakai's Zero had plunged down about 7000 feet before he finally brought it back under control. Apparently, the dive had extinguished the fire. The engine was running smoothly, but precious fuel had been consumed in his combats.... Sakai ... was permanently blinded in his right eye and had only partial vision in his left.... He was floating on the edge of unconsciousness, a bloodied mess, trying to find his way home. Over the

*Esders' aircraft also ditched, but Brazier died of his wounds shortly after.

expanse of ocean, he was never quite sure where he was. It was also questionable whether he had enough fuel left to make it back to Rabaul. However, in an epic struggle, which has become legendary, Sakai *did* make it back, covering almost 560 nautical miles in about four and a half hours.

On the airfield at Rabaul, pilots and groundcrew heard the familiar sound of a Zero's engine.... Scanning the sky with his binoculars, Lieut (jg) Junichi Sasai ... recognized the fighter. 'It's Sakai! It's Sakai! He's come back to us!' He could hardly contain his joy. Sakai had been ... given up for dead when he had not returned with the others.... [He] was unbuckled and helped out of the plane.... His face was unrecognizable even to his ... flying mates.

After a few days in the base hospital at Rabaul, Saburo Sakai was ordered back to Japan.... Sasai accompanied him to the pier where a flying boat was waiting. As they exchanged farewells, Sasai unbuckled his belt and handed Sakai his silver buckle, depicting a roaring tiger. Sasai's father had presented it to him with the thought that the tiger would protect him and bring him safely home. He wanted Sakai to have it as a remembrance of their friendship.* 'Come back to us when you have recovered,' said Sasai. 'We will meet again.'

But this was not to be. Sasai, known as the 'Richthofen of Rabaul', failed to return from another mission to Guadalcanal ... a victim of F4Fs from VMF-223 led by USMC ace, Marion Carl. He had been credited with 27 personal victories at the time of his death.

Arriving in Japan, Sakai spent over six months in various naval hospitals, but even the best surgeon could not restore the sight of his damaged eye. After recovering ... Sakai served as a test pilot and instructor. He was then transferred to Iwo Jima as part of the Yokosuka Kokutai and there (once more) engaged F6F Hellcats. He claimed two Hellcats shot down in one combat, an incredible feat for a one-eyed pilot in an ageing Zero fighter!

The war finally came to an end for Saburo Sakai on 18 August 1945 – three days after the official surrender – when he flew with a group of pilots who shot up a B-32 Dominator of the 312th Bombardment Group (USAAF) over Tokyo....

Sakai was promoted to Lieut. (jg) after the end of hostilities. He had engaged in more than 200 combats and claimed 64 enemy aircraft destroyed or damaged.

Dumbos in a 'Friendly' Sea

Air Commodore S.G. Quill, RNZAF (Retd), Newbury Line, Palmerston North, 1982

Almost all our flying in the Pacific was done over the sea. What little land there was, was pretty well all hostile – hostile for two reasons. It was enemy territory and, almost worse, it was all jungle. Simply, the land was to be avoided.

It was remarkable how quickly one became friendly with the sea.... But it

*The superstitious pilots would never give up a 'lucky' charm or emblem as the trusting Sasai did. (Ed.)

had its traps. Most of the current led away from where we were based and, more generally, towards enemy-held islands. Some of the fish weren't exactly friendly, either – barracuda, sharks. And you had to be prepared to work hard on your survival. But, for all that, it was better than the jungle.

A finely-tuned rescue system, using Catalinas – Dumbos we called them – gave us every confidence. Those chaps took some incredible risks to make their saves. They were mostly US Navy with a sprinkling of US Marines. They would take their Cats right in close to shore, often under fire, sometimes having to make their catch while still under way and taxiing well away before coming back and hauling in their man. They were remarkably brave and highly efficient people.

Dumbo escort could be quite exciting. Anything up to 12–15 fighters weaving over a slow Catalina, then performing all sorts of contortions to distract the opposition, while the rescue was made. Looking after those Dumbos was one of our most serious missions, for they were the chaps who gave us the confidence to fly so much over the sea. They had to be looked after.

'Pappy'

Colonel Gregory Boyington, USMC (Retd), *Baa Baa Black Sheep* (Wilson Press Inc., Fresno, USA, 1958)

Less lucky than some with the Catalina search-and-pick-up service was Colonel Gregory 'Pappy' Boyington, commander of the famed Black Sheep Squadron (VMF 214) in the Pacific, and the US Marine Corps' best-known fighter leader. Pappy Boyington, with 28 victories to his name, was shot down in a Corsair on 3 January 1944, right at the end of this third operational tour, leading a fighter sweep to Rabaul. Some evidence now suggests that his victor could well have been the Imperial Navy pilot, W.O. Takeo Tanimizu whose wartime total reached 32 and who was leading the fighter defence that day.*

After baling out, Boyington drifted on his 'little raft' for eight hours while the search went on. In the end a Japanese submarine found him first.

[As I drifted] . . I found I had something clutched in my hand. It was a small card that had been sent to me by a Catholic nun from Jersey City. The card was soaked with sea water. . . . Then I remembered how, a couple of years previously, when passing through Jersey City, I had given a talk to a Catholic orphanage, and a couple of little girl orphans took a fancy to me. So, when I was out in the south Pacific long afterwards I had sent this nun some money

*Sources: The press release of the Zero Pilots' Association of Tachikawa – Shi, Tokyo, dated 18 June 1982. Captain Robert A.G. Strickland, former US Marine Corps member of Boyington's Black Sheep Squadron, and founder and president of the Biwakokisen Steamship Company of Otsu City, now feels able to go further: 'After years of research,' he wrote to the editor on 25 September 1982, 'there is now documentary proof that Takeo Tanimizu shot down Pappy.' Thirty-eight years on there seemed to be no lingering feelings. Excusing himself from attending a Zero pilots' reunion, Boyington wrote to Bob Strickland on 2 July 1982: '. . . Above all . . . will you please give Tanimizu-San an affectionate hug for me? Tell him it is a great relief to learn I was shot down by one of Japan's top "aces" who is a warm and compassionate person. . . .' The generosity of a great pilot. (Ed.)

to buy dresses for these two little orphans, and the nun had mailed it to me. . . . I had paid little or no attention to the card when I received it, but absently had stuck it in my jacket pocket, the pocket that happens to be above the heart. But why now on the raft I had it in my hand, water-soaked though it was, I never will know. Yet, for some peculiar reason, I now looked at it more closely than I had before. It was a picture of a lady with a baby in her arms, and there was a boat on a stormy sea. On the back of it I could make out the blurred lettering of a lengthy prayer. I read it over time after time, while drifting there. I probably read it over forty or fifty times, and it seemed to give me a great deal of company.

. . . I had time to meditate. I wondered how it was possible for me to be saved from death so many times when people I considered so much better than myself had to die. I hated myself, but it was for the first time that I realized that a Higher Power than myself does with one as it wishes. I truthfully wondered why a Higher Power might be saving a bum like me.

I look back now and realize that this was the first time I had ever prayed without asking for something, the first time I had ever prayed honestly, or properly, in my entire lifetime. I wasn't asking for a deal, although I wasn't conscious that it is impossible to make a so-called deal with one's Higher Power.

I prayed: 'I don't know why I was saved, and I don't really care, but you have my permission to do anything with me you want. You take over. You've got the controls!' And, oddly enough, this seemed to help me through the next two hectic months.

After being released from the prison camp, at the end of the war, and returning to the United States, I sent a letter to the nun and told her about the card and that it had come to be in my hand at such a time, when there seemed no chance at all of getting out. I wrote all this to her as best I could, and when I was in New York City sometime later the little nun presented me with my medallion to replace the card. And this is why I always wear this medallion. It is about the size of a dime, or even smaller. On one side is the Virgin Mary, and the edge is bordered with stars.

McGuire's Last Mission

Carroll R. Anderson, reprinted by permission of *Air Force Magazine*, January 1975

Major Thomas McGuire of the 475th Fighter Group – Satan's Angels – was a stand-out in the Pacific War. On 7 January 1945, he was leading a section of P-38 Lightnings in a marauding sweep over the Philippines hunting for the kills that would take his total past Major Dick Bong's, America's leading exponent of the fighter game. He did not return. Thirty years on, Carroll R. 'Andy' Anderson, himself a seasoned member of the 475th, with 89 combat missions in his log book, told the story of the fateful, early morning sortie after he had researched the background with Mizonori Fukuda who was also engaged in the fighting that day, albeit on the opposing side. As was so often his way in his lifetime, Bob Anderson brought to his narrative the authority that first-hand experience alone can provide.

'... The battle for the skies over Mindoro Island [in the Philippines] was as vicious as any fought in the southwest Pacific area. One of the Fifth Air Force pilots drawn into those murderous skies was a slight, steely-eyed extrovert from the 431st Fighter Squadron of the 475th Fighter Group. His name was Major Thomas Buchanan McGuire, Jr.

En route to Mindoro in his Lockheed P-38 on 7 January 1945, he led a fighter sweep over Negros Island, seeking his thirty-ninth victory. This is the story of that sweep – the last mission of Thomas McGuire....

The dynamic young man from Ridgewood, NJ, was America's leading active ace with 38 confirmed victories. His score was second only to that of Major Richard Ira Bong, who had recently received the Medal of Honour and completed his second tour of duty with the Fifth Air Force. Bong had 40 Japanese aircraft to his credit. McGuire wanted to be No. 1, and he was due to go home in February. Time was running out.

The 7 January mission had originated the night before when a group of 431st fighter pilots gathered in one of the ramshackle tents that passed for home. They heard McGuire say, almost casually, "How about going out in a four-plane sweep tomorrow? Thropp, you want to go?"

"Hell, yes, Major," replied Thropp. "I'd like to go."

Within minutes, Lt. Douglas S. Thropp, Jr., Capt. Edwin R. Weaver and Major Jack B. Rittmayer had volunteered and the fighter sweep was organized.'

'With sure dispatch, McGuire finished the cockpit check of his twin-engined fighter the next morning. The big P-38 and the three others from "Daddy Flight" were sitting at the edge of the Marston strip at Dulag on Leyte Island. It was dawn. 0615.

[Very soon] the mission was underway, a hell-for-leather fighter sweep to Negros Island, and from there to Mindoro, where they expected the hunting to be excellent.

McGuire levelled the flight at 10,000 feet....

This was no game to be played in the cloud-flecked sky. You lived or died by how well you could see movement above and behind you, by how well you scanned a break in the clouds as your fighter skimmed beneath, or by how often you sensed a reflection from up sun.

West of Leyte Island, the weather thickened. The sunny upper reaches of the sky dissolved into masses of brooding blackness beneath the glistening anvil tops of the approaching thunderheads.... It became prematurely dark.

Throttling back, Mac gradually led Daddy Flight to 6000 feet, surrendering strategic height. To get through what was clearly a threatening storm, he would lead Weaver, Rittmayer, and Thropp under the lowest layer of clouds....

Apparently they had the sky to themselves. The VHF radios were silent. If anyone else was on patrol over Negros in this weather, he was saying nothing.

But there *were* other planes in the air.'

'Warrant Officer Akira Sugimoto of the 54th Sentai [squadron] was tired. Headquarters had detailed him and one fighter pilot from each of the sentais on Negros Island and Luzon to fly search missions for an American reinforcing convoy headed for Mindoro or Lingayen. The fleet was to be attacked with bombs or by whatever means were necessary to sink the ships.... But his search had been fruitless.

The weather was bad and had grown worse, and Sugimoto knew the despair of being unable to locate the Yankee ships, if indeed there were any nearby.... He dropped his green Ki-43, the Nakajima Hayabusa known to American pilots as "Oscar", below the cloud cover as he flew back to the strip at Fabrica.

Elsewhere in the dismal skies, 21-year-old Sgt. Mizunori Fukuda was heading home ... to Manapla strip on Negros Island. His search for the American convoy had also been unsuccessful....'

'McGuire led Daddy Flight directly to Fabrica arriving over the strip at 0700. The flight lazily circled for 5 minutes at 1400 feet, but none of the fifteen parked fighters gave any indication of taking off ... nor was there any flak.

McGuire, who had been ordered not to strafe, [now] altered course for the Japanese dromes on the western coast of the island. The radio remained silent....'

'Ten or fifteen miles from Fabrica strip, Capt. Edwin Weaver sighted a single green-coloured plane about 500 feet below Daddy Flight. It was approaching head-on and was no more than 1000 yards away.

Weaver's voice, filled with urgency, barked over the radio, "Daddy Leader! This is Daddy Two! Bandit, twelve o'clock low!"

In the brief moment it took McGuire to acknowledge the call, the enemy plane, piloted by Sugimoto, had flown directly beneath the American flight.

[The Japanese fighter] ... had suddenly sighted the Lightnings directly ahead and above....

At that moment, Sergeant Fukuda had settled his Ki-84* into its final approach to Manalpa with gear and flaps down. He glanced toward Fabrica just in time to see the Oscar attack four Lockheed P-38 Lightnings. Instantly, he retracted his landing gear and applied full throttle, carefully milking up the flaps. He could see the Oscar dogfighting the Yankee planes, darting in close as the Americans continued to bank sharply to the left, apparently trying to hem in his comrade.'

'Sugimoto drove his Oscar closer to Thropp. Rittmayer wracked his P-38 to the edge of a stall and fired one burst from his guns which temporarily drove off Sugimoto. Still, it did not take him out of the fight. Turning even more tightly, Sugimoto drew a bead on Weaver and fired.

"Daddy Leader! This is Weaver! He's on me now!"

The urgency in the voice was clearly evident. Weaver knew the Japanese was no ordinary pilot. He was a true "wild eagle" of Nippon and he damn well meant to kill him, Thropp, Rittmayer and McGuire if he could!

Weaver tightened his bank slightly, skidding at the same time to throw off the Oscar pilot's aim. The green-coloured Oscar clung tenaciously to the

*Ki-84 Hayate, or Frank.

jinking Lightning, Sugimoto firing frugally. He wasted no ammunition in long, undisciplined bursts.

McGuire used all his skill to bring his guns to bear on the enemy plane. Ordinarily, a Lufbery circle would have worked, but not this time. The Lightning shuddered at the edge of a stall. McGuire felt it. He had to feel it!

The P-38 struggled ... snap-rolled in one wild gyration and plunged, inverted, to the ground, 200 feet below.'

'Weaver did not see the impact, but he saw an explosion and fire on the ground. He knew someone in the flight had crashed.'

'With the P-38s scattered, Sugimoto broke off the attack and sped north, climbing fast for the base of the clouds.... His Oscar had taken a mauling from Rittmayer and from Thropp's last burst. Now he had to get the crippled fighter down. The mountains posed a problem but as he glided below the cloud base, well away from the fight, he found a flat spot and landed safely. He saw bedraggled Filippinos appear and run towards his plane. Sugimoto died from six bullets fired into his chest. Within minutes, the guerillas had stripped the body of its clothing and melted back into the jungle.'

'Sergeant Fukuda was approaching the fray at top speed when he saw the blazing crash on the ground and the Oscar zoom up into the clouds. He found the Lightnings beneath the overcast and attacked head-on.... His guns rumbled, sending a stream of slugs smashing into Jack Rittmayer's plane....

The big Lightning swerved and plunged straight down. A huge explosion mushroomed from a point approximately a mile and a half from the village of Pinanamaan, where McGuire had crashed....

Just as quickly as the "wild eagle" had seized the initiative, he relinquished it, leaving the sky to the two remaining Americans.'

'"McGuire, this is Thropp. McGuire, this is Thropp. Do you read? Do you read?" There was no answer, and it was then that both Thropp and Weaver knew who had crashed

"Daddy Four, this is Weaver. I'm right behind you. Let's go home!" Thropp now knew that Rittmayer was gone, too.'

'The struggle was not yet over for Sergeant Fukuda as he flew his badly damaged Frank back to Manapla. Weaver had gotten in some telling shots.

In the landing pattern, only one of the plane's main gears locked in the down position. The aircraft flared, stalled and touched down. With a horrendous crash, it cart-wheeled onto its back in a cloud of dust. Fukuda half-crawled and was half-dragged from the wreckage. His ground crew later counted twenty-three bullet holes in the aircraft.

Thropp landed at Dulag at 1755, followed by Weaver ten minutes later.'

'Word of McGuire's death swept through the 475th Fighter Group and the Fifth Air Force like wildfire. No one wanted to believe the Iron Major from Hades Squadron was gone.'

Writing to Bob Anderson three decades after the engagement, and signing himself the 'Former Flying Sergeant of the Japanese Army', Mizunori

Fukuda asked to have this 'message' sent to 'all members of the 475th Fighter Group'.

'We fought each other in the nightmare of war, but after thirty years it is my pleasure that we, Americans and Japanese, lead the world in each [our] own way, and I hope to help each other [*sic*] with mutual understandings ever after. I hope you, the former members of the 475th Fighter Group, will take an active part in world peace with good health.'

Satsuma-gun, Japan, 1975

Palembang

On 24 January 1945, naval aircraft from the British fleet – Avengers, covered by Corsairs – launched the attack on the massive oil refinery at Palembang, in eastern Sumatra. They called it Operation Meridian. Pladjoe, the most southwesterly of the two refineries which made up Palembang, was hit first, followed the next day by Soenjei Gerong to the north.

High-flying reconnaissance aircraft had established the targets. Royal Air Force Intelligence officers, interpreting the photographs, had confirmed the location of heavy ack-ack defences, but no balloon barrage. Newly-excavated pits were identified, but the assurance was given, 'no balloons'.

Japanese aircraft were known to be based both on Sumatra and western Java and the assault would embrace a land crossing of some 200 nautical miles. The four carriers, *Illustrious, Victorious, Indomitable* and *Indefatigable*, with the remainder of the fleet, would be operating as close to the southern shores as reasonable safety might allow. The risks, manifestly, were great.

'Doc' Stewart, from *Indomitable*, Admiral Vian's flagship, was to lead one Avenger bombing wing, while David Foster, CO of 849 (FAA) Squadron, would command the TBF wing from *Victorious* and *Indefatigable*.

The first attack – Meridian I – despite resolute resistance, was successful. Aircraft damage was considerable although actual losses were relatively light. The squadron commander of *Indefatigable*'s Avengers was a casualty after hitting a barrage balloon cable that RAF Intelligence had assured the crews would not be there.

The second attack – Meridian II – on Soenjei Gerong the next day, with the enemy now right up on his toes, was a very different affair. Foster, with only 12 of his 21 Avengers from 849 now serviceable, was again leading the TBF wing from *Victorious* and *Indefatigable*. His recollection of the assault remains sharp.

David R. Foster, 'The Foster Papers', Mission Hills, California, 1982

'. . . The carriers' fighters found few Japanese on their airfields, they were either airborne or deployed elsewhere. The bombers, which had not been attacked before their bombing run on the previous day, were "jumped" before reaching the target area, with the Japanese fighter planes coming out of the sun.

I was about to give the signal to dive when two Japanese fighters "jumped" the lead flight and shot down my No. 2 and 3. With both wingmen gone, I started my dive through the balloon barrage. The second, third and last flights followed suit, with the *Indefatigable* squadron close behind. The Soenjei Gerong refinery was left in flames. Our bombing had again been accurate.

On arriving back over the fleet two of 849 Squadron's aircraft could not lower their undercarriages and had to ditch in the sea alongside destroyers. Five others that landed safely were badly damaged . . . so the squadron toll in Operation Meridian II was 2 crews missing, 2 crews ditched, 5 aircraft damaged. From the 12 aircraft which had taken off early that morning, this did not bode well for Meridian III which was expected the next day.

Thankfully, the damage to the two refineries was considered great enough to put them out of action for several months, so Meridian III was not thought necessary. I don't know how many planes we could have mustered for the third attack on Palembang had it been called.

Sad news followed the operations. The survivors of the crews who were shot down – some twenty-five officers and rating air gunners – were imprisoned on an island off Singapore. After the Armistice had been signed in Tokyo Bay, they were ordered to dig their own graves and were then buried alive by their captors. . . . One cannot forget these things.'

So, with an act of primaeval cruelty, was concluded a martial success that will rank in Fleet Air Arm history alongside Taranto, Matapan, the Mediterranean convoy battles and *Bismarck*.

Reception Party

John A. Iverach, Winnipeg, Manitoba, 1982

'In the summer of 1945 Flight Lieutenant Cliff Edwards, a pilot of the Royal New Zealand Air Force, was operating with a Mosquito squadron from the Cocos Islands. Also stationed there were two Liberator squadrons, a Spitfire unit and a Javanese-Dutch Air/Sea Rescue (Canso) squadron. The Liberators were on bombing missions and the Mossies on photo reconnaissance covering the Sumatra, Java and Singapore areas.

After the atomic bombs were dropped on Nagasaki and Hiroshima we waited from day to day for the official Japanese surrender.

Meanwhile, operations continued. The Mossies' main task now was to locate the many POW camps the Japs had tucked away in the jungles. When they found them, the Liberators followed with supply-dropping sorties.

Cliff was on such a flight near Singapore when an engine quit. He thought he could make it back on the other, but, on the way, that also began to misfire. He had to make a choice – ditch in the heavy sea and hope the rescue boat would get him before the sharks did, or land at Japanese-held Singapore and take his chance as a prisoner of war.

With the war almost over, Singapore seemed to be the better bet – except for one question. Would the Japanese, defeated and about to suffer the humiliation of unconditional surrender, take this opportunity to wreak vengeance on one more hated enemy, probably their last prisoner of war? As Cliff headed onto his final approach to the Singapore runway, he knew he would soon find out. He wasn't much encouraged by the Japanese track record to date.

As the Mosquito rolled to a halt on the tarmac, a horde of little men in white uniforms rushed towards it.

"At least I'll take a couple of the bastards with me!" thought Cliff, taking his pistol from its holster and bracing himself.

But instead of attacking him, the men stopped a few yards away and, quickly forming up into a line, stood stiffly to attention. A heavily braided senior officer stepped forward, saluted, and presented his sword.

Unknown to Edwards, a British aircraft was expected that day bearing the official Allied party which was to accept the Japanese surrender. This they thought, was that aircraft. Although the mistake was soon discovered, Edwards was, nevertheless, treated like visiting royalty.

A message was promptly despatched to the Cocos Islands. The following day S./L. Newman, Cliff's CO and close friend, flew over and brought him back to the Cocos – and to reality.'

Retrospect

Don Charlwood, *No Moon Tonight*, (Angus and Robertson, London and Melbourne, 1956; paperback to be published by Goodall Publications Ltd, London)

The train left Aberdeen in the late afternoon and by the time we had reached Stonehaven it was dusk. As I watched the North Sea beating at the Scottish cliffs and heard the shriek of the wind about the train, I realized that I might never again see this bleak expanse of water; certainly never from the air, perhaps never from the shore. Often we had looked down on it as darkness fell and we sped eastward to a waiting Germany; we had only felt affection for it when I was able to say to the crew, 'We have crossed the enemy coast and are well out to sea.' I used to feel then that everyone had sighed deeply. Down would go our nose and within forty minutes we would see the searchlights of England.

This same sea they said had claimed Tom and Max and God only knew what others from among my friends. Now, I alone of all of them was going home.

A porter drew the blackout curtain, leaving only the wind to remind me of the scene outside....

Sometimes in later years that last glimpse of the North Sea returned to me, but the emotions that it represented, I almost forgot. I even forgot how we had felt as we had watched for the Ruhr, crouched ahead of us in the darkness. Then one day long after, without knowing why, I became afraid. The other life began crowding back, until suddenly I realized that I was listening to a song we had known ten years before. Even as I listened my surroundings dissolved and the song became a worn recording, an arrangement of 'Tristesse' played on a gramophone.

I knew then that we were in the mess at Elsham Wolds on a night in January 1943. About me were many men I had known, most of them now ten years dead. There were Laing and Webber; Newitt and Berry; Maddern, our own pilot, and with him Richards, our Welsh engineer. They wore heavy white sweaters under their battle dress and each carried on his collar a silver whistle. Tonight we were to go to Düsseldorf. They were sending only thirty bombers from the whole Command, for it was to be an experiment that might easily fail.

On the faces of the men about me there was a similarity of expression, an expression I had often called 'contemptuous serenity'. They spoke very little and when they did their voices were subdued, as though at the same time they were listening.

I wondered why it was that this recording happened to be played so often as we waited to leave. To me it as a song without hope, full of urgent pleadings we could never heed.

I sat at a corner table finishing a letter home, as I had often done while we waited to leave. Familiar sounds registered on my mind as they had done before: the clack of billiard balls; rain on the windows and the unrelenting song.

Soon the crews about me began leaving. In their hands they carried red packages of escape equipment – tabloid foods, maps printed on silk, a small compass – in case tonight they were shot down, but survived. It occurred to me how young they were and how foreign to their task. There was Morris, who at nineteen was so soon to die, and Syd Cook, now a sergeant, but in the year of life left to him to rise to squadron leader and to be doubly decorated. And Ian Robb, the first among those I knew who was not to return. Other men lay dozing in the long green chairs. As members of their crews roused them I noticed how child-like they appeared in the moment they woke. I saw them leaving the mess. Outside the rain had increased and as they passed through the door they put on their coats and turned up their collars.

With a feeling of urgency I turned back to my letter.

'It is time now to go. Geoff and the rest of the crew have already left and except for a few ground staff men I am alone in the mess. Once again it scarcely seems possible that we will leave this room, with its chairs and fire, for the grey miles of the North Sea, then Germany.'

I paused, not knowing how to end. The imploring crescendo of the song filled the room. In a moment of defiance I wrote quickly, 'Whatever happens, I feel that when all is known, all will be well.'

Between the inner and outer doors I paused. Outside the night was empty and very dark, the rain heavier. I shuddered and pulled on my coat. As I left the building the last words of the song followed me, as on other nights they had followed other men now no longer here,

'No moon tonight,
No moon tonight.'

'I Close My Bombing Years . . .'

Group Captain T.G. 'Hamish' Mahaddie, final extract from 'The Bombing Years', Winnipeg, Manitoba, 5 November 1982

. . . 'Mossies'* were sent on 36 consecutive nights to Berlin, carrying 4000-lb

*Mosquitoes.

'cookies'. There were the grand slams, destroying viaducts and bombing Hitler's lair at Berchtesgaden. Manna from heaven was dropped on Rotterdam in one of the most exciting episodes of the whole war. Sides of bacon, and half a carcase of beef, went thundering down in two bags on the city's racecourse. And the crowds were there, sitting in the stands, cheering each drop....

We come now to the end – and to the last sortie. *We had lost 7122 aircraft; 43,786 aircrew had been killed; 4000 had been very, very badly injured; 10,000 had been taken prisoner. This was the heavy toll of the Harris years. It takes no account of the earlier period which brought so little reward for so much effort and loss of life* [editor's italics].

The last sortie I led with a force of sixty bombers. I flew with one of our early Pathfinders, acting as his flight engineer. We flew to Lübeck to pick up prisoners of war. For me, it was the most satisfying, the most rewarding sortie of the war. After nearly six years of bombing, it brought, I dare say it, not a few tears. Here they were, these fine characters all sitting about on the ground, spent, exhausted.

We flew them back to the United Kingdom and to the receiving stations – thirty prisoners-of-war in each Lancaster. There, they were welcomed by lovely WAAFs who took them gently by the arm and led them through the processing routine and prepared them for leave and return to their homes.

I close my Bombing Years with a tribute to the girls, the wives, the WRAF, the mothers, that favourite aunt.... To the girls who obeyed the old naval command, 'pack and follow'. They threw the spaniel, and the kids, and what rations they could beg, into the back of the car and drove off, sometimes 500 miles, to support us in a way that made it all possible. They sustained us with their courage and their faith; they brought us fun and cheer, gaiety and humour; they embellished our exacting lives....

To do it, they had to endure the waiting, the anxieties and the worries. Many must have been the times when they recited to themselves those moving lines of Noël Coward's* with which I close:

Lie in the dark and listen.
It's clear tonight, so they're flying high –
Hundreds of them: thousands perhaps,
Riding the icy moonlit sky –
Men, machinery, bombs and maps,
Altimeters and guns and charts,
Coffee, sandwiches, fleece-lined boots,
Bones and muscles and minds and hearts,
English saplings with English roots
Deep in the earth they've left below.
Lie in the dark and let them go.
Lie in the dark and listen.

Lie in the dark and listen.
They're going over in waves and waves,
High above villages, hills and streams,
Country churches and little graves,
And little citizens' worried dreams.

*'Lie In the Dark and Listen', *The Collected Lyrics of Noël Coward* (Heinemann).

Very soon they'll have reached the bays
And cliffs and sands where they used to be
Taken for summer holidays.
Lie in the dark and let them go.
Theirs is a world we'll never know.
Lie in the dark and listen.

Lie in the dark and listen.
City magnates and steel contractors,
Factory workers and politicians,
Soft, hysterical little actors,
Ballet dancers, reserved musicians,
Safe in your warm, civilian beds,
Count your profits and count your sheep,
Life is passing above your heads.
Just turn over and try to sleep.
Lie in the dark and let them go.
There's one debt you'll forever owe.
Lie in the dark and listen.

Hamish Mahaddie also recited Noël Coward's poem on 4 September 1982, in Guildhall, at the dinner held in honour of Marshal of the Royal Air Force, Sir Arthur Harris which was attended by some 600 former aircrew as well as pilots who took part in the Falkland Islands fighting earlier in the year. It was on this occasion that Group Captain Sir Douglas Bader made his last speech. He died of a heart attack the same evening while being driven home by his wife after the dinner.

The Himalayan Men

Don Charlwood, *No Moon Tonight* (Angus and Robertson, London and Melbourne, 1956; paperback to be published by Goodall Publications Ltd, London)

When I saw him I felt that Hughie Edwards could almost have been awarded the VC on appearance. He looked the personification of courage.

Edwards and such other men as Gibson, Cheshire and Pickard were mountains, far above our heads. We, who had merely a tour behind us, were foothills; above us again were men of two and even three tours; but towering over us all were a few Himalayan names, men with a richer combination of skill and luck than comes to more than one in ten thousand. Below us were the broad plains, the average men of Bomber Command, for squadron life had convinced me that the average man, however great his skill, reached no more than ten operations before he was lost to sight. At Elsham the stream of average men flowed before us still. But for us, the anguish of seeing them vanish had lessened, for most of them we barely knew. In seven months we had reached old age and our contemporaries were dead. The public seldom heard of these missing thousands; rather they heard of those whose luck lasted sufficiently for their courage to be revealed. Probably no one realized this more clearly than did the Himalayan men themselves and yet, knowing the odds, men like Edwards strained their luck to the uttermost – and seldom did they win.

LAST MISSION

Perhaps anything we are doing for the last time we do with regret; or, if not regret, with sharp memories of all the times that have gone before. Perhaps thus we acknowledge our mortality....

Over Duisburg high cloud swirled continuously, hectically lighted by flak, but blotting out all view of the target. We bombed in silence, took our photograph and turned away.... Behind us the mysterious land of all our adventure was slipping into darkness; the land that held the secret of the missing; the most tragic land in the world. Except for shell bursts and the watchful beams of searchlights, it was soon enveloped by night. I fancied that in the plane there was an inarticulate spirit of humility and gratitude....

The Advanced Science

The Rt Hon. Lord Justice Kerr, London, 1982

'I only did about six operations against E-boats, patrolling up and down the Dutch and Belgian coasts for 6 hours at night. We were fitted with a new type of radar-controlled bomb release, always flying at 100 feet above the sea. *This worked perfectly in practices on a single buoy as a target, but since the E-boats came out in twos and threes, the bomb invariably fell in the geometrical middle between them!* [Editor's italics.] So we missed them, and fortunately, they also missed us with their rockets.

I only remember one attack ... when we seemed to be stationary in the middle of a Christmas tree with these rockets of various colours appearing to come past us quite slowly – something which I have never understood. Most of the "ops" were spent looking at one's watch every minute and chewing peanuts, wanting the time to go and to get back to Langham: after "ops" one got eggs and bacon for breakfast and a 48-hour pass!'

Good (Royal) Question

It was the professors of the flying art who gathered at the Empire Central Flying School at Hullavington, in the West Country. They were the alumni of the species – the chief flying instructors, the chief ground instructors of the various air forces, British, Canadian, South African, Australian, New Zealand. These, and picked officers of the US Navy, Army and Air Force, and their counterparts from the other Allied air forces, used to assemble for the three-months' courses in all their imposing strength.

Here was the Establishment and here was pooled the sum of the knowledge of wartime flying and of the aeroplane as a means of making modern war.

Next door to Hullavington, across the county boundary, in Gloucestershire, was Badminton, the seat of the Duke of Beaufort, Master of the Royal Horse. It was here, during a Wings for Victory week, that Her Majesty Queen Mary, resident, then, at Badminton, witnessed a specially mounted flying display by CFS's favourite sons.

A feature of the programme was a dive-bombing demonstration by Miles Master III aircraft [*sic*], the leading aeroplane being piloted by Squadron Leader Bristow, of the CFS directing staff, and Squadron Leader Jacklin, from Rhodesia, flying as No. 2. Queen Mary, sitting regally erect, was at the ringside to watch the results. Every detail is still retained in Squadron Leader Bristow's mind.

The Hon. Mr Justice Peter Bristow, London, 1982

'... After a visit from the US Navy's dive-bomber pilots (10 bombs into a 50-foot circle to graduate) we know all about this. If you dive the aircraft at 80 degrees you can't miss. Unhappily, we have no dive-bombers with dive brakes. But if you dive the Miles Master III at 80 degrees from 4000 feet to 1500 feet you should not pull the wings off as you pull out, and the engine and airscrew noise are impressive....

The great afternoon arrives, and we are airborne, each with two 4-lb practice bombs which make a nice puff of smoke. Corporal Sutcliffe, our keen aircraft-hand flight sweeper-upper, begs a ride in the back seat. It's a lovely day, with some cumulus at about 3000 feet, but clear when we arrive over Badminton at 4000 feet on time. There is the target. Some optimist has put it only 150 yards in front of the crowd. So....

"Eighty degrees and you can't miss ... 10 bombs in a 50-foot circle to graduate." Confidence supreme, and the reputation of the Royal Air Force at stake. Nose up, throttle back, 80 knots, and over we go into our half-roll and straight down on the target. Release, pull out low over the house with plenty of engine noise, and as we climb to repeat the performance confidence is justified; both bombs are practically direct hits. We repeat the exercise. Just as close. We hope the pull-out is just as spectacular. And so home, pleased with ourselves and all well – except that Corporal Sutcliffe's stomach, unaccustomed to such drastic manoeuvres, has let him down.

Our group captain has been sitting in the front row next to Queen Mary. Pleased with the performance, he asks Her Gracious Majesty what she thought of it. He gets a very cold look.

"Are those young men really *allowed* to do that?" she asks.'

Norway's 'Black Problem'

Major General Werner Christie, Oslo, 1982

One incident springs to mind from our happy international relationships. In 1945, I took over the Hunsdon Wing of Mustangs (P-51s). In one of the squadrons there were two black pilots from Jamaica. They were called Fifty-eight and Fifty-nine.

I was a little worried about how to handle Fifty-eight and Fifty-nine. I remembered having read quite a bit about 'Negro Problems' before the war. Now I was in the middle of them! Once, when I was on leave in London, I even bought a book on the subject. I read it in bed until late at night.

One evening in the bar, I wanted to test my knowledge, so I approached the two Jamaicans. How come, I asked, that their nicknames were Fifty-eight and Fifty-nine. They both smiled with their rows of faultless teeth. I had the impression that my opening was appropriate. I did, however, have to ask again.

Fifty-nine laughed. 'Can't you see, sir? Can't you see that I am blacker than Fifty-eight?'

I felt rather uncomfortable at having to verify the comparison. Yes, I said, hesitantly, I thought perhaps I could see that Fifty-nine was possibly the blacker.

'Well, there you are, sir,' said Fifty-eight. 'Fifty-nine is short for 2359 – one minute to midnight, whereas I am 2358 – two minutes to midnight!'

That night I put away my book about 'Negro Problems'!

New Zealand Contribution

James Hayter (Squadron Leader J.C.F. Hayter, RNZAF), known mostly as 'Spud', and a farmer by upbringing, carried the New Zealand flag throughout one of the longest operational runs of the war. Fairey Battles in France in 1940 ... Hurricanes in Fighter Command ... an extended and varied tour in the Middle East ... and finally back again to the UK and Spitfire IXs in time for D-Day, in June 1944, and the advance northwards through France and into Belgium and Holland.... In four years Jim Hayter's war had turned the full circle. He had come a long way.

How long, was told in an endearingly simple inscription he wrote in a well-worn copy of Cuthbert Orde's *Book of Fighter Pilots – Drawings and Written Biographies*. In it he had collected and kept press clippings, signals, messages, pictures and all the other items a fighter pilot tended to gather along the way. He gave it to his mother and father as a souvenir of his service and his safe return home.

'To Mum and Dad – A "line" book. Love James
Left NZ 14 July 1939
Returned 24 September 1945'

It was an individual measure of New Zealand's contribution to victory.

Feminine Assets

Katrine Christie, Oslo, 1982

'... Being quite an advanced amateur photographer, I soon found myself in charge of the photographic work in Little Norway.*

When taking the boys' pictures I naturally wondered whether they would ever live through the war. Maybe someone who loved them at home would eventually ask for these pictures. I learned to bear this in mind because my first assignment was to photograph a double funeral.

Nevertheless the daily work was so varied that it did not leave much time for sentimental speculations. When the last shots were made before embarkation for United Kingdom I was often taken into a boy's confidence.

A miniature picture for a locket or a big one in a frame was to be sent to a very special girl friend on condition that none of the other boys got hold of the address. Or it might be a small memento to be brought home in case....

The Norwegians were very popular with the Canadian girls. Overstaying their leave was common. I roomed in the city and drove into the camp every morning. I wonder how many boys I smuggled through the gate in the back of my car....

The selection procedures they had been through were so many and so severe – mentally and physically – that only the best reached their goal in Little Norway, where they all hoped that flying school would be open to them. Today, about forty years later, I am constantly reminded of their quality. Names from Little Norway appear regularly among our most notable citizens in politics, science, high professions, and business.

In the end, everybody was transferred back to Norway. On my way to the airport in Scotland I met some Polish officers. "For us the war is not over," was their tragic remark. Next to me in the plane was a Norwegian boy in a sailor's uniform, David Rubinstein. Needless to say I had guessed, even before he told me, that he had none of his family left in Norway.

Back home I was given charge of the Office of Missing Personnel. It was an awful challenge. I soon realized, however, that, even in military service there was a position where it was an asset to be a woman. A meeting with the parents or widows would necessarily be very personal. I found I was privileged because in most cases I knew the person concerned and could give a good deal of information beyond the written records. Moreover, I could provide whatever photos that had been taken. Their gratitude was pathetic.

Our women's service during the war was improvised and after the war it was disbanded. Nevertheless, experience had shown that many positions could adequately be filled by women.'

Epitaph

There is an epitaph, composed by the Norwegian poet, Nordal Grieg, inscribed on the air force memorial in Oslo. The Royal Norwegian Air Force preferred to remember the fallen collectively, with these moving lines, rather than by recording individual names.

Before the war, Nordal Grieg was said to have been a pacifist and a communist sympathizer. The occupation changed his beliefs and he came to

*Little Norway was the Royal Norwegian Air Force's training camp in Toronto, established under the general aegis of the Royal Canadian Air Force's Empire Training Scheme which contributed so much to victory. (Ed.)

England to join others among his countrymen who had escaped. From here he sought to rally the morale of his occupied country.

He was lost riding with a bomber crew into Germany. Just before this he had visited the Norwegian fighter wing at North Weald where Wilhelm Mohr (Lieut.-General W. Mohr), to whom the editor is indebted for this inscription, was then commanding No. 332 Squadron. Grieg composed the two couplets immediately after this visit. David Scott-Malden, himself a former leader of the North Weald wing, has provided the English translation.

> To name even one of them,
> Would break faith with those unknown:
> Above the ranks of the fallen
> Stands Heaven, serene, alone.

True to his Precepts

Those who questioned General Galland in 1945 recorded this note about him as he led the elite jet unit, JV 44, in the final days of the conflict.

... Most of the other Me 262 units in southern Germany gave up their aircraft to Galland in the last weeks of the war, and he was able to finish his career as a flyer leading the last effective unit of the Luftwaffe according to the tactical precepts which he himself had always fought for.

As American armour approached Salzburg/Maxglan, the last base of JV 44, Galland sent a message by air offering to keep his unit intact for capture, but when the American force demanded that he fly over with his whole force, he refused and blew up all aircraft. Weather would not have permitted the flight within the stipulated time limit, anyway....*

'Scientific Developments'

Air Chief Marshal Sir Arthur T. Harris, despatch on War Operations, 23 February 1942 to 8 May 1945. Addressed to Under-Secretary of State for Air on 18 December 1945

... We are ending this war on the threshold of tremendous scientific developments – radar, jet propulsion, rockets, and atomic bombing are all as yet in their infancy. Another war, if it comes, will be vastly different from the one which has just drawn to a close. While, therefore, *it is true to say that the heavy bomber did more than any other single weapon to win this war*, it will not hold the same place in the next. The principles will, however, hold true; *the quickest way of winning the war will still be to devastate the enemy's industry and thus destroy his war potential* [editor's italics]. But the means by which that end is achieved will certainly be different from those which

*It is almost incomprehensible that, with the Germans so far ahead of the Allies in the design and operation of jet aircraft, so dunderheaded a response could have been given to Galland's offer. The emotions of war can be bad counsellors. (Ed.)

were used, with such far-reaching effect, to destroy the most powerful enemy we have faced for centuries.

The Dangerous Sky

Desmond Scott, Christchurch, New Zealand, 1983

'Many times I would yearn for my native New Zealand. To stand again by some high country stream, as the gold of the day was deepening into purple. Such dreams were often short lived; shattered by the sound of bursting bombs or the staccato bark of ack-ack guns. The mad scramble to our aircraft and the crazy weaving climb into the heavens.... Messerschmitts shining high in the sun like wicked little daggers – the hair-raising whirl of combat and the screaming breakaways that pushed your head into your chest, and your eyes into the back of your neck.... Such days were common. They shattered the nerves, turned some boys into men, some into dithering idiots. It was usually the inexperienced who fell first. Most were recent graduates from the school-room. They silently dropped away, or blossomed earthwards into balls of orange fire.

The passing years have in no way erased the magnitude of each brief encounter, for death is a very personal thing for those who flew face to face with it. Is it any wonder, even to this day, that some of us avoid memorial services? Men are not supposed to cry, but when the bugle's "Last Post" rings down the years, its long-drawn-out note brings back a flood of memories, and the tears come with them.'

Desmond Scott is now writing the manuscript of his next work, Airfield Commander, *to follow his recently published* Typhoon Pilot. *The editor is grateful to the author for permission to publish this reflection from it.*

To Douglas Bader 1910–82

James Sanders, Matakana, North Auckland, October 1982

I like to think of him as flying on and on,
A phantom phoenix risen from a noble pyre.
Can Britain's heart and hope be gone
With all the heavens flaming with his fire?
Look upwards! Jets – above the cloud!
And in the thunder of their urgent way
Know evermore, his legacy endowed
Against all odds.... Courage from yesterday.

PRIMUS INTER PARES

The history of the air war is studded with magical names. Many find a place in this volume; many more have, of necessity, had to be passed by. If the selections reflect editorial judgement, so also is personal prejudice exposed. The collector must have his heroes – if he is human.

Two among the air's wartime legends stand together as a pair, a pace or two in front of the rest. Once they traversed the Allied sky; now they bestride the record as, perhaps, no other surviving couple has quite been able to do.... Lieut.-General James Doolittle of the United States and Group Captain Leonard Cheshire of Great Britain – each, in his own right, a first.

Doolittle

Douglas Bader finished this personal reflection on James Doolittle on 1 September 1982. It was only just in time; three days later Bader was dead. It was the last important piece that he wrote and he invested unusual effort in it. 'There you are,' he said to the editor, as he delivered his corrected copy, 'that'll do for Jimmy. I'm glad I was asked to do it.' His pleasure at having had the chance to set down his thoughts upon his old friend and colleague was manifest.

'General Doolittle, born in Alameda, California, on 14 December 1896, was the most remarkable and probably the most successful United States senior air force commander of World War II. The record supports the contention.

James Doolittle was forty-three when he returned to the Service in 1940 as a major, having spent the previous ten years with the Shell Petroleum Corporation following an initial spell of thirteen years – 1917 to 1930 – with the United States Army Air Corps. From 1940 onwards, the story is one of endeavour, outstanding success and promotion.

His leadership (at the age of forty-five, remember) of the extraordinary, carrier-borne Tokyo raid on 18 April 1942, was to become one of the war's epics. It made Doolittle internationally famous. But it was the next three years that saw his qualities as a commander – a fighting, air force commander – truly revealed. The catalogue of achievement is impressive: commanding general of the Twelfth Army Air Force in North Africa in 1942; commander of the Fifteenth in Italy a year later; then on to England in 1944 to take over the Eighth Army Air Force for the culminating stages of the crushing, daylight offensive against Germany.

With the end of the European war, it was then to the Pacific theatre – and while he was still at the head of the Eighth Army Air Force – that James Doolittle's special talents were transferred. The period of his final command in 1945 was, however, characteristically brisk and short. After the atomic attacks on Hiroshima and Nagasaki, and the subsequent surrender of Japan's Imperial forces, the General's time as a brilliant war leader was over. Having risen in four years of almost continuous action from major to lieutenant general, he returned once more to Shell, just as I did myself at the conclusion of the war in Europe.

What kind of a man, then, is this Jimmy Doolittle whom I have known as friend and colleague for more than forty-five years? What are the characteristics which contributed to his outstanding wartime success? My picture of his ability and attributes is quite clear.

I first met Jimmy Doolittle in 1936 or 1937, two or three years before the European war. He was then the Aviation Manager of Shell Petroleum in St Louis. I was working in the Aviation Department of Shell in London. We met and gossiped for only a short time, but my impression of that first meeting endures. Here was this little man, slim, standing absolutely upright – just as he has done all his life. There was no doubt, even then, that this was an exceptional individual of decisive and straightforward character. Of course, I knew his reputation as a pilot. He was already a considerable figure in aviation with all sorts of awards and distinctions to his name. I well remember him telling me how, in 1929, at the time of the Cleveland Air Races, he took to his parachute for the first time. He was flying a Hawk fighter very fast low down when the wings came off. He was left in the fuselage, hurtling earthwards from 700 or 800 feet. After breaking free, he pulled the rip-cord (remember, there were no ejector seats in those days) and the chute opened just in time. He picked himself up, dragged the canopy after him and went off to find another aeroplane. He flew a scheduled stunt display later the same day.

The next time he jumped, he told me, was in 1931. He had joined Shell by then and was testing an aeroplane he had himself built for another attack on the world air speed record standing then at 177 m.p.h. Approaching the "barrier" of 300 m.p.h., the aircraft ran into serious trouble. He pulled up, got whatever altitude he could and rolled over onto his back. The splendid telegram he sent to his chum, Les Irvin, the founder of the Irvin Air Chute Company, said it all.

"Aeroplane failed, chute worked."

It was typical Jimmy – but it made no difference to his attempt on the record, for the next year, in his famous Gee Bee Racer, he set the world mark at 196 m.p.h.

What, I think, has never been stressed sufficiently about Jimmy Doolittle is his technical knowledge. It has often surprised me whenever we have talked about aeroplanes. He *knows* flying and everything to do with it – with aircraft and their engines. Engineering-wise he is very sound. Like all first-class pilots, he has had a quick, precise and sure mind.

A combination of these attributes, plus his ability to fly and "feel" an aircraft, made him one of the best test pilots of his day and widely respected by that remarkable bunch of men. He was also a sales pilot pre-war, selling aeroplanes, I think, for the Curtiss Wright company. As such he used, on occasions, to visit South America.

I well recall an incident when we were together in Tangier in 1946. He suddenly burst into Spanish, talking to some official. "Jimmy," I said, "what on earth's going on? Wherever did you learn that language?"

"Oh, that's nothing, Doug," he replied, "it isn't proper Spanish. I used to travel one time down to South America when I was demonstrating aircraft. I had to pick up the lingo to get by."

Of course Jimmy will always primarily be remembered as a great wartime commander, a great war leader. The Tokyo raid had so much to do with it. I got the story – his story – of the attack out of him during a trip we did round

Europe and North Africa just after the war.

I shall always think of that mission as the bravest operation of the war. He led those 16 B-25 medium bombers – specially prepared, twin-engined, American Mitchell bombers – off the aircraft carrier, *Hornet*, on a cold, squally, windy morning, with high waves lashing against the ship. They were some 650 miles from the Japanese mainland and were prepared, after the bombing, to fly on into China and, if necessary, bale out or force land. It wasn't much of a prospect.

There was supposed to be a medium frequency radio beacon on the Chinese mainland which Chiang Kai-shek's boys were said to have put up. This was what the crews hoped to home onto after they had bombed the Japanese capital, and turned south. In the event, they took off, made the flight to Japan, did the bombing and turned south for the China coast. Getting no signals they flew on and on and I recall Jimmy saying that, in the poor visibility and darkness and with fuel getting low, he felt he must give the order to his four crewmen to bale out. "You know, Doug," he said, "that was the most awful command I ever had to give in my life."

Doolittle, having set the controls on automatic went out last. "I baled out," he said, "and landed right up to my ass in a foul-smelling paddy field. I remember standing up and saying to myself, 'Doolittle, when you get back home, you'll sure be dropped a rank – maybe several ranks.'"

The rest of the story is well known. Of the 16 aircraft which took off from the carrier, 11 crews baled out, 4 crash-landed and 1 made a wheels-down landing on an airstrip near Vladivostok in Russia, and was interned. Of all the crew members who survived this wonderfully courageous and hazardous operation, 8 fell into Japanese hands. In a dreadful act of retribution, 3 were executed.

It was very soon after we had both started again with Shell after the war that Jimmy came over to London. The Company had got the idea that he and I should do a goodwill tour together around the capitals of Europe, North Africa and Egypt where Doolittle was now a household name.

I had then a little single-engined Proctor aeroplane which Shell provided for my work. Sitting side by side in this tiny aircraft, the great four-star general and I flew from capital to capital. Everywhere there was a wonderful welcome.

After Paris, we flew down south to the Mediterranean. It was high summer and very hot. In Nice, we decided we'd like to bathe. So we went down the shopping street in search of a pair of swimming trunks. At the first shop we tried, a madame in a black dress produced two slips which were very, very narrow. The bit which goes through the legs was so narrow that I said to Jimmy, "These won't do. They aren't big enough." But madame hadn't got anything bigger.

"What are we going to do?" I asked.

Jimmy smiled. "Don't worry, Doug," he said. "I guess it's being worn outside this year."

Much has happened to Doolittle since those days and I've seen him many times, both in the United States and here in London. When he retired from Shell in 1967, he went on working at his other directorships, and often for the Pentagon, just as hard as ever. He would never let go.

The last time we met was in 1981 in Los Angeles. He was then eighty-four and my wife and I were lunching with him and his wife, Jo. There were only the four of us. He looked just the same – slim, back as straight as a ramrod. I don't suppose he weighed a pound more than when I first met him more than forty-five years ago. He was still as bright as a button. And yet what a life it had been in between.

"Jimmy," I said, "when are you going to retire?" The question seemed to take him by surprise.

"Doug," he replied, "I just don't have time to retire."

It summed the man up perfectly – a splendid and exhilarating companion.'

Cheshire

Leonard Cheshire was invested with the Victoria Cross in war and the Order of Merit in peace. There never was such a double; nor can one ever imagine it being repeated. The two honours are a measure of the man and tell the story of a life lived in courage and compassion, humility and Godliness – a life profoundly moved by experience yet totally untouched by success.

The brief extracts that follow disclose something of the character of one of the most exceptional men of war – and peace.

THE SCHOOLBOY

J.F. Roxburgh, first Headmaster of Stowe (1923–49), on Leonard Cheshire, quoted by Andrew Boyle in *No Passing Glory* (Collins, 1955)

'He was very successful as a schoolboy, in the ordinary sense.... He became head of his house, a prefect, a member of the sixth form ... and captain of lawn tennis. But any first-rate boy can expect a career of that kind at his public school, and I cannot say that he made more impression on what was certainly a distinguished generation than several others did.

Leonard could not have reached the position he did among boys like that if he had not been pretty good. But that does not mean that he showed as a schoolboy the extraordinary qualities that he developed later as a fighting man. I personally was much attached to him, and I felt for him not only affection but respect. There was something about him – was it perhaps a kind of moral dignity? – which made it inconceivable that he would think or do anything which was below top level. He knew how to make other people do the right thing, too, and his courteous, ever gentle manner covered a pretty tough will, which made him a strong ruler in his house and an effective prefect in the school.

But all this could be said of other boys who as men never became pre-eminent as Leonard did.... I should have expected him to do well when tested by war, but I should not have foreseen that he would become what he became or achieve what he achieved. I doubt if such development and such achievements could have been predicted of anyone....'

Andrew Boyle, *No Passing Glory* (Collins, London, 1955)

For the whole half-hour he spent in Harris's* presence, Cheshire's brain was in a whirl. A hard and belligerently practical man, the C-in-C had a 'soft spot' for pilots who did not realize when they had done more than their stint of operations. The gruff gentleness of his welcome was the first shock; the second was his statement, still in those velvety tones, that he wanted to congratulate him before anyone else on winning a VC. Cheshire was reduced to a stunned silence. The VC of all things! He had never thought of Harris as a 'fatherly man' with feelings and a heart. The interview was like a fantasy with wish-fulfilment, a triumph of hope over experience.

Harris did not seem to notice his bewilderment. He was used to the spectacle. ('On every occasion when I informed anyone that they had been awarded that very high honour, they have invariably been overcome with astonishment and given the impression that it was the last thing they expected or deserved,' he told me.

'I have never known any recipient of the Victoria Cross not to be astonished at the news of the award – but I have sometimes met others astonished that they have not been awarded the Victoria Cross.')

Generosity

Commanding Officer, No. 617 Squadron, Bomber Command, Royal Air Force

Leonard Cheshire on Air Marshal Sir Harold Martin – 'Mick' Martin, the Australian 'dambuster' – quoted by Peter Firkins in The Golden Eagles (St George Books of Perth, Western Australia, 1980)

'The backbone of the squadron were Martin, Munro, McCarthy and Shannon, and of all these by far the greatest was Martin. He was not a man to worry about administration then (though I think he is now), but as an operational pilot I consider him greater than Gibson, and indeed the greatest that the air force ever produced. I have seen him do things that I, for one, would never have looked at....

'I learned all I knew of this low-flying game from Mick. He showed me what you could do by coming in straight and hard and low, and I never saw him make a mistake....

'Mick Martin was the greatest bomber pilot of the war....'

Endeavour

Leonard Cheshire, *No Hidden World* (Collins, 1981)

... I know of no substitute in any field of human endeavour for hard work, for clear and realistic thinking and planning, and, most important of all, for

*Air Chief Marshal Sir Arthur Harris, Air Officer Commanding-in-Chief, Bomber Command.

perseverance. The person who ferrets away, who never lets go, who, when faced with an impasse or just cannot see what he should do next, is content to wait and relax until something happens to give him an opening, is the one who will usually achieve the most. Without in any way renouncing the need to set our sights high, to be satisfied with nothing less than the best, and to commit ourselves totally and unreservedly to participating in the struggle to build a more liveable world, I have come to believe that the important thing is to keep going and to appreciate that even one small improvement is infinitely worth making. It is the multiplication of many people, each working in their own chosen field and in their own individual way that brings about genuine change....

Tribute

On 23 October 1982, Leonard Cheshire gave the address at the Service of Thanksgiving for the life of Group Captain Sir Douglas Bader at St Clement Danes, the Royal Air Force's church, in the Strand. Spoken without a note, and with a compelling gentleness, it was a performance of rare quality. The packed congregation, which included representatives of the Royal Family, and the Prime Minister in person, was held by the simple and unaffected sincerity of the message.

There was a dignity about the peroration.

We are here today to honour Douglas Bader's memory and to commend his soul to Almighty God, from Whose hand comes all that is good and noble, and Who rewards each man as his works deserve. We would not have gathered in such numbers and from such distant parts but for two fundamental things, the crash that cost him his legs, and the RAF. If it was his disability that released the spirit that was in him and gave it wings with which to soar, it was the RAF that harnessed that spirit and gave it direction. It was there, in the RAF, that he learned what it is to be part of a Service tradition, what it is to be in succession to a long heritage of courage, self-sacrifice and compassion.

It was there that he received the torch from those who had gone before, carried it forward as far as he was able and has now handed it on to those who follow after.

Epilogue

On 5 October 1938, in a moment of gathering national euphoria ('peace for our time'), Winston Churchill, from the back benches, warned the Chamberlain Government and the House of Commons that the 'agreement' which the British prime minister had just made with Adolf Hitler over Czechoslovakia was a sell-out and that awful consequences would ensue (see page 33).

On 1 March 1955, nearly seventeen years and another world war later, Churchill, then eighty, rose again in his place in the House of Commons – this time at the dispatch box to deliver his last great speech at the head of his second administration. It was a moving and masterly defence of Britain's possession of the nuclear weapon as the best deterrent to World War III.

His closing sentences, beautifully delivered, and full of emotion, make, in the light of the peace that has followed, a salutary end to a book about war:

> ... What ought we to do? Which way shall we turn to save our lives and the future of the world? It does not matter so much to old people; they are going soon anyway, but I find it poignant to look at youth in all its activity and ardour and, most of all, to watch little children playing their merry games, and wonder what would lie before them if God wearied of mankind....
>
> ... There is time and hope if we combine patience and courage.... The day may dawn when fair play, love for one's fellow men, respect for justice and freedom, will enable tormented generations to march forth serene and triumphant from the hideous epoch in which we have to dwell. Meanwhile, never flinch, never weary, never despair.

Winston Churchill, Hansard, 1 March 1955

Index